Signaling and Communication in Plants

Series Editors

František Baluška
Department of Plant Cell Biology, IZMB, University of Bonn, Kirschallee 1, D-53115 Bonn, Germany

Jorge Vivanco
Center for Rhizosphere Biology, Colorado State University, 217 Shepardson Building, Fort Collins, CO 80523-1173, USA

František Baluška
Editor

Plant-Environment Interactions

From Sensory Plant Biology to Active Plant Behavior

Springer

Editor
František Baluška
Department of Plant Cell Biology
IZMB
University of Bonn
Kirschallee 1
D-53115 Bonn
Germany
email: baluska@uni-bonn.de

ISSN 1867-9048
ISBN 978-3-642-10037-6 e-ISBN 978-3-540-89230-4
DOI: 10.1007/978-3-540-89230-4

Cover design: WMXDesign GmbH, Heidelberg, Germany

Printed on acid-free paper

9 8 7 6 5 4 3 2 1

springer.com

František Baluška dedicates this book to Prof. Dieter Volkmann for introducing him to sensory plant biology and in recognition of his great support and friendship.

Preface

Plants are generally considered to be passive and insensitive organisms. One can trace this strong belief back to Aristoteles, who positioned immobile plants outside of the sensitive life domain. The millennia that have elapsed between time of Aristoteles and the present day highlight the fact that it will very difficult to change this almost dogmatic view. For instance, one of the first serious attempts to rehabilitate plants was performed by no less than Charles Darwin, in 1880. At the end of the book *The Power of Movement in Plants*, which he wrote together with his son Francis, they proposed that the root apex represents the brain-like anterior pole of the plant body.

This volume, in fact the whole series, documents a paradigm shift that is currently underway in the plant sciences. In the last two or three decades, plants have been unmasked as being very sensitive organisms that monitor and integrate large numbers of abiotic and biotic parameters from their environment. That plants react to electric stimuli in the same manner as animals was shown by Alexander von Humboldt a few years after Luigi Galvani discovered the electrical stimulation of animal muscles in frogs' legs. Later, when animal action potentials were discovered in animals, similar action potentials were soon recorded in plants too. Initially only "sensitive plants" were tested, but some 30 years ago it was found that all plants use action potentials to respond to environmental stimuli. This rather dramatic breakthrough went almost unnoticed in the mainstream plant sciences. Only recently, the emergence of plant neurobiology has highlighted this neglected aspect of biology. The obvious conservation that occurs throughout evolution means that action potentials provide both plants and animals with evolutionary advantages that are crucial to their adaptive behavior and survival. As plants evolved action potentials independently of animals, this phenomenon also holds the key to illuminating the mystery of convergent evolution, a phenomenon that does not conform to the classical Darwinian principles of biological evolution.

Recent advances in chemical and sensory ecology have revealed that plants communicate via volatile and allelochemical chemical messengers with other plants and insects. By using a wide variety of volatiles, plants are able to attract or repel diverse insects and animals, enabling them to shape actively their biotic niche. The number of volatile compounds released and received by plants for communication is immense, requiring complex signal-release machinery, as well as "neuronal" decoding apparatus to correctly interpret the received signals. These aspects of plant activity have not been studied yet. Plants integrate and memorize numerous sensory "experiences" in order to adapt effectively to an ever-changing environment.

Plants also show active behavior, including kin and self/nonself recognition, cognition, and a plant-specific form of intelligence. In order to find their prey, parasitic plants use sophisticated sensory detection systems, and after colonizing the prey tissues they conform to an animal-like heterotrophic lifestyle. Plants often apply deception as an effective strategy to manipulate other organisms, including insects, other animals, and perhaps even us humans. They use colors, forms and odors, as well as taste-stimulating, nutritional and neuroactive substances to manipulate insects, animals and humans in order to aid their spread around the globe. Crop plants like wheat, maize, barley and rice are the most successful species in this respect. New concepts are needed and new questions must be asked in order to advance our rather rudimentary understanding of the communicative nature of sensory plants.

One of the goals of current plant science is to improve the agricultural properties and stress adaptabilities of plants. However, we will not achieve this goal until we unravel the communicative, sensory, and cognitive aspects of these organisms. Moreover, our civilization still is—and will continue to be in the future—fully dependent on plants, since they (together with unicellar photosynthetic organisms) are the only primary source of oxygen and organic matter on this planet. Recently, humans have begun to use plants extensively to produce biofuels. Due to the continuing problems with hunger in underdeveloped countries, this presents our civilization with a dilemma: what proportion of plants should be grown for food and what proportion for energy? Our future depends on us gaining a complete understanding of plants in their full complexity.

Bonn, October 2008 František Baluška

Further Reading

Baluška F, Volkmann D, Mancuso S (2006) Communication in plants: neuronal aspects of plant life. Springer, Berlin

Baluška F, Mancuso S (2009) Signaling in plants. Springer, Berlin

Barlow PW (2008) Reflections on plant neurobiology. *Biosystems* 92:132–147

Brenner E, Stahlberg R, Mancuso S, Vivanco J, Baluška F, Van Volkenburgh E (2006) Plant neurobiology: an integrated view of plant signaling. *Trends Plant Sci* 11:413–419

Conway Morris S (2003) Life's solution: inevitable humans in a lonely universe. Cambridge University Press, Cambridge

Darwin CR (assisted by Darwin F) (1880) The power of movement in plants. Murray, London

Karban R (2008) Plant behaviour and communication. *Ecol Lett* 11:727–739

Mancuso S, Shabala S (2006) Rhythms in plants. Springer, Berlin

Pollan M (2002) The botany of desire: a plant's-eye view of the world. Random House, New York

Volkov AG (2006) Plant electrophysiology. Springer, Berlin

Scott P (2008) Physiology and behaviour of plants. Wiley, New York

Contents

Mechanical Integration of Plant Cells

Anna Kasprowicz, Daniel Kierzkowski, Michalina Maruniewicz, Marta Derba-Maceluch, Ewelina Rodakowska, Paweł Zawadzki, Agnieszka Szuba, and Przemysław Wojtaszek

1 Introduction

In order to function in changing environmental conditions, all living organisms need to be equipped with two sets of seemingly contradictory mechanisms; these enable them to (1) function as an integrated entity independent of the environment, and (2) sense and communicate with their immediate surrounding. During the course of evolution, several factors—both physical and chemical—have emerged as organismal integrators. Among these, gravity provides a major directional stimulus, while chemical compounds are usually used as internal integratory molecules (Bhalerao and Bennett 2003).

Although the same cellular toolkit of their common ancestor gave rise to present-day eukaryotes through evolution, it should be remembered that plant and animal lineages diverged about 1 billion years before they became multicellular organisms. As a consequence, plants and animals differ in their lifestyles, responses to stimuli, and adaptations to the environment. This distinction results from the adoption of two different strategies of coping with the regulation of intracellular water content, and is reflected in the properties and behavior of "naked" animal cells vs. "walled"

A. Kasprowicz, D. Kierzkowski, M. Maruniewicz, M. Derba-Maceluch, E. Rodakowska, and P. Zawadzki
Department of Molecular and Cellular Biology, Faculty of Biology, Adam Mickiewicz University, Umultowska 89, 61-614 Poznań, Poland

A. Szuba
Institute of Bioorganic Chemistry, Polish Academy of Sciences, Noskowskiego 12/14, 61-704 Poznań, Poland

P. Wojtaszek (✉)
Department of Molecular and Cellular Biology, Faculty of Biology, Adam Mickiewicz University, Umultowska 89, 61-614 Poznań, Poland and
Institute of Bioorganic Chemistry, Polish Academy of Sciences, Noskowskiego 12/14, 61-704 Poznań, Poland
e-mail: przemow@amu.edu.pl

F. Baluška (ed.), *Plant-Environment Interactions*, Signaling and Communication in Plants,
DOI: 10.1007/978-3-540-89230-4_1,

plant cells (Peters et al. 2000). Thus, while animals are able to move away when conditions are unfavorable, plants—since they are sessile organisms—must react and/or adapt to changes. As a result, much greater plasticity of plants and their cells is observed (Valladares et al. 2000).

All organisms have the ability to sense and respond to a variety of physical stimuli, such as radiation, temperature, and gravity (Volkmann and Baluška 2006). Although physical forces act in the same manner on different organisms, the effects of their actions depend on the organism's habitat. For example, the effect of gravitational force on an organism depends greatly on whether it lives in water or on land. On the other hand, the forces exerted on terrestrial plants by the movement of air are much lower than those exerted on aquatic ones by the movement of water (Niklas et al. 2000). Thus, although the overall construction of any particular plant or plant cell is generally similar to that of any other, the details of their biochemical and mechanical designs can vary considerably, as these are also shaped by the changing conditions in the cell's or organism's immediate surroundings.

2 Mechanical Organization of Plant Cells

From a mechanical point of view, the end product of the evolutionary transition to present-day plant cells could be considered a *tensegral hydrostat*. In normal plant cells, compression-resistant turgid protoplast is surrounded by and presses against tension-resistant and mechanically stable cell walls (Wojtaszek 2000; Zonia and Munnik 2007). This design principle has several important implications for the functioning of plant cells and plants: (1) functional cell walls become indispensable elements of plant cells; (2) the vast majority of plant cells do not move in relation to their neighbors; (3) both the cell walls and the steep gradient of hydrostatic pressure across the plasma membrane (which exceeds 2 MPa) can be used to mechanically stabilize plant bodies; (4) the interplay between the cell walls and turgor is the major determinant of cellular shape and organismal morphogenesis; (5) the presence of a hermetic matrix around protoplasts limits the ability to acquire energy and nutrients (Peters et al. 2000; Wojtaszek 2001). However, phragmoplast-based incomplete cytokinesis, which leads to the formation of the cell plate and enables a new type of intercellular communication through plasmodesmata (Lucas et al. 1993; Heinlein and Epel 2004), and the inclusion of newly synthesized cell walls into the supracellular structure of the apoplast (Wojtaszek 2000), have allowed plants to overcome at least some of the constraints of this mechanical design.

The major structural and functional organizer of plant cells is the continuum formed by cell walls, plasma membrane and cytoskeleton (WMC continuum; Wyatt and Carpita 1993; Kohorn 2000; Wojtaszek 2000; Baluška et al. 2003). Plant protoplasts are able to rapidly and reversibly retract from the cell wall in plasmolytic response to changing osmotic conditions (Lang-Pauluzzi and Gunning 2000) while maintaining localized membrane-wall attachments (Lang et al. 2004). The functioning of the cytoskeleton, which is anchored via plasma membrane to the walls, provides mechanisms for (1) the regulation of cellular volumes (Komis et al. 2003); (2) the directional transport and spatial distribution of cellular components (Sato et al. 2003; Chuong et al. 2006), and; (3) the rearrangement of cellular architecture in response to internal and external stimuli (e.g. Wojtaszek et al. 2005; Schmidt and Panstruga 2007). However, the wall-anchored cytoskeleton seems to function as not only a detector of physical forces but also a transmitter of mechanical signals as well as a transducer of those signals into biochemical messages (Forgacs 1995; Ingber 2003a, b). These processes are rather poorly recognized in plants, and important linker molecules within the WMC continuum are still not characterized (for review see Kohorn 2000; Baluška et al. 2003). However, from studies in animal systems, it is now becoming clear that proper ECM–cytoskeleton contacts are crucial to the determination of cellular shapes and thus cell fate (e.g., Nelson et al. 2005; Engler et al. 2006; Vogel and Sheetz 2006; Assoian and Klein 2008). This reinforces the idea that information stored in molecular and cellular structures is used during the generation of form, giving rise to new, emergent properties that are not directly deducible from the properties of the initial components (Harold 1995).

Our questions about the influence of physical forces on the functioning of cells and organisms are not yet fully answered. However, some general rules of mechanosensing and mechanotransduction are becoming apparent. According to the tensegral model of cellular architecture, microfilaments are tension-responsive elements, whereas microtubules serve as contraction-resisting structures, and the cell and tissue shape depends on a balance between the physical states of those prestressed filamentous networks (Ingber 2003a, b). Upon arrival at the cell surface, mechanical stimuli are recognized by specialized receptors. Those receptors—which are connected to both the ECM and the internal cytoskeleton spanning the whole cytoplasm—will be able to transmit these mechanical signals into cells, while other membrane receptors will fail (Ingber 2003a). At least two possible and nonexclusive ways of mechanotransduction can be envisioned. One of them involves the direct transduction of the mechanical stress imposed on the receptor into a chemical signal which can be propagated into the cell. The other one makes use of local conformational changes of proteins, at least within a portion of the signaling pathway (Kung 2005; Valle et al. 2007). The first path offers the versatility of secondary chemical messengers and the possibility of cross-talk with other signaling pathways, enabling the fine tuning of cellular reactions (Orr et al. 2006). The second provides the speed and fidelity of signal transmission, which is a unique feature of mechanotransduction (Na et al. 2008). Interestingly, if we assume that the same forces act on all elements of the tensegral structure (ignoring the size), the same rules of tensegrity will apply at not only the cellular level but also the tissue and organismal ones (Ingber 2003a).

There are many examples (in plants too) of cellular processes in which the transmission of mechanical force has been documented or is commonly assumed, starting with changes in the activities of enzymes or protein complexes (Aon et al. 2000), through organized movement of molecules, particles and organelles (van der

Honing et al. 2007), and ending with the reorganization of whole cells in response to external cues, such as osmotic stress (e.g., Wojtaszek et al. 2005, 2007) or pathogenic infection (Schmidt and Panstruga 2007; Hardham et al. 2008). Over ten years ago, a direct mechanical connection between the cell surface and the nucleus via the cytoskeleton was demonstrated in animal cells (Maniotis et al. 1997); this profoundly affects the organization of chromatin (Maniotis et al. 2005). Interestingly, it seems that the nucleolus is to some extent mechanically independent from the rest of the nucleus (Yang et al. 2008). In plant cells, nuclei are highly dynamic; they are able to undergo polymorphic shape changes and rapid, long-distance movements (Chytilova et al. 2000). Both the positioning and movements of nuclei are mediated by actin (Baluška et al. 2000). Importantly, mechanical stimulus seems to be the primary signal that induces nuclear repositioning (Hardham et al. 2008), and it has been demonstrated that isolated nuclei are also able to sense physical forces (Xiong et al. 2004). As the position of the nucleus is strictly correlated with the cell cycle progression, especially with the determination of the plane of cell division, the sensing and transduction of mechanical stimuli provide the mechanism for the coordinated development of supracellular plant structures (Lintilhac and Vesecky 1984; Qu and Sun 2007; see also below).

2.1 Constructing the Pathway for Mechanotransduction

In accordance with what was said above, at least two broad classes of mechanosensitive (MS) molecules can be distinguished. The first comprises proteins that sense the tension within the lipid bilayers of biological membranes (Martinac 2004). These can then open rapidly, allowing a large number of ions to enter, thus amplifying the signal. Examples include the bacterial MscS (mechanosensitive channel of small conductance) channels that regulate cellular responses to osmotic stress. In the *Arabidopsis* genome there are ten genes coding for MscS-like (MSL) proteins. Among them, MSL2 and MSL3 are involved in the control of plastid size and morphology (Haswell and Meyerowitz 2006), while MSL9 and MSL10, and possibly three other MSL proteins, are required for MS channel activities in root cells (Haswell et al. 2008). The regulation of cellular volumes has been ascribed to some MS anion channels (reviewed by Roberts 2006), while the gating of Ca^{2+} influx is thought to be a major function of the MS ion channels in lily pollen tubes (Dutta and Robinson 2004) and Mca1 protein from *Arabidopsis* roots (Nakagawa et al. 2007).

Proteins belonging to the second group are characterized by their ability to sense mechanical distortions in either cytoskeleton or extracellular matrices (ECM), such as cell walls in plants. Their ectodomains are usually embedded in ECM or they strongly interact with the ECM components. These domains are connected with transmembrane domains, and—if present—with cytoplasmic domains of various lengths and different activities. On the intracellular side, they interact with cytoskeleton either directly or via cytoskeleton-binding proteins. Further signal transmission can occur in several different and nonexclusive ways, depending on the design of the given protein. First, the distortion can be propagated as conformational changes within a chain of interacting proteins. Second, the stimulus can be transduced into an electrical signal via the activity of ion channels. Third, the mechanical signal can be transformed into a chemical message, e.g., through the phosphorylation of target proteins by the intracellular domain of a sensor with kinase activity or a specialized kinase interacting with a sensor (Ingber 2003b; Orr et al. 2006). Typical examples taken from animal systems include integrins, which are able to detect and transmit mechanical perturbations in both directions: inside–outside and from the ECM to the cytoskeleton, reacting to changes in the cellular neighborhood and stabilizing cell–ECM interactions. The extent and the quality of the interactions with the integrins are then recognized and transformed into various biochemical messages regulating metabolism and cellular behavior (Arnaout et al. 2007; Assoian and Klein 2008). In plants, the most diverse group of proteins are the protein kinases with specialized extracellular domains. These include receptor-like kinases (RLKs), such as wall-associated kinases (WAKs; Kohorn 2001) and proline-rich extensin-like receptor kinases (PERK), and other kinases with, say, carbohydrate-binding motifs (reviewed by Shiu and Bleecker 2001). Although WAKs (for example) have been shown to be embedded in the pectin matrix of the walls (Decreux and Messiaen 2005), the involvement of RLKs in mechanotransduction has rarely been demonstrated (Gouget et al. 2006). An interesting example is the specialized potassium channel KAT1, located in plasma membrane and probably associated with the surrounding cell walls of *Vicia faba* guard cells, although whether it transmits mechanical distortion into the cell is yet to be elucidated (Homann et al. 2007).

In animal cells, integrin activity can be directly modulated by peptides containing RGD (Arg–Gly–Asp) motifs that are characteristic of many of the extracellular proteins interacting with integrins. Although genes coding for integrins or integrin-interacting proteins have not been identified in the *Arabidopsis* genome (Hussey et al. 2002), the existence of proteins similar to integrins (e.g., those recognized by heterologous antibodies) has been demonstrated in many plant species. Moreover, the treatment of plant cells with RGD-containing peptides affects their functioning in processes such as gravisensing (Wayne et al. 1992), the plasmolytic cycle (Canut et al. 1998), the plant defense response to fungal infection (Mellersh and Heath 2001), as well as growth and differentiation (Schindler et al. 1989; Barthou et al. 1998). The application of RGD peptides also leads to the modulation of cytoplasmic streaming (Hayashi and Takagi 2003) and the formation of Hechtian strands (Canut et al. 1998; Mellersh and Heath 2001). As Hechtian strands contain both actin filaments and microtubules, these observations provide direct evidence of active linkages between plasma membrane proteins and the cytoskeleton, which play an important role in cell-to-cell communication and signal transduction from the cell wall into the protoplast.

Several other molecules has been proposed to function as linkers within the WMC continuum; myosin VIII and formins are thought to be the most probable adhesive molecules (reviewed by Kohorn 2000; Baluška et al. 2003). Their localization at the cross-walls of cells in the axial organs may be crucial to their functioning (Deeks et al. 2002; Baluška and Hlavačka 2005). Intriguingly, some of the proposed

WMC linkers, such as WAKs and arabinogalactan proteins (AGPs), were found to associate with plasma membrane-located MS calcium-selective channels in tobacco BY-2 cells, supporting the view that the WMC continuum is the sensor and transducer of mechanical signals (Gens et al. 2000). Interestingly, such an association enables the discrimination of various signals, as stretch-activated Ca^{2+} channels are involved in the sensing of both hypotonic and hypertonic conditions, whereas the WMC continuum is only involved in sensing a hypertonic environment (Hayashi et al. 2006).

3 Control of Cell Morphogenesis and Fate Determination

Cell organization and functioning takes place in four dimensions (Wojtaszek 2000). To understand these processes, we must, in the words of Frank Harold (1995), "ask how organisms produce successive shapes as they traverse their life cycles. This query focuses attention on structures, forces and flows that modulate form, rather than on molecules and genes." Research on various systems, but especially animal cells, has provided evidence that cellular shapes and the sensing of geometry and mechanical environment are tightly intertwined with cellular functions. For example, cell–ECM interactions are crucial in deciding the cellular fate (Engler et al. 2006) and the frequency of cell division within an organ (Nelson et al. 2005). The presence of turgor and the "walled" organization of plant cells (Peters et al. 2000) provide other mechanisms of shape determination. As turgor is a scalar quantity, its effects are isodiametric, and wall-less protoplasts are invariably spherical. The continuous interplay between turgor and the differentiated mechanics of wall domains surrounding individual cells provide the means to achieve the great diversity of cellular shapes (Panteris and Galatis 2005; Mathur 2006). Even more importantly, although the organized cytoskeleton carries out cytokinesis, it is the presence of the walls as well as the resulting shape of the cell that provide spatial cues that are indispensable when organizing the cytoskeleton and determining the plane of cell division (Meyer and Abel 1975; Niklas 1992; Green 1999; Cleary 2001). In growing plants, the characteristic mechanical environment of the cells in a given organ results in an ordered pattern of cell divisions. This is lost in regenerating tissues such as callus, but can be restored with the external application of directional forces (Lintilhac and Vesecky 1984), which are sensed by protoplasts (Lynch and Lintilhac 1997). Moreover, the mechanical environment of the maternal tissues has a crucial influence on the plane of first asymmetric division in fertilized zygotes (Kaplan and Cooke 1997). Mechanical patterns are also important in suspension-cultured cells, in which mechanical stimuli dictate the proper organization of cellular metabolic networks (Yahraus et al. 1995; Aon et al. 2000).

At the cellular level, the turgor is used to identify mechanically weaker domains of the cellular boundary in order to enable growth in that direction (Mathur 2006). As the cellulose–hemicellulose network constitutes the major tension-resistant element of the walls, it is commonly assumed that the orientation of newly deposited cellulose microfibrils restricts the possible growth directions of expanding cells. However, the question of what determines the orientation of cellulose microfibrils is still a matter of debate. The classical point of view is that the deposition of cellulose microfibrils is affected by the alignment of cortical microtubules (Wymer and Lloyd 1996). Experiments with tobacco suspension-cultured cells have demonstrated that spatial cues for the organization of microtubules might come from biophysical forces, and that microtubules themselves can respond to vectorial changes in such forces (Wymer et al. 1996). According to the geometrical model, new microfibrils are oriented by the cell geometry together with existing wall components, while the orientation of microtubules is a simple reflection of the directed delivery of cellulose synthase complexes to the plasma membrane (reviewed by Emons and Mulder 2000). However, recent biochemical and genetic data suggest the existence of a bidirectional flow of information between cortical microtubules and cellulose microfibrils, with the latter providing spatial cues for the internal organization of microtubules, most probably through the cellulose synthesis machinery (Fisher and Cyr 1998; Paredez et al. 2006, 2008). Microtubules aside, filamentous actin is also essential for cell elongation during plant development (Baluška et al. 2001) and for the directed delivery of cellulose synthase complexes to the sites of wall synthesis (Wightman and Turner 2008).

In many cases, tissue geometry has a crucial influence on cell fate. In axial plant organs, the pressure exerted by external epidermal cell walls allows inner cells of the root to perceive the mechanical environment nearby and adjust properly to it (Kutschera 2008). Externally applied pressure can lead to an ordering of the cell division planes in callus (Lintilhac and Vesecky 1984), and to an altered developmental pattern, combined with changes in organ identity (Hernández and Green 1993). The laser removal of cells from *Arabidopsis* root meristem reorients the emerging division planes in remaining cells to fill in the empty space. Moreover, daughter cells are able to change their directions of development and differentiate according to their new positions in the root (van den Berg et al. 1995, 1997). These changes can be coupled with the remodeling of the structure and composition of the cell wall in order to reinforce and stabilize the mechanical message. This was first demonstrated in fucoid algae, where zygote differentiation into thallus and rhizoid cells depends on asymmetric division and the formation of cell-specific cell walls (Berger et al. 1994). Similarly, during zygotic embryogenesis in tobacco, the original zygotic cell wall is crucial for the maintenance of apical–basal polarity and for determining the fates of daughter cells (He et al. 2007).

4 Responses of Plants and Plant Cells to Mechanical Stimuli

In the classical view, for an organism to be able to respond to a given stimulus, it should be equipped with a complete signaling pathway that ends with the modulation of the activities of regulatory elements that affect the expression of stimulation-dependent genes. The activities of the gene products would eventually lead to changes at the cell level and at the organismal level. As mentioned above, sensing and transduction of mechanical stimuli are among the oldest evolutionary mechanisms that enable plant cells to respond to external cues. It should be noted, however, that although reactions to osmoticum, touch, and gravity are all responses to physical signals, they can be and are differentiated according to their "directionality." Touch stimuli arrive from the outside of the cell and are signaled into the cell. In contrast, the reaction to a gravitational stimulus is initiated through its sensing inside the cell. Finally, the reaction to osmotic changes is most probably bidirectional, as it involves sensing the stimulus at both the plasma membrane and the tonoplast.

4.1 Osmoregulation in Plant Cells

Water availability is crucial to the proper functioning of the plant cell, as a hypotonic environment causes an influx of water into the protoplast, causing it to swell, whereas hypertonic conditions draw the water out of the cell, decreasing turgor and inducing a plasmolytic response. Stresses such as drought and high salinity result in effects similar to those evoked by a hyperosmotic environment, leading to a loss of mechanical strength and a wilting of soft, nonlignified plant tissues (Boudsocq and Laurière 2005). Osmotic conditions are carefully sensed by all cells, and their changes induce active responses, mainly mechanisms regulating the cell volume (Zonia and Munnik 2007). In walled cells such as yeast, osmotic stress sensing depends on cell wall integrity (Hohmann 2002), and this is also postulated for plant cells (Marshall and Dumbroff 1999; Nakagawa and Sakurai 2001).

Sensing and signaling systems for osmotic conditions occur in all groups of organisms. Relatively little is known about osmosensors in plants (Grefen and Harter 2004). Plasma membrane protein AHK1 has been identified in *Arabidopsis* and was shown to be a homolog of yeast osmosensory two-component histidine kinase SLN1 (Urao et al. 1999). Its involvement in water stress responses in plants has been demonstrated (Wohlbach et al. 2008). There are eight genes coding for two-component histidine kinases in total in the *Arabidopsis* genome, some of which are also potential receptors for cytokinins and ethylene (Grefen and Harter 2004). Interestingly, one of the cytokinin receptors—CRE1—is also regulated by changes in turgor pressure (Reiser et al. 2003). A close homolog of another cytokinin receptor (AHK3 from *Medicago sativa*) has been shown to be transcriptionally activated in response to high salinity, which suggests that MsHK1 can also function as an osmosensor (Coba de la Peña et al. 2008). On the other hand, the overexpression of the transmembrane protein NtC7, which is most similar to RLKs, provides tobacco with a tolerance to the osmotic stress evoked by 500 mM mannitol. Interestingly, this tolerance appears to be stress-specific, as seeds were not able to germinate on media containing high salt concentrations (Tamura et al. 2003). Such stress-specific response phenomena have also been demonstrated in other osmotic signaling steps (Zonia and Munnik 2004). During the drought, the detection of the cell's turgor state forms part of a hydraulic signaling pathway that allows for a rapid stomatal response in cooperation with the abscisic acid signaling pathway that is activated in roots (Comstock 2002). The tight control over stomatal aperture size depends on osmotically induced rapid shrinking–swelling cycles of guard cells (Blatt 2000). These plasmolytic cycles also involve continuous membrane turnover (Shope et al. 2003; Meckel et al. 2005).

Changes in hydrostatic pressure across the plasma membrane generate stretch and compression forces that induce rapid responses in plant cells. The hydrodynamic condition of the plant cell and oscillations between different osmotic states have recently been postulated to affect cell shape, structure and growth as well as vesicle trafficking (Proseus et al. 2000; Shope et al. 2003; Meckel et al. 2005; Mathur 2006; Proseus and Boyer 2006a, b; Zonia et al. 2006). The cell walls and cytosol are highly anisotropic. Inside the cell, organelles and cytoskeleton are organized and distributed nonrandomly (e.g., Wojtaszek et al. 2005; Chuong et al. 2006). These features allow for a local response to the vector of mechanical force. The anisotropic tip growth of pollen tubes and root hairs is strictly controlled by the local weakening of cell walls and cortical cytoskeleton arrays (Mathur 2006). Following the appearance of the bulge, tip growth is still maintained due to the weaker cortical arrays at the tip than in the distal regions. Modulation of culture medium osmolality causes changes in apical volume, cell wall composition and expansion, and this affects pollen tube growth rates (Zonia et al. 2002, 2006; Zonia and Munnik 2004). The mechanical properties of cell walls can thus be tuned precisely, using either enzymatic or nonenzymatic mechanisms, to withstand dynamic changes in extra- and intracellular pressure.

4.2 Reactions to Touch

All plants sense and respond to mechanical perturbations in their environment, such as wind, rain, snow and sound waves, as well as to contact with other organisms or elements of the physical environment, like soil. These reactions are collectively termed touch responses, and are usually divided into thigmotropic or thigmonastic reactions, depending on the influence of the stimulus vector on the direction of movement. The former usually occur in the direction determined by the arriving stimulus, while nastic movements are largely independent of the direction of the stimulus. Touch responses can be extremely quick, as in carnivorous plants or *Mimosa pudica*, or very slow, eventually resulting in changes to the morphology of the plant in a process called thigmomorphogenesis (Braam 2005; Esmon et al. 2005; Telewski 2006). An interesting example is the growth of roots in the soil, as it combines responses to both touch and gravity (Fasano et al. 2002; see also below). Under normal conditions, plant roots grow along the gravitational vector. However, when a root approaches an obstacle, it seems that gravitropic behavior is compromised and touch responses take place (Okada and Shimura 1990; Massa and Gilroy 2003).

Although responses of plants to mechanical stimuli are usually observed at an organismal level, there are changes at the cellular and subcellular levels that are crucial to the selection and modulation of those responses. It has been demonstrated in yeast that cell walls exhibit local temperature-dependent nanomechanical motion with an amplitude of ca. 3 nm (Pelling et al. 2004). If the situation is similar in plant cells, this may suggest that touching such an oscillator will immediately induce not only a slight perturbation of the surface of the wall but also changes in either the frequency or amplitude of the wall's oscillations. Thus, even a very small stimulus could be recognized and transduced into a cellular response. This response can be further amplified by the activities of cellular machinery and maintained over time, giving rise to all kinds of responses. At the cellular level, touching the cell surface induces very rapid changes in both cellular metabolism and intracellular organization, like chloroplast movement (Sato et al. 2003) or nuclear and cytoplasmic migration towards the contact site (Hardham et al. 2008). The cell returns to it previous state as soon as stimulus is removed. Examples include the reactions of plant cells to physical forces exerted by fungal or oomycete pathogens infecting plant epidermal cells. In many cases, fungi use mechanical force to break through the physical barriers of plant cell walls, and these attempts can be detected in a mechanosensitive way (Gus-Mayer et al. 1998). Such reactions can also be induced experimentally, by applying gentle pressure to the epidermal cell surface using a microneedle. Interestingly, the changed cell morphology tracks the needle tip as it moves along the plant cell surface (Hardham et al. 2008).

Several genes that are upregulated in response to touch stimulation (TCH) have been identified and characterized (Braam and Davis 1990). Interestingly, the expression of *TCH* genes is also regulated in response to other environmental stimuli (reviewed by Braam et al. 1997), and at least some of them also seem to be under the phytohormonal control of, e.g., auxin and brassinosteroids (Antosiewicz et al. 1995; Xu et al. 1995). Touch stimulation leads to the rapid and transient elevation of $[Ca^{2+}]_{cyt}$ in plants (Knight et al. 1991), while the exogenous addition of Ca^{2+} to suspension-cultured cells upregulates the expression of *TCH* genes (Braam 1992). These findings strongly support the idea that Ca^{2+} acts as a second messenger in touch responses (Braam et al. 1997), and probably also as a stimulus-specific signal that allows touch and gravitational stimulation to be discriminated (Legué et al. 1997). Thus, it is not a surprise to discover that three out of four of the initially identified *TCH* genes are in fact calcium-binding proteins. TCH1 is a plant calmodulin, while TCH2 and TCH3 belong to a family of calmodulin-like proteins that are also able to bind Ca^{2+}, but their exact role is unknown (Braam et al. 1997; McCormack and Braam 2003). An interesting suggestion derives from the finding that TCH3 interacts with PINOID—a serine/threonine kinase involved in auxin signaling—to regulate its activity in response to changes in calcium levels (Benjamins et al. 2003). Finally, the product of *TCH4* is xyloglucan endotransglycosylase/hydrolase (XTH), one of the major wall-modifying enzymes. The *TCH4* expression pattern is also touch- and Ca^{2+}-dependent, and changes in localization are also observed (Xu et al. 1995; Antosiewicz et al. 1997).

In the years following the description of the *TCH* genes, many other genes were found to be induced by touch. Genome-wide analysis of expression patterns in touch-stimulated *Arabidopsis* plants revealed that expression of 589 genes was upregulated within 30 min of touch stimulation, while 171 genes were downregulated (Lee et al. 2005). As expected, a relatively high proportion of the upregulated genes coded for proteins involved in cellular calcium binding as well as cell wall synthesis and modification. Interestingly, among seven genes coding for calmodulins, only *TCH1* was upregulated by touch stimulus. Importantly, genes implicated in disease resistance formed the third biggest functional group of upregulated genes (Lee et al. 2005).

4.3 Responses to Gravity

As mentioned above, gravity is a major relatively constant physical force on Earth and is thus considered to be one of the major driving forces in evolution (Volkmann and Baluška 2006). At the organismal level, gravity is the most important integratory physical factor, and it is also a source of mechanical stress that must be accommodated (Kern et al. 2005). Gravity affects plant body architecture via two mechanisms: gravitropism and gravity resistance. Gravitropism is the orientation of the growth of plant organs along (e.g., roots) or against (e.g., shoots) the gravitational vector (Blancaflor and Masson 2003). On the other hand, gravity resistance comprises there are also a set of mechanisms that allow plants to support their own weight, e.g., by strengthening their cell walls (Ko et al. 2004; Hoson et al. 2005). Graviperception is the first step in a series of events leading to various graviresponses. Its major element is a translation of an internal mechanical stimulus, usually caused by the displacement of some mass, into biophysical and biochemical signals (Perbal and Driss-Ecole 2003). Although graviperception in plants is now understood in quite some detail, the precise mechanisms involved are still a matter of debate. It seems also that mechanisms of graviperception utilized in gravitropism and in resistance to gravity are at least partially different (Hoson et al. 2005).

Different cells are specialized in order to detect the gravitational vector in gravitropism. In roots, these cells are statocytes, which are located in the root-cap columella; in hypocotyls, these are dedicated cells within the endodermal cell layer; in shoots, they are cells within the bundle sheath parenchyma (Blancaflor et al. 1998; Fukaki et al. 1998). Some data indicate, however, that the orientation of the gravitational vector can also be perceived outside of those regions (Wolverton et al. 2002). Following transduction, the perceived gravitational signal is transferred from the statocytes to the responding tissues (Perbal and Driss-Ecole 2003; Perrin et al. 2005). Statocytes are polarized cells containing starch-filled amyloplasts (statoliths; Driss-Ecole et al. 2003). Because they are quite different in density from the cytoplasm, statocytes are able to sediment. This notion gave rise to the commonly accepted explanation of gravity sensing: the starch–statolith hypothesis, according to which the sedimentation of statoliths provides the vectorial information required to orient the direction of organ growth (reviewed by, e.g., Sack 1997; Blancaflor and Masson 2003). Observations of the gravitropic responses of starch-deficient mutants and starch-overproducing mutants as well as experiments utilizing high-gradient magnetic fields generally support this model (Kuznetsov and Hasenstein

1997; Kiss et al. 1997; Vitha et al. 2007). However, some data indicate that starch-deficient mutants still exhibit some degree of gravitropic response (Caspar and Pickard 1989).

The question how the displacement of starch-filled amyloplasts is sensed in statocytes is still debatable. One possibility is that statoliths act as ligands that activate receptors located in the cellular membrane system (Braun 2002; Limbach et al. 2005). However, not all of the experimental data fit into such a model (Wendt et al. 1987). The sensing of statolith movement by MS ion channels is another possibility (Yoder et al. 2001; Pickard 2007). Over the last two decades, various MS ion channel activities have been identified in plant membranes (see above). It has been shown that gravitational stimulation of roots is correlated with the rapid alkalinization of the cytosol and the transient influx of Ca^{2+} into protoplasts (Fasano et al. 2001; Plieth and Trewavas 2002). The question of how statolith movement activates the MS channels remains, however. At the moment it appears that the tensegral concept of cellular organization provides the answer, and that the mechanical signal is sensed within the WMC continuum (Blancaflor 2002; Baluška et al. 2003). The statoliths' trajectories indicated that they usually move along cellular channels located at the interface between the ER-less central region and the ER-dense cortical region of columella cells. These regions are pervaded by the prestressed actin network, which is denser in the ER-less region. Statolith movement can then disturb the mechanical balance of the cytoskeleton, and this (through the connection to the plasma membrane) can activate the MS ion channels (Yoder et al. 2001). In accordance with this, pharmacological disruption of the microfilaments affects the distribution and sedimentation of amyloplasts (Baluška and Hasenstein 1997; Palmieri and Kiss 2005). At the same time, such disruption does not usually abolish gravitropic response (Staves et al. 1997; Yamamoto and Kiss 2002; Hou et al. 2004). This may indicate that other cytoskeletal components are also important, and a role for microtubules has indeed already been suggested (Himmelspach et al. 1999). It is also important to note that the sedimentation of statoliths is probably not a free, passive precipitation, as their positions are precisely controlled by the actomyosin system (Braun et al. 2002; Wojtaszek et al. 2005). Finally, the precise spatial organization of the actin filaments and the way that they are anchored to the walls via polysaccharides and proteins are also important for gravisensing (Wayne et al. 1992; Wojtaszek et al. 2005, 2007).

Another hypothesis for gravisensing has been proposed by Staves (1997). The hydrostatic pressure model postulates that what is sensed is not the disturbance in the balance of forces within intracellular structures, but rather the difference in the tension/compression forces exerted by the entire protoplast between the apical and basal sites of attachment to the walls in axial organs (Staves et al. 1992). The tension exerted by an entire protoplast could also locally activate MS ion channels (Pickard 2007), triggering the transduction of the gravitational signal. It has also been proposed that such differences can result in a shift in the positions of symplastic domains that secrete auxin, resulting in the accumulation of auxin at the bottoms of the cells (Friml et al. 2002). Such a shift could result from differential membrane trafficking in domains subjected to variable tensile forces (Morris and Homann 2001). Finally, the possibility that several gravisensing mechanisms operate together cannot be excluded (Barlow 1995; LaMotte and Pickard 2004).

In contrast to gravitropism, gravity resistance can occur in virtually all cells, so there is probably no signal transmission between perceiving and responding cells (Hoson et al. 2005). In this case, gravity produces tensile and compressive forces in some regions of the plant body. The gravisensing that occurs in resistance to gravity is independent of statolith sedimentation, since mutants that have abolished gravitropism and lack sedimentable amyloplasts still exhibit full gravity resistance reactions (Tasaka et al. 2001). Also, the removal of the root cap does not influence gravity resistance (Soga et al. 2005a). On the other hand, MS ion channels have been shown to be a crucial element here (Soga et al. 2002, 2005b), as has the composition of the cellular membranes, with sterols being particularly important (Koizumi et al. 2007). Moreover, upregulation of tubulin gene expression is involved in gravity-induced modification of microtubule dynamics, which may play an important role in the resistance of plant organs to gravity (Soga et al. 2006; Matsumoto et al. 2007). However, further elaboration of the molecular mechanisms of gravity resistance is strongly needed.

References

Antosiewicz D, Polisensky D, Braam J (1995) Cellular localization of the Ca^{2+} binding CH3 protein of *Arabidopsis*. Plant J 8:623–636

Antosiewicz D, Purugganan M, Polisensky D, Braam J (1997) Cellular localization of *Arabidopsis* xyloglucan endotransglycosylase-related proteins during development and after wind stimulation. Plant Physiol 115:1319–1328

Aon MA, Cortassa S, Gomez Casati DF, Iglesias AA (2000) Effects of stress on cellular infrastructure and metabolic organization in plant cells. Int Rev Cytol 194:239–273

Arnaout MA, Goodman SL, Xiong J-P (2007) Structure and mechanics of integrin-based cell adhesion. Curr Opin Cell Biol 19:495–507

Assoian RK, Klein EA (2008) Growth control by intracellular tension and extracellular stiffness. Trands Cell Biol 18:347–352

Baluška F, Hasenstein KH (1997) Root cytoskeleton: its role in perception of and response to gravity. Planta 203:S69–S78

Baluška F, Hlavačka A (2005) Plant formins come of age: something special about cross-walls. New Phytol 168:499–503

Baluška F, Salaj J, Mathur J, Braun M, Jasper F, Šamaj J, Chua N-H, Barlow PW, Volkmann D (2000) Root hair formation: F-actin-dependent tip growth is initiated by local assembly of profilin-supported F-actin meshworks accumulated within expansin-enriched bulges. Dev Biol 227:618–632

Baluška F, Jasik J, Edelmann HG, Salajova T, Volkmann D (2001) Latrunculin B-induced plant dwarfism: plant cell elongation is F-actin-dependent. Dev Biol 231:113–124

Baluška F, Šamaj J, Wojtaszek P, Volkmann D, Menzel D (2003) Cytoskeleton–plasma membrane–cell wall continuum in plants. Emerging links revisited. Plant Physiol 133:482–491

Barlow P (1995) Gravity perception in plants—a multiplicity of systems derived by evolution. Plant Cell Environ 18:951–962

Barthou H, Petitprez M, Brière C, Souvré A, Alibert G (1998) RGD-mediated membrane-matrix adhesion triggers agarose-induced embryoid formation in sunflower protoplasts. Protoplasma 206:143–151

Benjamins R, Ampudia C, Hooykaas P, Offringa R (2003) PINOID-mediated signaling involves calcium-binding proteins. Plant Physiol 132:1623–1630

Berger F, Taylor A, Brownlee C (1994) Cell fate determination by the cell wall in early *Fucus* development. Science 263:1421–1423

Bhalerao RP, Bennett MJ (2003) The case for morphogens in plants. Nat Cell Biol 5:939–943

Blancaflor EB (2002) The cytoskeleton and gravitropism in higher plants. J Plant Growth Regul 21:120–136

Blancaflor EB, Masson PH (2003) Plant gravitropism. Unraveling the up and downs of a complex process. Plant Physiol 133:1677–1690

Blancaflor EB, Fasano JM, Gilroy S (1998) Mapping the functional roles of cap cells in the response of *Arabidopsis* primary roots to gravity. Plant Physiol 116:213–222

Blatt MR (2000) Cellular signaling and volume control in stomatal movements in plants. Annu Rev Cell Dev Biol 16:221–241

Boudsocq M, Laurière C (2005) Osmotic signalling in plants. Multiple pathways mediated by emerging kinase families. Plant Physiol 138:1185–1194

Braam J (1992) Regulated expression of the calmodulin-related TCH genes in cultured *Arabidopsis* cells: Induction by calcium and heat shock. Proc Natl Acad Sci USA 89:3213–3216

Braam J(2005)In touch: plant responses to mechanical stimuli. New Phytol 165:373–389

Braam J, Davis R(1990)Rain-, wind-, and touch-induced expression of calmodulin and calmodulin-related genes in *Arabidopsis*. Cell 60:357–364

Braam J, Sistrunk M, Polisensky D, Xu W, Purugganan M, Antosiewicz D, Campbell P, Johnson K (1997) Plant responses to environmental stress: regulation and functions of the *Arabidopsis* TCH genes. Planta 203:S35–S41

Braun M (2002) Gravity perception requires statoliths settles on specific plasma membrane areas in characean rhisoids and protonemata. Protoplasma 219:150–159

Braun M, Buchen B, Sievers A (2002) Actomyosin-mediated statolith positioning in gravisensing plant cells studied in microgravity. J Plant Growth Regul 21:137–145

Canut H, Carrasco A, Galaud JP, Cassan C, Bouyssou H, Vita N, Ferrara P, Pont-Lezica R (1998) High affinity RGD-binding sites at the plasma membrane of *Arabidopsis thaliana* links the cell wall. Plant J 16:63–71

Caspar T, Pickard BG (1989) Gravitropism in starchless mutant of *Arabidopsis*—implication for the starch-statolith theory of gravity sensing. Planta 177:185–197

Chuong SDX, Franceschi VR, Edwards GE (2006) The cytoskeleton maintains organelle partitioning required for single-cell C_4 photosynthesis in Chenopodiaceae species. Plant Cell 18:2207–2223

Chytilova E, Macas J, Śliwinska E, Rafelski SM, Lambert GM, Galbraith DW (2000) Nuclear dynamics in *Arabidopsis thaliana*. Mol Biol Cell 11:2733–2741

Cleary AL (2001) Plasma membrane-cell wall connections: roles in mitosis and cytokinesis revealed by plasmolysis of *Tradescantia virginiana* leaf epidermal cells. Protoplasma 215:21–34

Coba de la Peña T, Cárcamo CB, Almonacid L, Zaballos A, Lucas MM, Balomenos D, Pueyo JJ (2008) A salt stress-responsive cytokinin receptor homologue isolated from *Medicago sativa* nodules. Planta 227:769–779

Comstock JP (2002) Hydraulic and chemical signalling in the control of stomatal conductance and transpiration. J Exp Bot 53:195–200

Decreux A, Messiaen J (2005) Wall-associated kinase WAK1 interacts with cell wall pectins in a calcium-induced conformation. Plant Cell Physiol 46:268–278

Deeks MJ, Hussey PJ, Davies B (2002) Formins: intermediates in signal-transduction cascades that affect cytoskeletal reorganization. Trends Plant Sci 7:492–498

Driss-Ecole D, Lefranc A, Perbal G (2003) A polarized cell: root statocyte. Physiol Plant 118:305–312

Dutta R, Robinson KR (2004) Identification and characterization of stretch-activated ion channels in pollen protoplasts. Plant Physiol 135:1398–1406

Emons AMC, Mulder BM (2000) How the deposition of cellulose microfibrils builds cell wall architecture. Trends Plant Sci 5:35–40

Engler AJ, Sen S, Sweeney HL, Discher DE (2006) Matrix elasticity directs stem cell lineage specification. Cell 126:677–689

Esmon C, Pedmale U, Liscum E (2005) Plant tropisms: providing the power of movement to a sessile organism. Int J Dev Biol 49:665–674

Fasano JM, Swanson SJ, Blancaflor EB, Dowd PE, Kao TH, Gilroy S (2001) Changes in root cap pH are required for gravity response of the *Arabidopsis* root. Plant Cell 13:907–921

Fasano JM, Massa GD, Gilroy S (2002) Ionic signaling in plant responses to gravity and touch. J Plant Growth Regul 21:71–88

Fisher DD, Cyr RJ (1998) Extending the microtubule/microfibril paradigm. Cellulose synthesis is required for normal cortical microtubule alignment in elongating cells. Plant Physiol 116:1043–1051

Forgacs G (1995) On the possible role of cytoskeletal filamentous networks in intracellular signaling: an approach based on percolation. J Cell Sci 108:2131–2143

Friml J, Wiśniewska J, Benkova G, Mendgen K, Palme K (2002) Lateral relocation of auxin efflux regulator PIN3 mediates tropism in *Arabidopsis*. Nature 415:806–809

Fukaki H, Wysocka-Diller J, Kato T, Fujisawa H, Benfey PN, Tasaka M (1998) Genetic evidence that the endodermis is essential for shoot gravitropism in *Arabidopsis thaliana*. Plant J 14:425–430

Gens JS, Fujiki M, Pickard BG (2000) Arabinogalactan protein and wall-associated kinase in a plasmalemmal reticulum with specialized vertices. Protoplasma 212:115–134

Gouget A, Senchou V, Govers F, Sanson A, Barre A, Rouge P, Pont-Lezica R, Canut H (2006) Lectin receptor kinases participate in protein–protein interactions to mediate plasma membrane–cell wall adhesions in *Arabidopsis*. Plant Physiol 140:81–90

Green PB (1999) Expression of pattern in plants: combining molecular and calculus-based biophysical paradigms. Am J Bot 86:1059–1076

Grefen C, Harter K (2004) Plant two-component systems: principles, functions, complexity and cross talk. Planta 219:733–742

Gus-Mayer S, Naton B, Hahlbrock K, Schmelzer E (1998) Local mechanical stimulation induces components of the pathogen defense response in parsley. Proc Natl Acad Sci USA 95:8398–8403

Hardham AR, Takemoto D, White RG (2008) Rapid and dynamic subcellular reorganization following mechanical stimulation of *Arabidopsis* epidermal cells mimics responses to fungal and oomycete attack. BMC Plant Biol 8:63

Harold FM (1995) From morphogenes to morphogenesis. Microbiology 141:2765–2778.

Haswell ES, Meyerowitz E (2006) MscS-like proteins control plastid size and shape in *Arabidopsis thaliana*. Curr Biol 16:1–11

Haswell ES, Peyronnet R, Barbier-Brygoo H, Meyerowitz E, Frachisse J-M (2008) Two MscS homologs provide mechanosensitive channel activities in the *Arabidopsis* root. Curr Biol 18:730–734

Hayashi T, Takagi S (2003) Ca^{2+}-dependent cessation of cytoplasmic streaming induced by hypertonic treatment in *Vallisneria* mesophyll cells: possible role of cell wall–plasma membrane adhesion. Plant Cell Physiol 44:1027–1036

Hayashi T, Harada A, Sakai T, Takagi S (2006) Ca^{2+} transient induced by extracellular changes in osmotic pressure in *Arabidopsis* leaves: differential involvement of cell wall–plasma membrane adhesion. Plant Cell Environ 29:661–672

He YC, He YQ, Qu LH, Sun MX, Yang HY (2007) Tobacco zygotic embryogenesis in vitro: the original cell wall of the zygote is essential for maintenance of cell polarity, the apical–basal axis and typical suspensor formation. Plant J 49:515–527

Heinlein M, Epel BL (2004) Macromolecular transport and signaling through plasmodesmata. Int Rev Cytol 235:93–164

Hernández LF, Green PB (1993) Transductions for the expression of structural pattern: analysis in sunflower. Plant Cell 5:1725–1738

Himmelspach R, Wymer CL, Lloyd CW, Nick P (1999) Gravity-induced reorientation of cortical microtubules observed in vivo. Plant J 18:449–453

Hohmann S (2002) Osmotic stress signaling and osmoadaptation in yeasts. Microbiol Mol Biol Rev 66:300–372

Homann U, Meckel T, Hewing J, Hütt M-T, Hurst AC (2007) Distinct fluorescent pattern of KAT1::GFP in the plasma membrane of *Vicia faba* guard cells. Eur J Cell Biol 86:489–500

Hoson T, Saito Y, Soga K, Wakabayashi K (2005) Signal perception, transduction, and response in gravity resistance. Another graviresponse in plants. Adv Space Res 36:1196–1202

Hou G, Kramer VL, Wang YS, Chen R, Perbal G, Gilroy S, Blancaflor EB (2004) The promotion of gravitropism in *Arabidopsis* roots upon disruption is coupled with the extended alkalization of the columella cytoplasm and a persistent lateral auxin gradient. Plant J 39:113–125

Hussey P, Allwood E, Smertenko A (2002) Actin-binding proteins in the *Arabidopsis* genome database: properties of functionally distinct plant actin-depolymerizing factors/cofilins. Philos Trans R Soc Lond B Biol Sci 357:791–798

Ingber DE (2003a) Tensegrity I. Cell structure and hierarchical systems biology. J Cell Sci 116:1157–1173

Ingber DE (2003b) Tensegrity II. How structural network influence cellular information processing network. J Cell Sci 116:1397–1408

Kaplan DR, Cooke TJ (1997) Fundamental concepts in the embryogenesis of dicotyledons: a morphological interpretation of embryo mutants. Plant Cell 9:1903–1919

Kern VD, Schwuchow JM, Reed DW, Nadeau JA, Lucas J, Skripnikov A, Sack FD (2005) Gravitropic moss cells default to spiral growth on the clinostat and in microgravity during spaceflight. Planta 221:149–157

Kiss JZ, Guisinger M, Miller A, Stackhouse K (1997) Reduced gravitropism in hypocotyls of starch-deficient mutants of *Arabidopsis*. Plant Cell Physiol 38:518–525

Knight M, Campbell A, Smith S, Trewavas A (1991) Transgenic plant aequorin reports the effects of touch and cold-shock and elicitors on cytoplasmic calcium. Nature 352:524–526

Ko JH, Han KH, Park S, Yang J (2004) Plant body weight-induced secondary growth in *Arabidopsis* and its transcription phenotype revealed by whole-transcriptome profiling. Plant Physiol 135:1069–1083

Kohorn BD (2000) Plasma membrane–cell wall contacts. Plant Physiol 124:31–38

Kohorn BD (2001) WAKs: cell wall associated kinases. Curr Opin Cell Biol 13:529–533

Koizumi T, Sakaki T, Usui S, Soga K, Wakabayashi K, Hoson T (2007) Changes in a membrane lipid composition in adzuki bean epicotyls under hypergravity conditions: possible role of membrane sterols in gravity resistance. Adv Space Res 39:1198–1203

Komis G, Apostolakos P, Galatis B (2003) Actomyosin is involved in the plasmolytic cycle: gliding movement of the deplasmolyzing protoplast. Protoplasma 221:245–256

Kung C (2005) A possible unifying principle of mechanosensation. Nature 436:647–654

Kutschera U (2008) The growing outer epidermal wall: design and physiological role of a composite structure. Ann Bot 101:615–621

Kuznetsov OA, Hasenstein KH (1997) Magnetophoretic induction of curvature in coleoptiles and hypocotyls. J Exp Bot 48:1951–1957

LaMotte CE, Pickard BG (2004) Control of gravitropic orientation. II. Dual receptor model for gravitropism. Funct Plant Biol 31:109–120

Lang I, Barton DA, Overall RL (2004) Membrane-wall attachments in plasmolysed plant cells. Protoplasma 224:231–243

Lang-Pauluzzi I, Gunning BES (2000) A plasmolytic cycle: the fate of cytoskeletal elements. Protoplasma 212:174–185

Lee D, Polisensky D, Braam J (2005) Genome-wide identification of touch- and darkness- regulated *Arabidopsis* gene: a focus on calmodulin-like and XTH genes. New Phytol 165:429–444

Legué V, Blancaflor E, Wymer C, Perbal G, Fantin D, Gilroy S (1997) Cytoplasmic free Ca^{2+} in *Arabidopsis* roots changes in response to touch but not gravity. Plant Physiol 114:789–800

Limbach C, Hauslage J, Schöfer C, Braun M (2005) How to activate a plant gravireceptor. Early mechanisms of gravity sensing studied in Characean rhizoids during parabolic flights. Plant Physiol 139:1030–1040

Lintilhac PM, Vesecky TB (1984). Stress-induced alignment of division plane in plant tissues grown in vitro. Nature 307:363–364

Lucas WJ, Ding B, van der Schoot C (1993) Plasmodesmata and the supracellular nature of plants. New Phytol 125:435–476

Lynch TM, Lintilhac PM (1997) Mechanical signals in plant development: a new method for single cell studies. Dev Biol 181:246–256

Maniotis AJ, Chen CS, Ingber DE (1997) Demonstration of mechanical connections between integrins, cytoskeletal filaments, and nucleoplasm that stabilize nuclear structure. Proc Natl Acad Sci USA 94:849–854

Maniotis AJ, Valyi-Nagy K, Karavitis J, Moses J, Boddipali V, Wang Y, Nuñez R, Setty S, Arbieva Z, Bissell MJ, Folberg R (2005) Chromatin organization measured by *Alu*I restriction enzyme changes with malignancy and is regulated by the extracellular matrix and the cytoskeleton. Am J Pathol 166:1187–1203

Marshall JG, Dumbroff EB (1999) Turgor regulation via cell wall adjustment in white spruce. Plant Physiol 119:313–319

Martinac B (2004) Mechanosensitive ion channels: molecules of mechanotransduction. J Cell Sci 117:2449–2460

Massa GD, Gilroy S (2003) Touch modulates gravity sensing to regulate the growth of primary roots of *Arabidopsis thaliana*. Plant J 33:435–445

Mathur J (2006) Local interactions shape plant cells. Curr Opin Cell Biol 18:40–46

Matsumoto S, Saito Y, Kamasaki S, Soga K, Wakabayashi K, Hoson T (2007) Up-regulation of expression of tubulin genes and roles of microtubules in hypergravity-induced growth modification in *Arabidopsis* hypocotyls. Adv Space Res 39:1179–1181

McCormack E, Braam J (2003) Calmodulins and related potential calcium sensors of *Arabidopsis*. New Phytol 159:585–598

Meckel T, Hurst AC, Thiel G, Homann U (2005) Guard cells undergo constitutive and pressure driven membrane turnover. Protoplasma 226:23–29

Mellersh DG, Heath MC (2001) Plasma membrane–cell wall adhesion is required for expression of plant defense responses during fungal penetration. Plant Cell 13:413–424

Meyer Y, Abel WO (1975) Importance of the wall for cell division and in the activity of the cytoplasm in cultured tobacco protoplasts. Planta 123:33–40

Morris CE, Homann U (2001) Cell surface area regulation and membrane tension. J Membrane Biol 179:79–102

Na S, Collin O, Chowdhury F, Tay B, Ouyang M, Wang Y, Wang N (2008) Rapid signal transduction in living cells is a unique feature of mechanotransduction. Proc Natl Acad Sci USA 105:6626–6631

Nakagawa N, Sakurai N (2001) Cell wall integrity controls expression of endoxyloglucan transferase in tobacco BY2 cells. Plant Cell Physiol 42:240–244

Nakagawa Y, Katagiri T, Shinozaki K, Qi Z, Tatsumi H, Furuichi T, Kishigami A, Sokabe M, Kojima I, Sato S, Kato T, Tabata S, Lida K, Terashima A, Nakano M, Ikeda M, Yamanaka T, Lida H (2007) *Arabidopsis* plasma membrane protein crucial for Ca^{2+} influx and touch sensing in roots. Proc Natl Acad Sci USA 104:3639–3644

Nelson CM, Jean RP, Tan JL, Liu WF, Sniadecki NJ, Spector AA, Chen CS (2005) Emergent patterns of growth controlled by multicellular form and mechanics. Proc Natl Acad Sci USA 102:11594–11599

Niklas KJ (1992) Plant biomechanics: an engineering approach to plant form and function. University of Chicago Press, Chicago

Okada K, Shimura Y (1990) Reversible root-tip rotation in *Arabidopsis* seedlings induced by obstacle-touching stimulus. Science 250:274–276

Orr AW, Helmke BP, Blackman BR, Schwartz MA (2006) Mechanisms of mechanotransduction. Dev Cell 10:11–20

Palmieri M, Kiss JZ (2005) Disruption of F-actin cytoskeleton limits statolith movement in *Arabidopsis* hypocotyls. J Exp Bot 56:2539–2550

Panteris E, Galatis B (2005) The morphogenesis of lobed plant cells in the mesophyll and epidermis: organization and distinct roles of cortical microtubules and actin filaments. New Phytol 167:721–732

Paredez AR, Somerville CR, Ehrhardt DW (2006) Visualization of cellulose synthase demonstrates functional association with microtubules. Science 312:1491–1495
Paredez AR, Persson S, Ehrhardt DW, Somerville CR (2008) Genetic evidence that cellulose synthase activity influences microtubule cortical array organization. Plant Physiol 147(4):1723–1734
Pelling AE, Sehati S, Gralla EB, Valentine JS, Gimzewski JK (2004) Local nanomechanical motion of the cell wall of *Saccharomyces cerevisiae*. Science 305:1147–1150
Perbal G, Driss-Ecole D (2003) Mechanotransduction in gravisensing cells. Trends Plant Sci 8:498–504
Perrin R, Young L, Murthy N, Harrison B, Wang Y, Will A, Masson PH (2005) Gravity signal transduction in primary roots. Ann Bot 96:737–743
Peters WS, Hagemann W, Tomos DA (2000) What makes plants different? Principles of extracellular matrix function in "soft" plant tissues. Comp Biochem Physiol A 125:151–167
Pickard BG (2007) Delivering force and amplifying signals in plant mechanosensing. In: Mechanosensitive ion channels, Part A. Elsevier, New York, pp 361–392
Plieth C,Trewavas AJ (2002) Reorientation of seedlings in the Earth's gravitational field induces cytosolic calcium transiently. Plant Physiol 129:789–796
Proseus TE, Boyer JS (2006a) Identifying cytoplasmic input to the cell wall of growing *Chara corallina*. J Exp Bot 57:3231–3242
Proseus TE, Boyer JS (2006b) Periplasm turgor pressure controls wall deposition and assembly in growing *Chara corallina* cells. Ann Bot 98:93–105
Proseus TE, Zhu GL, Boyer JS (2000) Turgor, temperature and the growth of plant cells: using *Chara corallina* as a model system. J Exp Bot 51:1481–1494
Qu L-H, Sun M-X (2007) The plant cell nucleus is constantly alert and highly sensitive to repetitive local mechanical stimulation. Plant Cell Rep 26:1187–1193
Reiser V, Raitt DC, Saito H (2003) Yeast osmosensor Sln1 and plant cytokinin receptor Cre1 respond to changes in turgor pressure. J Cell Biol 161:1035–1040
Roberts SK (2006) Plasma membrane anion channels in higher plants and their putative functions in roots. New Phytol 169:647–666
Sack FD (1997) Plastids and gravity sensing. Planta 203:S63–S68
Sato Y, Kadota A, Wada M (2003) Chloroplast movement: dissection of events downstream of photo- and mechano-perception. J Plant Res 116:1–5
Schindler M, Meiners S, Cheresh DA(1989)RGD-dependent linkage between plant cell wall and plasma membrane: Consequences for growth. J Cell Biol 108:1955–1965
Schmidt SM, Panstruga R (2007) Cytoskeleton functions in plant-microbe interactions. Physiol Mol Plant Pathol 71:135–148
Shiu SH, Bleecker AB (2001) Receptor-like kinases from *Arabidopsis* form a monophyletic gene family related to animal receptor kinases. Proc Natl Acad Sci USA 98:10763–10768
Shope JC, DeWald DB, Mott KA (2003) Changes in surface area of intact guard cells are correlated with membrane internalization. Plant Physiol 133:1314–1321
Soga K, Wakabayashi K, Kamisaka S, Hoson T (2002) Perception of gravity stimuli by mechanosensitive ion channels in plant seedlings. Plant Cell Physiol 43:182
Soga K, Wakabayashi K, Kamisaka S, Hoson T (2005a) Mechanoreceptors rather than sedimentable amyloplasts perceive the gravity signal in hypergravity-induced inhibition of root growth in adzuki bean. Funct Plant Biol 32:175–179
Soga K, Wakabayashi K, Kamisaka S, Hoson T (2005b) Hypergravity inhibits elongation growth of adzuki bean epicotyls independently of the direction of stimuli. Adv Space Res 36:1269–1279
Soga K, Wakabayashi K, Kamisaka S, Hoson T (2006) Hypergravity induces reorientation of cortical microtubules and modifies growth anisotropy in adzuki bean epicotyls. Planta 224:1485–1494
Staves MP (1997) Cytoplasmic streaming and gravity sensing in Chara internodal cells. Planta 203(Suppl 1):S79–S84
Staves MP, Wayne R, Leopold AC (1992) Hydrostatic pressure mimics gravitational pressure in characean cells. Protoplasma 168:141–152

Staves MP, Wayne R, Leopold AC (1997) Cytochalasin D does not inhibit gravitropism in roots. Am J Bot 84:1530–1535
Tamura T, Hara K, Yamaguchi Y, Koizumi N, Sano H (2003) Osmotic stress tolerance of transgenic tobacco expressing a gene encoding a membrane-located receptor-like protein from tobacco plants. Plant Physiol 131:454–462
Tasaka M, Kato T, Fukaki H (2001) Genetic regulation of gravitropism in higher plants. Int Rev Cytol 206:135–154
Telewski F (2006) A unified hypothesis of mechanoperception in plants. Am J Bot 93:1466–1476
Urao T, Yakubov B, Satoh R, Yamaguchi-Shinozaki K, Seki M, Hirayama T, Shinozaki K (1999) A transmembrane hybrid-type histidine kinase in *Arabidopsis* functions as an osmosensor. Plant Cell 11:1743–1754
Valladares F, Gianoli E, Gómez JM (2007) Ecological limits to plant phenotypic plasticity. New Phytol 176:749–763
Valle F, Sandal M, Samorì B (2007) The interplay between chemistry and mechanics in the transduction of a mechanical signal into a biochemical function. Physics Life Rev 4:157–188
van den Berg C, Willemsen V, Hage W, Weisbeek P, Scheres B (1995) Cell fate in the *Arabidopsis* root meristem determined by directional signalling. Nature 378:62–65
van den Berg C, Willemsen V, Hendriks G, Weisbeek P, Scheres B (1997) Short-range control of cell differentiation in the *Arabidopsis* root meristem. Nature 390:287–289
van der Honing HS, Emons AMC, Ketelaar T (2007) Actin based processes that could determine the cytoplasmic architecture of plant cells. Biochim Biophys Acta 1773:604–614
Vitha S, Yang M, Sack FD, Kiss JZ (2007) Gravitropism in the starch excess mutant of *Arabidopsis thaliana*. Am J Bot 94:590–598
Vogel V, Sheetz M (2006) Local force and geometry sensing regulate cell functions. Nat Rev Mol Cell Biol 7:265–275
Volkmann D, Baluška F (2006) Gravity: one of the driving forces in evolution. Protoplasma 229:143–148
Wayne R, Staves MP, Leopold AC (1992) The contribution of the extracellular matrix to gravisensing in characean cells. J Cell Sci 101:611–623
Wendt M, Kuo-Huang L, Sievers A (1987) Gravitropic bending of cress roots without contact between amyloplasts and complexes of endoplasmic reticulum. Planta 172:321–329
Wightman R, Turner SR (2008) The roles of the cytoskeleton during cellulose deposition at the secondary cell wall. Plant J 54:794–805
Wohlbach DJ, Quirino BF, Sussman MR (2008) Analysis of the *Arabidopsis* histidine kinase ATHK1 reveals a connection between vegetative osmotic stress sensing and seed maturation. Plant Cell 20:1101–1117
Wojtaszek P (2000) Genes and plant cell walls: a difficult relationship. Biol Rev 75:437–475
Wojtaszek P (2001) Organismal view of a plant and a plant cell. Acta Biochim Polon 48:443–451
Wojtaszek P, Anielska-Mazur A, Gabry H, Baluška F, Volkmann D (2005) Rapid relocation of myosin VIII between cell periphery and plastid surfaces are root-specific and provide the evidence for actomyosin involvement in osmosensing. Funct Plant Biol 32:721–736
Wojtaszek P, Baluška F, Kasprowicz A, Łuczak M, Volkmann D (2007) Domain-specific mechanosensory transmission of osmotic and enzymatic cell wall disturbances to the actin cytoskeleton. Protoplasma 230:217–230
Wolverton C, Mullen JL, Ishikawa H, Evens ML (2002) Root gravitropism in response to a signal originating outside of the cap. Planta 215:153–157
Wyatt SE, Carpita NC (1993) The plant cytoskeleton–cell wall continuum. Trends Cell Biol 3:413–417
Wymer C, Lloyd C (1996) Dynamic microtubules: implications for cell wall patterns. Trends Plant Sci 1:222–228
Wymer CL, Wymer SA, Cosgrove DJ, Cyr RJ (1996) Plant cell growth responds to external forces and the response requires intact microtubules. Plant Physiol 110:425–430

Xiong TC, Jauneau A, Ranjeva R, Mazars C (2004) Isolated plant nuclei as mechanical and thermal sensors involved in calcium signalling. Plant J 40:12–21

Xu W, Purugganan M, Polisensky D, Antosiewicz D, Fry S, Braam J (1995) *Arabidopsis* TCH4, regulated by hormones and the environment, encodes a xyloglucan endotransglycosylase. Plant Cell 7:1555–1567

Yahraus T, Chandra S, Legendre L, Low PS (1995) Evidence for a mechanically induced oxidative burst. Plant Physiol 109:1259–1266

Yamamoto K, Kiss JZ (2002) Disruption of the actin cytoskeleton results in the promotion of gravitropism in inflorescence stems and hypocotyls of *Arabidopsis*. Plant Physiol 128:669–681

Yang L, Reece JM, Cho J, Bortner CD, Shears SB (2008) The nucleolus exhibits an osmotically regulated gatekeeping activity that controls the spatial dynamics and functions of nucleolin. J Biol Chem 283:11823–11831

Yoder TL, Zheng HQ, Todd P, Staehelin A (2001) Amyloplast sedimentation dynamics in maize columella cells support a new model for the gravity-sensing apparatus of root. Plant Physiol 125:1045–1060

Zonia L, Munnik T(2004) Osmotically induced cell swelling versus cell shrinking elicits specific changes in phospholipid signals in tobacco pollen tubes. Plant Physiol 134: 813–823

Zonia L, Munnik T (2007) Life under pressure: hydrostatic pressure in cell growth and function. Trends Plant Sci 12:90–97

Zonia L, Cordeiro S, Tupý J, Feijó JA (2002) Oscillatory chloride efflux at the pollen tube apex has a role in growth and cell volume regulation and is targeted by inositol 3,4,5,6-tetrakisphosphate. Plant Cell 14:2233–2249

Zonia L, Müller M, Munnik T (2006) Hydrodynamics and cell volume oscillations in the pollen tube apical region are integral components of the biomechanics of *Nicotiana tabacum* pollen tube growth. Cell Biochem Biophys 46:209–232

Root Behavior in Response to Aluminum Toxicity

Charlotte Poschenrieder, Montse Amenós, Isabel Corrales, Snezhana Doncheva, and Juan Barceló

Abstract Roots have an extraordinary capacity for adaptive growth which allows them to avoid toxic soil patches or layers and grow into fertile sites. The response of roots to aluminum toxicity, a widespread problem in acid soils, is an excellent model system for investigating the mechanisms that govern this root behavior. In this review, after a short introduction to root growth movement in response to chemical factors in the soil, we explore the basic mechanisms of Al-induced inhibition of root growth. The actinomyosin network and endocytic vesicle trafficking are highlighted as common targets for Al toxicity in cell types with quite different origins: root tip transition zone cells, tip-growing cells like root hairs or pollen tubes, and astrocytes of the animal or human brain. In the roots of sensitive plants, the perception of toxic Al leads to a change in root tip cell patterning. The disturbance of polar auxin transport by Al seems to be a major factor in these developmental changes. In contrast, Al activates organic acid efflux and the binding of Al in a nontoxic form in Al-resistant genotypes.

1 Introduction

Individual terrestrial higher plants are sessile, living anchored to the substrate by their roots. Migration to better, more fertile soil conditions is only possible for their genetic information (pollen) or their offspring (seeds), which have different mechanisms of

C. Pochenrieder (✉)
Lab. Fisiología Vegetal, Facultad Biociencias, Universidad Autónoma de Barcelona E-08193 Bellaterra, Spain
e-mail: charlotte.poschenrieder@uab.es

M. Amenós, I. Corrales, and J. Barceló
Plant Physiology, Bioscience Faculty, Autonomous University of Barcelona, Universidad Autónoma de Barcelona, E-08193, Bellaterra, Spain

S. Doncheva
Popov Institute of Plant Physiology, Bulgarian Academy of Sciences, Sofia, Bulgaria

F. Baluška (ed.), *Plant-Environment Interactions*, Signaling and Communication in Plants,
DOI: 10.1007/978-3-540-89230-4_2, © Springer-Verlag Berlin Heidelberg 2009

dissemination. Slow movement away from the original placement is also possible as clones by vegetative propagation, e.g., through the formation of stolons or rhizomal growth (Hart 1990).

Investigations into plant movements have so far mainly focused on aerial plant parts. Different mechanisms can be distinguished: those based on turgor changes (e.g., nyctinasty and thigmonasty), or those based on differential growth (such as phototropism and epinasty). An exception is gravitropism, another growth-based movement, which has mainly been investigated in roots. However, bending in response to gravitational stimulus is far from being the only movement available to roots (Barlow 1994). Hydrotropism, the directed growth of roots in relation to the gradient of soil water potential, is a well-established growth-based movement of roots in response to an essential chemical soil factor (water) (Ponce et al. 2008). The availability of other essential nutrients can also induce changes in the orientation of root growth in order to improve acquisition. Phosphorus and nitrogen are the best-studied examples (Desnos 2008). The movement of roots into nutrient-rich soil patches implies complex morphogenetic events, such as root hair formation, the induction of new laterals, or—in certain species—proteoid root formation. These trophomorphogenetic responses are controlled directly by the nutrient concentration in the external medium or indirectly by the nutrient status of the plant, or by both (Forde and Lorenzo 2001).

Avoiding toxic soil conditions by altering root growth patterns is a further mechanism that allows plants to move away and try to escape from inadequate growth conditions. Two different scenarios can be envisaged: (1) heterogeneous soil contamination with small hotspots of high toxicant concentrations embedded in less toxic soil, and (2) extended toxic layers in the subsoil.

A heterogeneous distribution of potentially toxic concentrations of metal ions is frequently observed in soils polluted by mining activities. The observation that less Cd was taken up by *Brassica juncea* from soil with a heterogeneous Cd distribution than from uniformly polluted soil supports the view that plants are able to sense the spot contamination and avoid growth into contaminated sites (Manciulea and Ramsey 2006). Contrastingly, *Thlaspi caerulescens*, a metal hyperaccumulating species with unusually high Zn requirements (Tolrà et al. 1996), exhibits zincophilic root foraging patterns, i.e., preferential growth into hot spots with high Zn concentrations (Haines 2002). The efficiencies of both avoidance and foraging responses seem to depend on the root system size of the species. While a negative correlation between species root biomass and precision of placement has been observed in foraging studies on nutrient-rich patches (Wijesinghe et al. 2001), larger root systems seem to be more effective at avoiding toxic spots than small ones (Manciuela and Ramsey 2006). A well-developed tap root system can also be very useful for avoiding the relatively uniform topsoil contamination produced by (for example) smelting activities or after years of applying copper sulfate to vines or hopyards.

In contrast, subsoil acidity is a typical scenario where the extension of roots into the deep soil is hampered by the presence of a layer of soil with high metal availability extending from several decimeters below the soil surface. Crop plants used in tropical and subtropical agriculture and forest stands affected by natural acidification or that due to acid rain are the plants of most concern in this context

(Jentschke et al. 2001; Kochian et al. 2004). Aluminum is considered to be the main toxic factor in acid soils with pH values of less than 4.5. More than 50% of the world's arable land is acidic, so Al toxicity should be considered one of the most important ion toxicity stressors in crop production worldwide. Intensive research into the mechanisms of Al toxicity and Al tolerance mechanisms has been carried out over the last few decades in order to provide the scientific background needed to speed up breeding programs in order to improve crop productivity in acid soils. Aside from this evident practical reason, the responses of plants to Al toxicity are also being used as highly informative model systems. The Al-induced alterations allow fundamental aspects of root stress perception and transduction to be investigated, as well as basic mechanisms of adaptative growth in roots, which are characterized by an enormous capacity for plastic responses to changing physical and chemical conditions in the soil.

2 Aluminum-Induced Inhibition of Root Growth

Root growth is a primary target for Al toxicity in plants. Maintenance of root elongation rate under Al stress is frequently used for Al tolerance screening purposes in hydroponics (Llugany et al. 1994; Ma et al. 2005; Narasihmamoorthy et al. 2007). Monitoring root elongation rates of maize varieties during the first minutes and hours upon exposure (Llugany et al. 1995) reveals various response patterns (Fig. 1): (1) The *threshold of toxicity* model, where a threshold time of 15–45 min and a threshold concentration (usually of a few μM) is required before Al-induced inhibition of elongation is detectable; (2) the *hormesis* response, where a transient Al-induced stimulation of root elongation followed by inhibition is observed, and; (3) the *threshold of tolerance* response, where a fast inhibition of elongation is followed by a recovery in the growth rate (Barceló and Poschenrieder 2002).

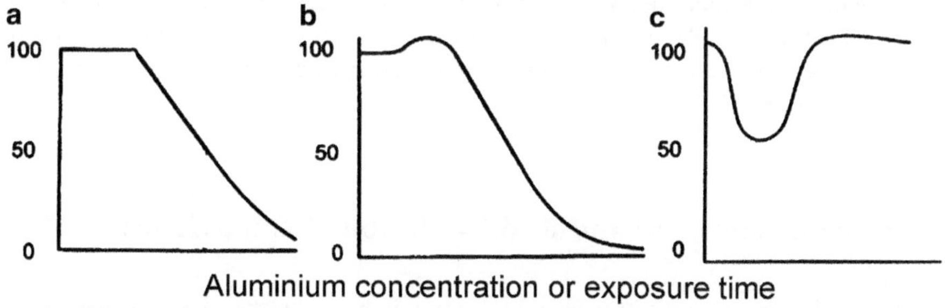

Fig. 1a–c Models for relative root elongation response (%) as a function of Al concentration or exposure time. **a** Toxicity threshold response: the greatest Al concentration or exposure time that does not have an observable effect on root elongation is an indicator of plant Al resistance. **b** Hormesis: growth stimulation by low Al concentrations or short exposure times due to the alleviation of another stress factor (e.g., proton toxicity). **c** Threshold of tolerance response: after the perception of Al-induced stress and the inhibition of elongation, defense mechanisms are activated (e.g., pattern 2 of organic acid efflux)

In the first response pattern, the threshold concentration and the time needed for growth inhibition are indicators of the Al tolerance of the plant. The need for a lag time of usually more than 15 min before elongation inhibition is detectable in sensitive plants (Llugany et al. 1995; Blamey et al. 2004) does not imply that key processes governing root growth cannot be affected even more rapidly (see Sects. 4 and 5).

The second pattern, a transient Al-induced stimulation of root elongation, is a clear hormetic effect, i.e., a positive response to a potentially toxic factor due to the alleviation of another stress suffered by the target organism. In experimental systems where plants are exposed to Al in nutrient solutions with low pH in order to maintain high Al^{3+} activity, proton toxicity is most probably the additional stress factor alleviated by Al (Llugany et al. 1994). The ameliorating effect of the trivalent Al^{3+} on the toxicity of monovalent H^+ can be attributed to competition among these cations in binding to the cell wall and plasma membrane surface, leading to site-specific amelioration at biological ligand targets and to alterations of the plasma membrane surface potential. Effects on the plasma membrane surface potential, in turn, influence the bioavailability of the intoxicating and ameliorating cations (Kinraide 2006; Kinraide and Yermiyahu 2007).

A threshold for tolerance response is observed in species with an inducible Al resistance mechanism, e.g., Al-induced secretion of organic acid anions following pattern II behavior (Ma 2000) (see Sect. 5). This response implies that the initial inhibition of root elongation is reversible upon the activation of the resistance mechanisms leading to the removal of the toxic Al species from the early targets that were responsible for the inhibition of elongation. In fact, even in sensitive plants, the initial inhibition of root elongation after short-term exposure to Al can be completely reversed by transferring the plants to Al-free medium (Kataoka and Nakanishi 2001). The duration of Al treatment after which full recovery of growth can be achieved in Al-sensitive plants varies between 15 and 120 min according to species and experimental conditions (Kataoka and Nakanishi 2001; Amenós 2007; Kikui et al. 2007). The observation that recovery is accelerated in solutions containing organic acids or high Ca concentrations (Alva et al. 1986) supports the view that lowering the Al concentration in the tips is crucial to the resumption of root elongation (Rangel et al. 2007). Recent investigations, however, suggest that malate secretion can stimulate regrowth in roots of sensitive wheat, even without decreasing root-tip Al concentrations (Kikui et al. 2007).

3 Mechanisms of Al-Induced Inhibition of Root Growth

Root growth is a complex process which implies not only the maintenance of cell viability, the production of new cells, and their enlargement, but also cell patterning, morphogenetic processes and coordination by hormonal signals (Barlow 2002; Osmont et al. 2007). As the Al-induced inhibition of root elongation is observable within minutes upon exposure, mechanistic research has mainly focused on the processes of cell enlargement. Cell division makes a negligible contribution to the root length in the short term, and Al-induced morphogenetic alterations are visible after prolonged exposure. Therefore, these processes have warranted less attention. However, recent investigations have demonstrated the relevance of alterations in cell patterning, morphogenetic processes and hormonal regulation in the primary responses of roots to Al toxicity (Doncheva et al. 2005).

3.1 Al-Induced Inhibition of Cell Expansion

Expansion growth of root cells occurs in the elongation zone, located in the subapical root zone a few millimeters from the apex. Turgor-driven expansion requires loosened and extensible primary cell walls, intact plasma membrane, and an adequate water supply to maintain the water potential gradient (Barceló et al. 1996). Cell integrity is a prerequisite for cell expansion. This begs the question of whether Al-induced cell death can account for fast inhibition of root elongation.

Aluminum is not a Fenton-type metal, but it clearly exhibits prooxidant activity (Exley 2004). Aluminum-induced oxidative stress in roots has been found in many investigations (Cakmak and Horst 1991). Aluminum-induced cell death has been observed after hours of exposure to extremely high Al concentrations (Pan et al. 2001; Šimonovičová et al. 2004). Such lethal distress treatments, however, provide scarce information on the dynamics of Al-induced inhibition of root growth. Vital staining of root tips of plants suffering from Al-induced inhibition of root elongation under less drastic conditions has revealed that massive cell death due to loss of cell compartmentation is not a primary cause of the inhibition of root elongation (Corrales et al. 2008). As an example, Fig. 2 shows root tips of a maize (Fig. 2a) and a cucumber plant (Fig. 2b) suffering from a 30–40% inhibition of relative root elongation rate in comparison to the untreated control (Fig. 2c). Note that only a few cells stain with propidium iodide, i.e., have damaged plasma membranes (Fig. 2). Time-dependent studies also demonstrated that cell death and protein oxidation in Al-exposed maize plants occurred later than inhibition of root elongation (Boscolo et al. 2003). Fast, locally induced formation of reactive oxygen species (ROS) can, however, play a crucial role in both stress signaling and cell wall alterations, leading to cell wall stiffening and inhibition of cell expansion.

3.1.1 Cell Wall Expansion and Al Binding

Large amounts of Al accumulate in the cell walls and intercellular spaces of root tips. This apoplastic Al comprises between 85 and 99.9% of the total Al fraction in roots (Ma 2007). Besides Al precipitation on the root surface and in intercellular spaces, an exchangeable form of Al bound to the negative charges of the pectin substances can be identified (Blamey et al. 1993), or it can be found in a more tightly bound form (Eticha et al. 2005).

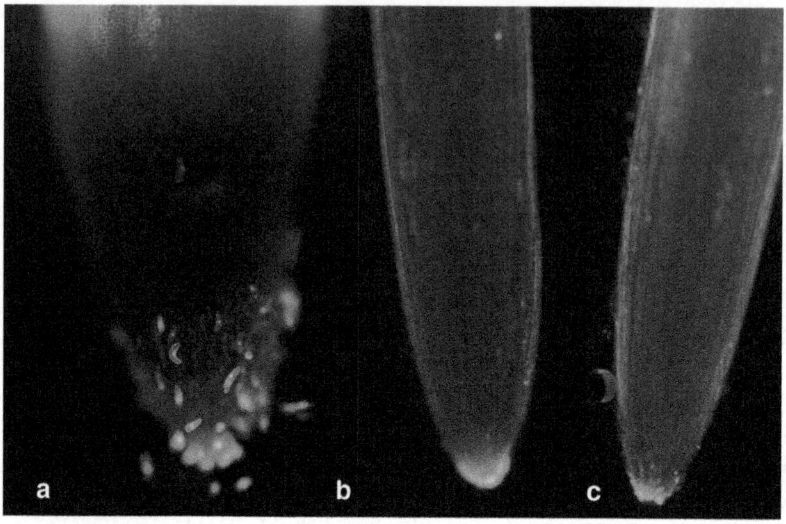

Fig. 2a–c Root tips double stained with fluorescein diacetate (green fluorescence of intact living cells) and propidium iodide (orange fluorescence of cells with damaged plasma membranes). **a** Root tip of maize plant exposed to 50 μM Al and suffering a relative root elongation inhibition of 40%. Only a few cells in the meristem and the transition zone are damaged. **b** Root of a cucumber plant exposed to 7 μM Al suffering a 30% inhibition of relative root elongation. Only a few cells in the elongation zone are damaged. **c** Control cucumber plant without damage. (Unpublished data and modified from Corrales et al. 2008)

Al-induced stiffening of cell walls has been observed in different experimental systems (Gunsé et al. 1997; Tabuchi and Matsumoto 2001; Ma et al. 2004). In vitro studies with maize coleoptiles floating on Al solutions (Llugany et al. 1992; Barceló et al. 1996) or dead root tips treated with Al (Ma et al. 2004) did not reveal Al-induced cell wall stiffening. This supports the view that Al-induced stiffening of cell walls is a biochemical process and not merely physical crosslinking of pectin substances by trivalent Al^{3+}. Cell wall expansion requires both the loosening of the wall matrix and the synthesis of new wall components. The binding of Al to the newly formed material, which is required for the elongation process, may lead to a deterioration in the mechanical properties of the walls, hampering cell elongation (Ma et al. 2004; Ma 2007).

Other polar wall constituents, such as the hydroxyproline-rich glycoprotein (HRGP), have received scant attention in Al toxicity research. Higher extensin concentrations were observed in Al-sensitive than in Al-resistant wheat (Kenzhebaeva et al. 2001). The binding of Al to extensin was observed both in vitro and in vivo (Kenjebaeva et al. 2001). The crosslinking of HRGPs by reactive oxygen species in combination with callose deposition induced by the ethylene precursor ACC has been shown to be an important mechanism for inhibiting cell expansion (de Cnodder et al. 2005). Aluminum-induced enhancement of ethylene evolution clearly precedes the inhibition of root growth in bean seedlings (Massot et al. 2002). Taken together, these results suggest that crosslinking of HRGPs—either directly by Al or indirectly through Al-induced enhancement of ethylene-derived, apoplastic ROS—plays an important role in the inhibition of root cell elongation (Laohavisit and Davies 2007). Therefore, reactive oxygen species participate in the Al-induced inhibition of elongation by inducing crosslinking reactions in proteins or cell wall phenolics rather than through a general breakdown of membrane integrity due to lipid peroxidation reactions. The inner cortical cell layers (Pritchard 1994) drive root elongation. However, cell wall rigidification of the epidermal cell layers could hamper this expansion process (Jones et al. 2006). Cracks in the epidermal layer (frequently observed after a few hours of Al exposure) are the visible consequence. Furthermore, Al-induced ROS can disturb Ca homeostasis through ROS-activated Ca channels (Kawano et al. 2004)

3.1.2 Plasma Membrane, Cytoplasm, and Tonoplast

Although cell walls make the initial contact with high Al concentrations in the soil solution, and most root-tip Al is localized in the apoplast, the primary toxic effects of Al on cell expansion are not restricted to impaired cell wall extensibility. Aluminum-induced impairment of the hydraulic conductivity (Gunsé et al. 1997) of the plasma membranes (PMs) and the tonoplasts of root cells have severe consequences for cell expansion. The importance of this toxic effect of Al on hydraulic conductance is reflected in the prominent changes in aquaporin gene transcription induced by Al within both plant roots and animal cells (Milla et al. 2002; Mathieu et al. 2006; Kumari et al. 2008). The PM responds very quickly to Al toxicity. Depolarization of PM has been observed immediately upon exposure to Al in root cells and Characeae (Sivaguru et al. 1999; Kisnierienë and Sakalauskas 2005). The cell membrane provides potential binding sites for Al, such as carboxyl and phosphate groups. The affinity of Al for the surfaces of phosphatidylcholine (PC) vesicles is 500 times higher than that of Ca (Akeson et al. 1989). The binding of Al to the plasma membrane can account for changes in key properties of this membrane, such as fluidity and lateral lipid phase separation. Decreased hydraulic conductivity of PM (Gunsé et al. 1997), changes in membrane potential and ion channel activity, alteration of Ca homeostasis (Rengel and Zhang 2003), and inhibition of H^+–ATPase (Ahn et al. 2001) are rapid consequences. All of these effects are characteristics of Al toxicity syndrome (Ma 2007; Poschenrieder et al. 2008). The exact sequence of events signaling the presence of Al at the plasma membrane, leading to adaptive root growth responses or inducible resistance mechanisms or both, is still not clearly established (see Sect. 4).

Classically, the plasma membrane was considered impermeable to trivalent cations. Aluminum was thought to penetrate into the symplast only after long-term exposure. As the inhibition of root elongation is a fast process, most research efforts have focused on the apoplast and membrane surface binding. In fact, studies with Al^{3+} or Ga^{3+} (used as an Al analog) have shown that the influx rate of these trivalent cations is slow. Rates on the order of 20–250 pmol m^{-2} s^{-1} have been reported (Reid et al. 1996; Ritchie and Raghupati 2008). However, even these slow rates allow small amounts of potentially toxic Al to enter the symplasm within minutes. This has now been clearly demonstrated by several investigations (Lazof et al. 1996; Vázquez et al. 1999; Taylor et al. 2000; Silva et al. 2000). The mechanisms and the chemical species that enable Al to pass through the plasma membrane are still unknown. Based on results with Al-tolerant accumulator species like *Fagopyrum* and *Melastoma* (Ma and Hiradate 2000; Watanabe et al. 2001), it was postulated that ionic Al^{3+} is taken up by a passive mechanism facilitated by an as-yet unidentified transporter and driven by a favorable electrochemical gradient. The gradient is maintained due to the immediate chelation of the incoming Al^{3+} by citrate or oxalate (Ma 2007).

Membrane transport of Al via endocytosis appears to be another path for Al intake. Internalization of aluminum into endosomal/vacuolar vesicles in cells of the distal transition zone of *Arabidopsis* roots has been visualized by fluorescence microscopy (Illéš et al. 2006). The presence of Al in the distal transition zone of maize and *Arabidopsis* was detected approximately 3 h after Al was supplied to the small root tip vacuoles (Vázquez et al. 1999; Illéš et al. 2006). This implies Al transport across the tonoplast. In *Arabidopsis*, chelated Al can be transported through the tonoplast by a half-type ABC transporter (Larsen et al. 2007).

Due to the low uptake rates of Al across the plasma membrane and the compartmentation of Al into the vacuole, combined with the close-to-neutral pH of symplastic solutions, it can be expected that the free activity of Al^{3+} in the cytoplasm is extremely low. However, even subnanomolar concentrations of Al can efficiently compete with Mg for binding to ATP (Ma 2007). In fact, the toxicity of symplastic Al would largely depend on the relative affinity for Al of toxicity targets and of protective ligands that are able to detoxify Al. Symplastic toxicity targets include (among others) ATP, GTP, nucleic acids, glutamate, endosomal vesicle transport and the cytoskeleton (Sect. 5). Organic acids, especially citrate and oxalate, are well-identified organic ligands that can prevent Al binding to these targets.

3.2 Effects of Aluminum on Cell Division

Pioneering work by Clarkson (1965) demonstrated that Al toxicity strongly affects root developmental features, and he pointed to the inhibition of cell division as a primary cause of Al-induced inhibition of root growth. The binding of Al to nucleic acid in root tips was demonstrated more than 40 years ago (Matsumoto et al. 1976; Morimura et al. 1978). More recent investigations revealed severe toxic effects of Al on root tip cell nuclei and cell division. Chromosome bridges, breaks and nuclear dissolution have been described in maize or onion roots (de Campos and Viccini 2003). Most of the early investigations were performed after several days of exposure to Al. As Al was thought to enter the symplasm only after long-term exposure, while Al-induced inhibition of root elongation can be observed after less than 1 h under Al stress, further investigations focused mainly on cell walls and root cell elongation (Horst 1995).

In recent years there has been a renewed interest in Al-induced alteration of the cell cycle for several reasons. On the one hand it is now well established that small amounts (at least) of Al can penetrate into the symplast quite rapidly (see Sect. 3.1.2). On the other hand, alterations of the cell cycle could be induced by Al in an indirect way, through a signaling cascade, without the need for Al to reach the nuclei of meristematic cells directly. Moreover, the strong influence of Al is not restricted to inhibition of the main root length. The fast developmental changes in response to Al seem to imply a complex coordination of cell patterning events that include inhibition of root cell elongation, inhibition of root cell division, and even stimulation of root cell division (Doncheva et al. 2005).

Lumogallion, a highly specific fluorescence stain for Al, revealed the presence of Al in root tip nuclei after only 30 min of exposure to low Al concentrations (Silva et al. 2000). Aluminum-induced inhibition of the cell cycle in root tips has been observed to occur even more quickly than this. Figure 3 shows the effects of Al in different zones (Fig. 3a) of root tips of maize plants. After only 5 min of exposure to Al followed by a 2-h labeling period, strong inhibition of the incorporation of fluorescent-labeled desoxybromouridine into the cells of the apical meristem is observable (Fig. 3b). Confocal microscopy of the apical meristems of control and Al-treated plants revealed a high number of S-phase cells in controls (Fig. 3d) and a virtual halting of cell cycle activity in the Al-treated plant (Fig. 3e).

This rapid negative effect on cell cycling in the apical meristem of maize root is not due to a general caryotoxic effect of Al in the root tips (Doncheva et al. 2005). On the contrary, the Al treatment quickly stimulated cell cycle activity in the subapical part of the root, in the transition zone (Fig. 3b). After 30 min an incipient protuberance with many dividing cells was observable. After longer Al exposure (3 h) the initial of a new lateral at a short distance from the apex of the main root was distinguished (Fig. 3c). This sequence of events shows that the plant is able to detect excess Al and react to it by adaptive root growth within minutes.

Stimulation of cell division by low Al concentrations has mainly been described in cell culture experiments. Cell cycle activity and cyclin-dependent kinase type A activity were stimulated in the Al-tolerant cell lines of *Coffea arabica*, while inhibition was observed in an Al-sensitive line (Valadez-Gonzalez et al. 2007). Aluminum-induced enhancement of cell division has also been described in human or animal osteoblasts and blood cells (Quarles et al. 1991; Yao et al. 1994) and in yeast (Zheng et al. 2007). The response is concentration-dependent and exposure to higher Al levels causes inhibition of mitosis and cell death. The Al-induced cell activation has been related to Al binding to an extracellular cation-sensing G-protein-coupled receptor (CaR) that is responsible for the perception of extracellular Ca^{2+} (Pi et al. 2005). The expression of a plasma membrane protein for extracellular Ca^{2+} sensing (CAS) has also been described in stems, leaves and stomata of *Arabidopsis* (Han et al. 2003). No ortholog exists in animal species. However, CAS apparently uses the same mechanism to increase intracellular Ca^{2+} by the phosphoinositide/Ca^{2+} pathway (Hofer 2005).

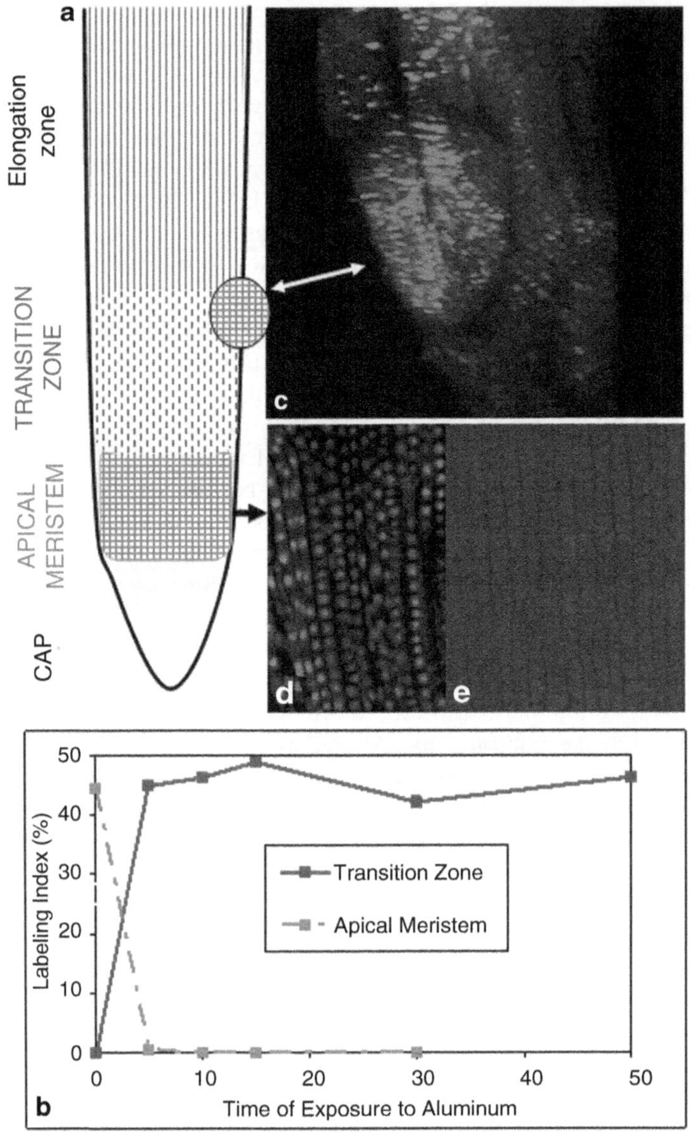

Fig. 3a Model of a maize root tip showing different developmental zones. **b** Labeling index (% of cells with S-phase nuclei) in the apical meristem and the transition zone cells of root tips of maize plants exposed to Al for different times followed by a 2 h bromodeoxyuridine (BrDU) labeling period. **c** Confocal image showing the formation of a lateral root initial close to the transition zone in a maize root exposed to Al for 3 h; S-phase nuclei exhibit green fluorescence due to BrDU labeling. **d** Apical meristem of a control root tip. **e** Apical meristem of a root tip exposed to Al for 30 min; no S-phase nuclei are detectable. (Unpublished data and modified after Doncheva et al. 2005)

3.3 *Root Transition Zone: Site for Al Perception and Al Signal Transduction*

Investigations on the spatial sensitivity to Al in different root tip zones revealed the transition zone (1–2 mm from the tips of maize roots) to be the main target of Al toxicity (Sivaguru and Horst 1998; Rangel et al. 2007). The transition zone is located between the meristem and the elongation zone (Fig. 3a). Distinctive features of the cells in the transition zone should be responsible for the perception of Al. Transition zone cells have a specific architecture that has been related to their exceptional capacity for sensing environmental factors (Baluška et al. 2001b, 2004).

Studies into the gravitropic responses of maize roots revealed a high sensitivity to extracellular Ca in the transition zone (Ishikawa and Evans 1992). Different membrane proteins are responsible for Ca binding and Ca transport in plant cells: the abovementioned CAS (Han et al. 2003); Mca1, a plasma membrane protein from *Arabidopsis* that enhances Ca influx into the cytoplasm upon distortion of the plasma membrane (Nakagawa et al. 2007); ROS-activated Ca channels (Mori and Schroeder 2004); other voltage dependent and independent Ca channels, and Ca efflux transporters (White and Broadley 2003). However, it is still unclear whether the high environmental sensitivity of the transition zone is related to a site-specific distribution of Ca receptors and/or Ca channels. The interference of Al with Ca homeostasis is well established (Rengel and Zhang 2003). Aluminum causes an increase in cytosolic Ca. This can be due to enhanced entry from the apoplast or enhanced release from intracellular storage sites, or both (Ma 2007). Aluminum-induced disturbance of Ca homeostasis can also be brought about by the interference of Al with the phosphoinositide cascade (Jones and Kochian 1995; Ramos-Diaz et al. 2007). Aluminum inhibits phospholipase C, which in turn affects the synthesis of phosphatidic acid.

The cytoskeleton plays a crucial role in driving the impressive changes in cell architecture that occur during the transition from mitotic to elongating cells. The fast impact of Al on the actin cytoskeleton has been documented in detail (Grabski and Schindler 1995; Blancaflor et al. 1998; Ahad and Nick 2007). Using high Al concentrations, Sivaguru et al. (1999) reported the most conspicuous effects of Al on the cytoskeleton in the epidermal and outer cortex cells of the distal transition zone in maize root tips. Under less severe toxicity, we have scored the most prominent Al-induced alterations on F-actin in the central, stelar part of the transition zone and, to a lesser extent, in the central part of the meristem zone (Amenós et al., unpublished). Actin filaments were also an early target of Al in the meristem cells of *Triticum turgidum* roots (Frantzios et al. 2005).

4 Al Toxicity Mechanisms: Common Features in Plant and Animal Cells?

The characterization of the structural and functional differences between transition zone cells and cells in less sensitive root zones is of fundamental interest when assessing primary mechanisms of Al toxicity in roots. Another approach arises from the question: what are the common features shared by the different highly Al-sensitive cell types? Besides root transition zone cells, examples of highly Al-sensitive cell types include plant cells that experience tip growth, like root hairs (Jones et al. 1995; Care 1995), pollen tubes (Konishi and Miyamoto 1983; Zhang et al. 1999) and filamentous algae (Alessa and Oliveira 2001), as well as astrocytes of the animal and human nervous systems (Suarez-Fernandez et al. 1999).

4.1 Actin–Myosin Network and Vesicle Trafficking: Common Targets for Al Toxicity in Plant and Brain Cells

Effects of Al on polar growing cells can be extremely fast. In *Vaucheria longicaulis*, a filamentous alga, cytoplasmic streaming was inhibited by more than 50% after 30 s of Al exposure, and the movement of cell organelles was completely inhibited after only 3 min (Alessa and Oliveira 2001). The movement of cell organelles should not be considered a passive flow movement but rather an active organelle translocation due to the actomyosin transport network (Peremyslov et al. 2008). Rigor has also been observed in the actin filament network as a fast Al-induced alteration in suspension-grown soybean cells (Grabski and Schindler 1995). In this system, the fast Al effects were not related to alterations in ion fluxes, and it was hypothesized that the formation of nonhydrolyzable Al–ATP or Al–ADP complexes and its binding to actin/myosin could be responsible for the stiffness of the network. Knocking out myosin genes XI-2 and XI-K severely affects Golgi-derived vesicle trafficking and root hair development (Peremyslov et al. 2008). Class VIII myosins play the role of endocytic motors in plants, and endocytosis is a fundamental process in cell tip growth (Šamaj et al. 2004, 2005).

Astrocytes in the brain are specific targets for Al toxicity (Levesque et al. 2000; Aremu and Meshitsuka 2005). Astrocytes play a crucial role in the functioning of neurons (Aremu and Meshitsuka 2006). Among others, clathrin-dependent endocytosis of GLT-1, a glutamate transporter that is predominantly expressed in astrocytes, seems to be important for the maintenance of local glutamate concentrations in synapsis. Impaired astroglial function leads to inhibition of glutamate clearance and excitotoxicity. Astroglia can respond to external stimuli by generating Ca waves that release the neurotransmitter glutamate, enhancing the activity in the synapses of nearby neurons. The signal can also be spread across distances through gap junctions. By altering the organization of the actin network, aluminum disturbs connexin trafficking and therefore the formation of gap junctions of two hemichannels in adjacent cells (Theiss and Meller 2002). In root tips of plants, Al also inhibits cell-to-cell transport via plasmodesmatal connections (Sivaguru et al. 2000). Plasmodesmata are located in the actomyosin-enriched domain of the cell periphery (Baluška et al. 2000). As Ca waves regulate fast changes in plasmodesmatal permeability (Baluška et al. 2001a), an Al-induced rise in intracellular Ca can be expected to account for plasmodesmata closure and inhibition of cell-to-cell trafficking.

Glutamate also plays a role in the response to Al in plant cells. Effects of glutamate on membrane depolarization, depolymerization of microtubules and root growth inhibition are similar to those of Al. However, the effects of glutamate occurred more rapidly than those of Al, and Al did not further enhance glutamate action. These observations suggest that glutamate or a glutamate-like substance is involved in the early signaling response to Al toxicity in plants (Sivaguru et al. 2003). The glutamate receptor GLR3.3 is required for Ca^{2+} transport into *Arabidopsis* cells in response to glutamate by a mechanism that can be considered homologous to the fundamental component of neuronal signaling (Qui et al. 2006). This glutamate-receptor-mediated Ca^{2+} influx also seems to be responsible for the glutamate-specific alterations in root branching (Walch-Liu et al. 2006). These root architectural changes are similar to those observed in Al-stressed plants.

Altogether, these observations reveal striking similarities in the responses to Al between Al-sensitive plant and animal cells. Tip-growing plant cells, such as root hairs, pollen tubes or filamentous algae, transition zone cells in plant root tips, and astrocytes are very different in terms of origin, morphology and function. However, a common characteristic of all of them is a high activity of vesicle trafficking. In both the quickly expanding tip-growing cells (Ishida et al. 2008) and the transition zone cells, intense vesicle trafficking is required to provide the new components for the expanding cell walls, among other reasons (Illéš et al. 2006). Vesicle trafficking in astrocytes is essential for the astrocyte-to-neuron communication in the brain (Potokar et al. 2007). Actomyosin network integrity is crucial to the correct functioning of this endocytic and exocytic transport. The fast impact of Al on this network can be considered the common toxicity target in both plant and animal cells. Moreover, in both root transition zone cells (Illéš et al. 2006) and astrocytes (Levesque et al. 2000), endocytosis appears to be an important mechanism for the entry of Al into cells. Therefore, the high Al sensitivities of cells with high endocytic activity may be due to the fact that the actomyosin network is a primary target for Al toxicity, as well as the preferential accumulation of Al in these cells.

5 Coordination of Root Developmental Features Under Al Stress

From this brief glance into the mechanisms of Al toxicity mechanisms, it has become clear that the response of plant roots to this important stress factor is not simply a disruption of cell elongation and a cessation of root growth due to the loss of cell viability. The perception of Al by transition zone cells induces a signaling cascade that can lead to changes in root architecture. The inhibition of main root extension and the induction of lateral roots are key processes in this adaptive growth response.

Inhibition of cell cycle activity in the root apical meristem and activation of cell division for lateral initiation are coordinated events in determinate root growth (Shishikova et al. 2008). Determinate root growth can be constitutive or inducible.

Constitutive determinate root growth is characteristic of certain species like Cactaceae. In these species, the apical meristem function is lost with age, and root hairs and laterals emerge very close to the tip. Exhaustion of the root apical meristem is temporally related to the onset of lateral development. This loss of meristem function has been described as being a physiological root decapitation (Dubrovsky 1997). Phosphorus deficiency (Sánchez-Calderon et al. 2005) and glutamate (Walch-Liu et al. 2006) have been found to induce determinate root growth. The exhaustion of the apical meristem induced by these factors requires several days and is reversible at the beginning. A stimulation of lateral root development close to the tip has also been observed in roots suffering from Cu^{2+} or Al^{3+} toxicity after a few days of exposure to the toxic factor (Llugany et al. 2003; Doncheva et al. 2005). However, the inhibition of the cell cycle in the apical meristem and stimulation of cell division in the subapical region can be observed after only a few minutes of exposure to Al. Similar effects can be induced when NPA (naphthylphthalamic acid), a auxin transport inhibitor, is locally applied to the transition zone of maize root tips (Doncheva et al. 2005).

Lateral roots originate from pericycle cells at a variable distance from the main root apex. Usually laterals emerge from the root zone, where a clearly differentiated vascular cylinder can be distinguished. However, early lateral root primordia initiation can arise close to the root tip (Dubrovsky et al. 2000). Cell division activity in the pericycle cells is restricted by the E2F–RB pathway. Auxin triggers cell division in these stem cells. In addition, an auxin-derived signal seems to be required for the proliferation of a new lateral (Vanneste et al. 2007). The role of polar auxin transport and its relation to differential gene expression in the patterning of morphogenetic events has mainly been investigated in plant shoots (Bowman and Floyd 2008). However, there is increasing evidence for a similar role of polar auxin transport in the development of the roots (Vanneste et al. 2007). In *Arabidopsis*, the patterning of root stem cells is mediated by PLETHORA genes (PLT) (Aida et al. 2004). The expression of PLT can be induced by maximum auxin concentrations.

Based on this, the plastic response of roots to environmental factors could be regulated by direct or indirect interactions between the environmental factor and the mechanism of polar auxin transport, leading to changes in the local auxin gradients and therefore to changes in developmental patterns; e.g., the induction of lateral root formation. It is now well established that polar auxin transport is mediated by a polar distribution of the auxin efflux transporter protein (PIN) (Wisniewska et al. 2006). Endocytotic cycling is considered a highly regulated mechanism for polar PIN localization (Benjamin and Scheres 2008).

Within this scenario, the mechanism responsible for the strong influence of Al on root architecture could be directly related to the toxic action of Al on the actomyosin network that governs vesicle trafficking required for polar auxin transport. The potential key molecule for this toxic action of Al could be small GTPases that are involved in vesicle trafficking and PIN localization (Molendijk et al. 2004). Aluminum fluoride (AlF^{4-}) is a well-known activator of trimeric G proteins, while it inhibits small GTPases.

6 Aluminum Tolerance

Plants adapted to grow in soils with high Al^{3+} activity must have efficient mechanisms for either Al exclusion or tolerance to high Al tissue concentrations (Barceló and Poschenrieder 2002; Ma 2007). Figure 4 summarizes some of these mechanisms. Internal detoxification of Al can be achieved by binding Al to strong chelators like oxalate, citrate, or phenolic substances and Al compartmentation in vacuoles (Vázquez et al. 1999). A constitutively expressed gene (*AlS1*) coding for a half-type ABC transporter protein has been identified in *Arabidopsis*. Located at the tonoplast, this transporter could be important for the compartmentation of chelated Al into the vacuoles (Larsen et al. 2007). It has been suggested that a phloem-located PM transporter protein that is inducible by Al removes the potentially toxic Al from sensitive parts of the root (Larsen et al. 2005). In rice, a gene coding for a possible Al efflux protein (Als1) located in the PM of root tip cells has recently been identified (Ma 2007). Rice mutants defective in this PM protein have higher cytoplasmic Al concentrations than the wild type. Even plants that can withstand the hyperaccumulation of Al in their shoots, such as members of the Melastomataceae or tea plants, must prevent the access of phytotoxic Al species to the sensitive cells in the transition zone. Different mechanisms have been proposed to operate in Al exclusion: plant-induced pH changes in the rhizosphere, production of mucilage and border cells, fewer binding sites in root tip cell walls, lower PM permeability, or enhanced Al efflux. The best-characterized mechanism, however, is the root-tip-located exudation of low molecular weight organic substances with a high affinity for Al (Kidd et al. 2001; Ryan et al. 2001; Kochian et al. 2005). Organic acid exudation seems to be the most widespread mechanism. Two exudation patterns in response to Al can be distinguished: pattern 1 exudation which is activated by Al almost immediately, and pattern 2, where a lag time of several hours is required before the Al-stimulated exudation of organic acids is detectable (Ma et al. 2001). The presence of an efficient, Al-activable, organic acid efflux system in root tips is responsible for the Al resistance (Fig. 4). In contrast, organic acid metabolism seems of minor importance (Ma 2007). Aluminum-activated malate efflux in wheat (*TaALMT1*) (Saski et al. 2006), in *Arabidopsis thaliana* (*AtALMT1*) (Hoekenga et al. 2006), and in *Secale cereale* (*ScALMT1*) (Fontecha et al. 2007; Collins et al. 2008) is related to Al resistance. Reversible phosphorylation is important in the transcriptional and posttranscriptional regulation of *ALMT1* (Kobayashi et al. 2007). In maize, *ZmALMT1* is not, however, involved in the specific Al-activated efflux of citrate (Piñeros et al. 2008). Aluminum-activated citrate efflux in barley and in sorghum is mediated by a protein of the MATE (Multidrug And Toxic compound Extrusion) efflux pump family (Furukawa et al. 2007; Magalhaes et al. 2007; Wang et al. 2007).

How Al activates these organic acid efflux systems has not yet been clearly established. Delhaize et al. (2007) recently proposed two hypothetical models for Al^{3+}-activated organic acid efflux by ALMT and MATE family proteins: model 1, where a direct interaction of Al with the membrane transporter occurs (e.g., TaALMT1

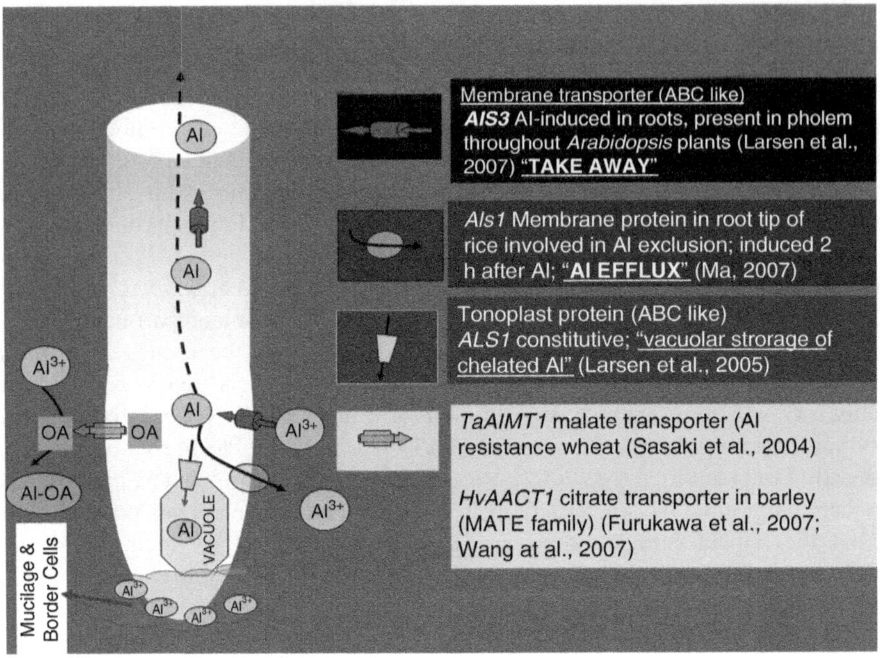

Fig. 4 Mechanisms for Al exclusion and compartmentation in root tips. Distribution of membrane transporter proteins involved in Al efflux, Al phloem transport and Al vacuolar transport are shown along with transporters for organic acid anions. Mucilage and border cells help to stop Al^{3+} from reaching the sensitive root tip (modified after Ma 2007)

in wheat), and model 2, which implies an Al-activated signal transduction cascade. This second model corresponds to Al-activated malate efflux in *Arabidopsis* and *Brassica* and to Al-activated citrate efflux in sorghum. In this pattern 2 response, Al induces the expression of the proteins either by binding to specific PM receptors or by activating a nonspecific stress response. Interaction of Al with these new proteins would then promote the organic acid efflux (Delhaize et al. 2007).

7 Conclusions and Outlook

During the last decades of intense research, substantial advances have been made in our understanding of the molecular mechanisms that are responsible for the resistance of plants to Al toxicity. The identification of Al resistance genes has provided new strategies for improving the breeding of crops adapted to acid soils with Al toxicity problems.

Besides this evident practical progress, the plant response to Al toxicity is becoming a very illustrative model system for basic research—not only in the field of membrane transport systems, but also in the area of studies into the mechanisms governing root developmental features. The information summarized in this review highlights the endocytic process as a common target for Al toxicity in very different cellular systems: tip-growing plant cells like pollen tubes, root hairs and filamentous algae, cells in the transition zones of plant roots, and astrocytes in the brain. Taken together, this information suggests the hypothesis that cells with high endocytotic activity are especially vulnerable to Al. Future research should clarify if his high Al sensitivity is due to enhanced Al entry into these cells via an endocytic uptake mechanism. Investigations into the differences in the Al-activated signal transduction cascades that can lead to adaptive root growth in Al-sensitive genotypes, while activation of anion efflux is induced in resistant genotypes of pattern 2 species, will help to establish the primary mechanism of Al perception in plant roots.

Acknowledgements Supported by the Spanish and the Catalonian Governments (BFU2007-60332/BFI and Grup de Recerca, expedient 2005R 00785).

References

Ahad A, Nick P (2007) Actin is bundled in activated-tagged tobacco mutants that tolerate aluminum. Planta 225:451–468

Ahn SJ, Sivaguru M, Chung GC, Rengel Z, Matsumoto H (2001) Aluminium-induced growth inhibition is associated with impaired efflux and influx of H^+ across the plasma membrane in root apices of squah (*Cucurbita pepo*). J Exp Bot 53:1959–1966

Aida M, Beis D, Heidstra R, Willemsen V, Billou I, Galinha C, Nussaume L, Noh YS, Amsino R, Scheres B (2004) The PLETHORA genes mediate patterning of the *Arabidopsis* root stem cell niche. Cell 119:109–120

Akeson MA, Munns DN, Burau RG (1989) Adsorption of Al^{3+} to phosphatidylcholine vesicles. Biochim Biophys Acta 986:33–40

Alessa L, Oliveira L (2001) Aluminum toxicity studies in *Vaucheria longicaulis* var. macouni (Xanthophyta, Tribophyceae). I. Effects on cytoplasmic organization. Environ Exp Bot 45:205–222

Alva AK, Asher CJ, Edwards DG (1986) The role of calcium in alleviating aluminum toxicity. Austr J Agric Res 37:375–382

Amenós M (2007) Respuestas primarias a la toxicidad por aluminio en raíces de plantas de maíz con diferente resistencia. Dissertion, Autonomous University of Barcelona.

Aremu DA, Meshitsuka S (2005) Accumulation of aluminum by primary cultures astrocytes from aluminum amino acid complex and its apoptotic effect. Brain Res 1031:284–296

Aremu DA, Meshitsuka S (2006) Some aspects of astroglial functions and aluminum implications for neurodegeneration. Brain Res Rev 52:193–200

Baluška F, Volkmann, D, Barlow PW (2000) Actin-based domains of the "cell periphery complex" and their associations with polarized "cell bodies" in higher plants. Plant Biol 2:253–267

Baluška F, Cvrcková F, Kendrick-Jones J, Volkmann D (2001a) Sink plasmodesmata as gateways for phloem unloading. Myosin VIII and calreticulin as molecular determinants of sink strength. Plant Physiol 126:39–46

Baluška F, Volkmann D, Barlow PW (2001b) A polarity crossroad in the transition growth zone of maize root apices: cytoskeletal and developmental implications. J Plant Growth Regul 20:170–181
Baluška F, Mancuso S, Volkmann D, Barlow P (2004) Root apices as plant command canters: the unique "brain-like" status of the root apex transition zone. Biologia 59(Suppl 13):7–19
Barceló J, Poschenrieder C (2002) Fast root growth responses, root exudates, and internal detoxification as clues to the mechanisms of aluminium toxicity and resistance: a review. Environ Exp Bot 48:75–92
Barceló J., Poschenrieder C, Vázquez MD, Gunsé B (1996) Aluminium phytotoxicity: a challenge for plant scientists. Fertil Res 43:217–223
Barlow PW (1994) Root movements: towards understanding through models the mechanisms involved. Plant Soil 165:293–300
Barlow PW (2002) Cellular patterning in root meristems: its origin and significance. In: Waisel Y, Eshel A, Kafkafi U(eds)Plant roots: the hidden half, 3rd edn. Marcel Dekker, New York, pp 49–82
Benjamin R, Scheres B (2008) The looping star in plant development. Annu Rev Plant Biol 59:443–465
Blamey FPC, Asher CJ, Kerven GL, Edwards DG (1993) Factors affecting sorption by calcium pectate. Plant Soil 192:269–275
Blamey FPC, Nishizawa NK, Yoshimura E (2004) Timing, magnitude, and location of initial soluble aluminum injuries to mungbean roots. Soil Sci Plant Nutr 50:67–76
Blancaflor BE, Jones DL, Gilroy S (1998) Alterations in the cytoskeleton accompany aluminum-induced growth inhibition and morphological changes in primary roots of maize. Plant Physiol 118:159–172
Boscolo PRS, Menossi M, Jorge RA (2003) Aluminum-induced oxidative stress in maize. Phytochemistry 62:181–189
Bowman JL, Floyd SK (2008) Patterning and polarity in seed plant shoots. Annu Rev Plant Biol 59:67–88
Cakmak I, Horst WJ (1991) Effect of aluminum on lipid peroxidation, superoxide-dismutase, catalase, and peroxidase activities in root tips of soybean (*Glycine max*). Physiol Plant 83:463–468
Care DA (1995) The effect of aluminium concentration on root hairs in white clover (*Trifolium repens* L.). Plant Soil 171:159–162
Clarkson DT (1965) The effect of aluminium and some trivalent metal cations on cell division in the root apices of *Allium cepa*. Ann Bot 29:309–315
Collins NC, Shirley NJ, Saeed M, Pallotta M, Gustafson JP (2008) An ALMT1 gene cluster controlling aluminum tolerance at the Alt4 locus of rye (*Secale cereale* L.). Genetics 179:669–682
Corrales I, Poschenrieder C, Barceló J (2008) Boron-induced amelioration of aluminium toxicity in a monocot and a dicot species. J Plant Physiol 165:504–513
de Campos JMS, Viccini LF (2003) Cytotoxicity of aluminum on meristematic cells of *Zea mays* and *Allium cepa*. Caryologia 56:65–73
De Cnodder T, Vissenberg K, Van der Straeten D, Verbelen JP (2005) Regulation of cell length in the *Arabidopsis thaliana* root by ethylene precursor 1-aminocyclopropane-1-carboxylic acid: a matter of apoplastic rections. New Phytol 168:541–550
Delhaize E, Gruber BD, Ryan PR (2007) The roles of organic anion permeases in aluminium resistance and mineral nutrition. FEBS Lett 581:2255–2262
Desnos T (2008) Root branching responses to phosphate and nitrate. Curr Opin Plant Biol 11:82–87
Doncheva S, Amenós M, Poschenrieder C, Barceló J (2005) Root cell patterning—a primary target for aluminum toxicity in maize. J Exp Bot 56:1213–1220
Dubrovsky JG (1997) Determinate primary-root growth in seedlings of Sonora desert Cactaceae; its organization, cellular basis, and ecological significance. Planta 203:85–92
Dubrovsky JG, Doerner PW, Colón-Carmona A, Rost TL (2000) Pericycle cell proliferation and lateral root initiation in *Arabidopsis*. Plant Physiol 124:1648–1657
釘icha D, Stass A, Horst WJ (2005) Localization of aluminium in the maize root apex: can morin detect cell wall-bound aluminium? J Exp Bot 56:1351–1357
Exley C (2004) The pro-oxidant activity of aluminum. Free Radical Biol Med 36:380–387

Fontecha G, Silva-Navas J, Benito C, Mestres MA, Espino FJ, Hernandez-Riquer, MV, Gallego FJ (2007) Candidate gene identification of an aluminum-activated organic acid transporter gene at the Alt4 locus for aluminum tolertance in rye (*Secale cereale* L.). Theor Appl Genet 114:249–260

Forde B, Lorenzo H (2001) The nutritional control of root development. Plant Soil 232:51–68

Frantzios G, Galatis B, Apostolakos P (2005) Aluminium causes variable responses in actin filament cytoskeleton of the root tip cells of *Triticum turgidum*. Protoplasma 225:129–140

Furukawa J, Yamaji N, Wang H, Mitani N, Murata Y, Sato K, Katsuhara M, Takeda K, Ma FJ (2007) An aluminum-activated citrate transporter in barley. Plant Cell Physiol 48:1081–1091

Grabski S, Schindler M (1995) Aluminum induces rigor within the actin network of soybean cells. Plant Phsyiol 108:897–901

Gunsé B, Poschenrieder C, Barceló J (1997) Water transport properties of roots and root cortical cells in proton and Al-stressed maize varieties. Plant Physiol 113:595–602

Haines, BJ (2002) Zincophilic root foraging in *Thlaspi caerulescens*. New Phytol 155:363–372

Han SC, Tang RH, Anderson LK, Woerner TE, Pei ZM (2003) A cell surface receptor mediates extracellular Ca^{2+} sensing in guard cells. Nature 425:196–200

Hart JW (1990) Plant tropisms and other growth movements. Unwin Hyman, London

Hoekenga OA, Maron LG, Piñeros MA, Cancado GMA, Shaff J, Kobayashi Y, Ryan PR, Dong B, DElhaize E, Sasaki T, Matsumoto H, Yamamoto Y, Koyama H, Kochian LV (2006) *AtALMT1*, which encodes a malate transporter, is identified as one of several genes critical for aluminum tolerance in Arabidopsis. Proc Natl Acad Sci USA 103:9738–9743

Hofer A (2005) Another dimension to calcium signaling: a look at extracellular calcium. J Cell Sci 118:855–862

Horst WJ (1995) The role of the apoplast in aluminium toxicity and resistance of higher plants: a review. Z Pflanzenrnähr Bodenk 158:419–428

Illéš P, Schlicht M, Pavlovkin J, Lichtscheidl I, Baluška F, Ovečka M (2006) Aluminium toxicity in plants: internalization of aluminium into cells of the transition zone in *Arabidopsis* root apices related to changes in plasma membrane potential, endosomal behaviour, and nitric acid production. J Exp Bot 57:4201–4213

Ishida T, Kurata T, Okada K, Wada T (2008) A genetic regulatory network in the development of trichomes and root hairs. Annu Rev Plant Biol 59:365–386

Ishikawa H, Evans ML (1992) Induction of curvature in maize roots by calcium or by thigmostimulation. Role of the postmitotic isodiametric growth zone. Plant Physiol 100:762–768

Jentschke G, Drexhage M, Fritz HW, Fritz E, Schella B, Lee DH, Heiman J, Kuhr M, Schmidt J, Schmidt S, Zimmermann R, Godbold DL (2001) Does soil acidity reduce subsoil rooting in Norway spruce (*Picea abies*)? Plant Soil 237:91–108

Jones DL, Kochian LV (1995) Aluminum inhibition of the inositol 1,4,5-triphosphate signal transduction pathway in wheat rots: a role in aluminum toxicity. Plant Cell 7:1913–1922

Jones DL, Shaff JE, Kochian LV (1995) Role of calcium and other ions in detecting root hair tip growth in *Linnobium stoloniferum*. I: Inhibition of tip growth by aluminum. Planta 197:672–680

Jones DL, Blancaflor EB, Kochian LV, Gilroy S (2006) Spatial coordination of aluminium uptake, production of reactive oxygen species, callose production and wall rigidification in maize roots. Plant Cell Environ 29:1309–1318

Kataoka, T, Nakanishi TM (2001) Aluminium distribution in soybean root tip for a short time Al treatment. J Plant Physiol 158:731–736

Kawano T, Kadono T, Fumoto K, Lapeyrie F, Kuse M, Isobe M, Furuichi T, Muto S (2004) Aluminium as a specific inhibitor of plant TPC1 Ca^{2+} channels. Biochem Biophys Res Commu 324:40–45

Kenjebaeva S, Yamamoto Y, Matsumoto H (2001) The impact of aluminium on the distribution off cell wall glycoproteins of pea root tip and their Al binding capcity. Soil Sci Plant Nutr 47:629–63

Kenzhebaeva SS, Yamamoto Y, Matsumoto H (2001) Aluminum-induced changes in cell wall glycoproteins in the root tips of Al-tolerant and Al-sensitive wheat lines. Russ J Plant Physiol 48:441–447

Kidd PS, Llugany M, Poschenrieder C, Gunsé B, Barceló J (2001) The role of root exudates in aluminium resistance and silicon-induced amelioration of aluminium toxicity in three varieties of maize (*Zea mays* L). J Exp Bot 52:1339–1352

Kikui S, Sasaki T, Osawa H, Matsumoto H, Yamamoto Y (2007) Malate enhances recovery from aluminum-caused inhibition of root elongation in wheat. Plant Soil 290:1–15

Kinraide TB (2006) Plasma membrane surface potential (Ψ_{PM}) as a determinant of ion bioavailability: a critical analysis of new and published toxicological studies and a simplified method for the computation of plant (Ψ_{PM}). Environ Toxicol Chem 25:3188–3198

Kinraide TB, Yermiyahu U (2007) A scale of metal ion binding strength correlating with ionic charge, Pauling electronegativity, toxicity, and other physiological effects. J Inorg Biochem 101:1201–1213

Kisnierienė V, Sakalauskas V (2005) Al^{3+} induced membrane potential changes in *Nitellopsis obtusa* cells. Biologija 1:31–34

Kobayashi Y, Hoekenga OA, Itoh H, Nakashima M, Saito S, Shaff JE, Maron LG, Piñeros MA, Kochian LV, Koyama H (2007) Characterization of AtALMT1 expression in aluminum-inducible malate release and its role for rhizotoxic stress in *Arabidopsis*. Plant Physiol 145:843–852

Kochian LV, Hoekenga OA, Piñeros MA (2004) How do crop plants tolerate acid soils? Mechanisms of aluminum tolerance an phosphorous efficiency. Annu Rev Plant Biol 55:459–493

Kochian LV, Piñeros MA, Hoekenga OA (2005) The physiology, genetics and molecular biology of plant aluminum resistance in plants. Plant Soil 274:175–195

Konishi S, Miyamoto S (1983) Alleviation of aluminum stress and stimulation of tea pollen tube growth by fluorine. Plant Cell Physiol 24:857–862

Kumari M., Taylor GJ, Deyholos MK (2008) Transcription responses to aluminum stress in roots of *Arabidopsis thaliana*. Mol Genet Genomics 279:339–357

Laohavisit A, Davies JM (2007) The gas that opens gates: calcium channel activation by ethylene. New Phytol 174:470–473

Larsen PB, Geisler MJB, Jones CA, Williams KM, Cancel JD (2005) ALS3 encodes a phloem-localized ABC transporter-like protein that is required for aluminum tolerance in *Arabidopsis*. Plant J 41:353–363

Larsen PB, Cancel J, Rounds M, Ochoa V (2007) Arabidopsis ALS1 encodes a root tip and stele localized half type ABC transporter required for root growth in an aluminum toxic environment. Planta 225:1447–1458

Lazof DB, Goldsmith JG, Rufty TW, Linton RW (1996) The early entry of Al into root cells of intact soybean roots. A comparison of three developmental root regions using secondary ion mass spectrometry imaging. Plant Physiol 108:152–160

Levesque L, Mizzen C, McLachlan DR, Fraser PE (2000) Ligand specific effects on aluminum incorporation and toxicity in neurons and astrocytes. Brain Res 877:191–202

Llugany M, Gunsé B, Poschenrieder C, Barceló J (1992) Total, plastic and elastic extensibility of *Zea mays* coleoptiles exposed to aluminum in vitro. Physiol Plant 85(part II):A76

Llugany M, Massot N, Wissemeier H, Poschenrieder C, Horst WJ, Barceló J (1994) Aluminium tolerance of maize cultivars as assessed by callose production and root elongation. Z Pflanzenernähr Bodenk 157:447–451

Llugany M, Poschenrieder C, Barceló J (1995) Monitoring of aluminium-induced inhibition of root elongation in four maize cultivars differing in tolerance to aluminium and proton toxicity. Physiol Plant 93:265–271

Llugany M, Lombini A, Poschenrieder C, Dinelli E, Barceló J (2003) Different mechanisms account for enhanced copper resistance in *Silene armeria* ecotypes from mine spoil and serpentine sites. Plant Soil 251:55–63

Ma JF (2000) The role of organic acids in detoxification of aluminum in higher plants. Plant Cell Physiol 41:383–390

Ma JF (2007) Syndrome of aluminum toxicity and diversity of aluminum resistance in higher plants. Int Rev Cytol 264:225–252

Ma JF, Hiradate S (2000) Form of aluminium for uptake and translocation in buckwheat (*Fagopyrum esculentum* Moench). Planta 211:355–360

Ma JF, Ryan PR, Delhaize E (2001) Aluminum tolerance in plants and the complexing role of organic acids. Trends Plant Sci 6:273–278

Ma JF, Shen RF, Nagao S, Tanimoto E (2004)Aluminum targets elongating cells by reducing cell wall extensibility in wheat roots. Plant Cell Physiol 45:583–589

Ma JF, Nagao S, Huang CF, Nishimura M (2005) Isolation and characterization of a rice mutant hypersensitive to Al. Plant Cell Physiol 46:1054–1061

Magalhaes JV, Liu J, Guimaraes CT, Lana UGP, Alves VMC, Wang YH, Schaffert RE, Hoekenga OA, Piñeros MA, Shaff JE, Klein PE, Carneiro NP, Coelho CM, Trick HN, Kochian LV (2007) A gene in the multidrug and toxic compound extrusion (MATE) family confers aluminum tolerance in sorghum. Nat Genet 39:1156–1161

Manciulea A, Ramsey MH (2006) Effect of scale of Cd heterogeneity and timing of exposure on the Cd uptake and shoot biomass, of plants with a contrasting root morphology. Sci Total Environ 367:958–967

Massot N, Nicander B, Barceló J, Poschenrieder C, Tillberg E (2002) A rapid increase in cytokinin levels and enhanced ethylene evolution precede Al^{3+} -induced inhibition of root growth in bean seedlings (*Phaseolus vulgaris* L.). Plant Growth Regul 37:105–112

Mathieu S, Millen N, Contini MD, Gonzalez M, Molinas SM, Elias NM (2006) Urinary concentrating mechanism and Aquaporin-2 abundance in rats chronically treated with aluminum lactate. Toxicology 223:209–218

Matsumoto H, Hirasawa E, Morimura S, Takahashi E (1976) Localization of absorbed aluminium in pea root and its binding to nucleic acid. Plant Cell Physiol 17:627–631

Milla MAR, Butler E, Huete AR, Wilson CF, Anderson O, Gustafson JP (2002) Expressed sequence tag-based gene expression analysis under aluminum stress in rye. Plant Physiol 130:1706–1716

Molendijk AJ, Ruperti B, Palme K (2004) Small GTPases in vesicle trafficking. Curr Opin Plant Biol 7:694–700

Mori IC, Schroeder JI (2004) Reactive oxygen species activation of plant Ca^{2+} channels. A signaling mechanism in polar growth, hormone transduction, stress signaling, and hypothetically mechanotransduction. Plant Physiol 135:702–708

Morimura S, Takahashi E, Matsumoto H (1978) Association of aluminium with nuclei and inhibition of cell division in onion (*Allium cepa*) roots. J Plant Physiol 88:395–401

Nakagawa Y, Katagiri T, Shinozaki K, Qui Z, Tatsumi H, Furuichi T, Kishigami A, Sokabe M, Kojima I, Sato S, Kato T, Tabata S, Iida K, Terashima A, Nakano M, Ikeda M, Yamanaka T, Iida H (2007) *Arabidopsis* plasma membrane protein crucial for Ca^{2+} influx and touch sensing in roots. Proc Natl Acad Sci USA 104:3639–3644

Narasimhamoorthy B, Blancaflor EB, Bouton JH, Payton ME, Sledge MK (2007) A comparison of hydroponics, soil, and root staining methods for evaluation of aluminum tolerance in *Medicago trunculata* (barrel medic) germplasm. Crop Sci 47:321–328

Osmont KS, Sibout R, Hardtke S (2007) Hidden breanches: developments in root system architecture. Annu Rev Plant Biol 58:93–113

Pan JW, Zhu MY, Chen H (2001) Aluminum-induced cell death in root-tip cells of barley. Environ Exp Bot 46:71–79

Peremyslov VV, Proknevsky Ai, Avisar D, Dolja VV (2008) Two class XI myosins function in organelle trafficking and root hair development in *Arabidopsis*. Plant Physiol 146:1109–1116

Pi M, Faber P, Ekema G, Jackson PD, Ting A, Wang N, Fontilla-Poole M, Mays RW, Brunden KR, Harrington JJ, Quarles LD (2005) Identification of a novel extracellular cation-sensing G-protein-coupled receptor. J Biol Chem 280:40201–40209

Piñeros MA, Cancado GMA, Maron LG, Lyi SM, Sangborn M, Menossi M, Kochian LV (2008) Not all ALMT1-type transporters mediate aluminum-activated organic acid responses: the case of ZmALMT1—an anion-selctive transporter. Plant J 53:352–367

Ponce G, Rasgado FA, Cassab GI (2008) Roles of amyloplasts and water deficit in root tropisms. Plant Cell Environ 31:205–217

Poschenrieder C, Gunsé B, Corrales I, Barceló J (2008) A glance into aluminium toxicity and resistance in plants. Sci Total Environ 400(1–3):356–268

Potokar M, Kreft M, Li L, Andersson JD, Pangršic, Chowdhury HH, Pekny M, Zorec R (2007) Cytoskeleton and vesicle mobility in astrocytes. Traffic 8:12–20

Pritchard J (1994) The control of cell expansion in roots. New Phytol 127:2–26

Quarles LD, Wenstrup RJ, Castillo SA, Drezner MK (1991) Aluminum-induced mitogenesis in MC373-E1 osteoblasts: potential mechanism underlying neoosteogenesis. Endocrinology 128:3144–3151
Qui Z, Stephens NR, Spalding EP (2006) Calcium entry mediated by GLR3.3, and *Arabidopsis* glutamate receptor with broad agonist profile. Plant Physiol 142:963–971
Ramos-Díaz A, Brito L, Munnik T, Hernandez-Sotomayor SMT (2007) Aluminum inhibits phosphatidic acid formation by locking the phospholipase C pathway. Planta 225:393–401
Rangel AF, Rao I, Horst WJ (2007) Spatial aluminium sensitivity of root apices of two common bean (*Phaseolus vulgaris* L.) genotypes with contrasting aluminium tolerance. J Exp Bot 58:3859–3904
Reid RJ, Rengel Z, Smith FA (1996) Membrane fluxes and comparative toxicities of aluminium, scandium and gallium. J Exp Bot 47:1881–1888
Rengel Z, Zhang WH (2003) Role of dynamics of intracellular calcium in aluminium-toxicity syndrome. New Phytol 159:295–314
Ritchie RJ, Raghupathi SS (2008) Al-toxicity studies in yeast using gallium as an aluminum analogue. Biometals 21:379–393
Ryan PR, Delhaize E, Jones DL (2001) Function and mechanism of organic anion exudation from plant roots. Annu Rev Plant Mol Biol 52:527–560
Šamaj J, Baluška F, Voigt B, Schlicht M, Volkmann D, Menzel D (2004) Endocytosis, actin cytoskeleton and signaling. Plant Physiol 135:1150–1161
Šamaj J, Read ND, Volkmann D, Menzel D, Baluška F (2005) The endocytic network in plants. Trends Cell Biol 15:425–433
Sánchez-Calderón L, López-Bucio J, Chacón-López A, Cruz-Ramírez A, Nieto Jacobo F, Dubrovsky JG, Herrera-Estrella L (2005) Phosphate starvation induces a determinate developmental program in the roots of *Arabidopsis thaliana*. Plant Cell Physiol 46:174–184
Saski T, Ryan PR, Delhaize E, Hebb DM, Ogihara Y, Kawaura K, Noda K, Kojima T, Toyoda A, Matsumoto H, Yamamoto Y (2006) Sequence upstream of the wheat (*Triticum aestivum* L.) *ALMT1* gene and its relationship to aluminum resistance. Plant Cell Physiol 47:1343–1354
Shishikova S, Rost TL, Dubrovsky (2008) Determinate root growth and meristem maintenance in angiosperms. Ann Bot 101:319–340
Silva IR, Smyth TJ, Moxley DF, Carter TE, Allen NS, Rufty TW (2000) Aluminium accumulation at nuclei of cells in the root tip. Fluorescence detection using lumogallion and confocal laser scanning microscopy. Plant Physiol 123:543–552
Šimonovičová M, Huttová J, Mistrík I, Široká B, Tamás L (2004) Root inhibition by aluminum is probably caused by cell death due to peroxidase-mediated hydrogen peroxide production. Protoplasma 224:91–98
Sivaguru M, Horst WJ (1998) The distal part of the transition zone is the most aluminum-sensitive apical root zone of maize. Plant Physiol 116:155–163
Sivaguru M, Baluška F, Volkmann D, Felle HH, Horst WJ (1999) Impacts of aluminum on the cytoskeleton of the maize root apex. Short-term effects on the distal part of the transition zone. Plant Physiol 119:1073–1082
Sivaguru M, Fujiwara T, Samaj J, Baluška F, Yang ZM, Osawa H, MAeda T, Mori T, Volkmann D, Matsumoto H (2000) Aluminum-induced 1-3-beta-D-glucan inhibits cell-to-cell trafficking of molecules through plasmodesmata. A new mechanism of aluminum toxicity in plants. Plant Physiol 124:991–1005
Sivaguru M, Pike S, Gassmann W, Baskin TI (2003) Aluminum rapidly depolymerizes cortical microtubules and depolarizes tha plasma membrane: evidence that these responses are mediated by a glutamate receptor. Plant Cell Physiol 44:667–675
Suarez-Fernandez MB, Soldado AB, Sainz-Medel A, Vega JA, Novelli A, Fernandez-Sánchez MT (1999) Aluminum-induced degeneration of astrocytes occurs via apoptosis and results in neuronal death. Brain Res 835:125–136
Tabuchi A, Matsumoto H (2001) Changes in cell wall properties of wheat (*Triticum aestivum*) roots during aluminum-induced growth inhibition. Physiol Plant 112:353–358

Taylor GJ, Stephens JL, Hunte DB, Bertsch PM, Elmore D, Rengel Z, Reid R (2000) Direct measurement of aluminium uptake and distribution in single cells of *Chara corallina*. Plant Physiol 123:987–996

Theiss C, Meller K (2002) Aluminum impairs gap junctional intercellular communications between astroglial cells in vitro. Cell Tissue Res 310:143–154

Tolrà R, Poschenrieder C, Barceló J (1996) Zinc hyperaccumulation in *Thlaspi caerulescens*. I. Influence on growth and mineral nutrition. J Plant Nutr 19:1531–1540

Valadez-Gonzalez N, Colli-Mull JG, Brito-Argaez L, Muñoz-Sanchez JA, Aguilar JJZ, Castano E, Hernandez-Sotomayor SMT (2007) Differential effect of aluminum on DNA synthesis and CDKA activity in two *Coffea arabica* cell lines. J Plant Growth Regul 26:69–77

Vanneste S, Inzé D, Beeckman T (2007) Auxin fuels the cell cycle engine during lateral root initiation. In: Inzé D (ed)Cell cycle control and plant development (Annals of Plant Reviews vol 32) . Blackwell, Oxford, pp 187–202

Vázquez MD, Poschenrieder C, Corrales I, Barceló J (1999) Change in apoplastic aluminium during the initial growth response to aluminium by roots of a tolerant maize variety. Plant Physiol 119:435–444

Walch-Liu P, Liu LH, Remans T, Tester M, Forde BG (2006) Evidence that L-glutamate can act as an exogenous signal to modulate root growth and branching in *Arabidopsis thaliana*. Plant Cell Physiol 47:1045–1057

Wang JP, Raman H, Zhou MX, Ryan PR, Delhaize E, Hebb DM, Coombes N Mendham N (2007) High-resolution mapping of the Alp locus and identification of a candidate gene HvMATE controlling aluminium tolerance in barley (*Hordeum vulgare* L.). Theor Appl Gen 115:265–276

Watanabe T, Osaki M, Tadano T (2001) Al uptake kinetics in roots of *Melastoma malabathricum* L.—an Al accumulator plant. Plant Soil 231:283–291

White PJ, Broadley MR (2003) Calcium in plants. Ann Bot 92:487–511

Wijesinghe DK, John EA, Beurskens S, Hutchings MJ (2001) Root system size and precision in nutrient foraging: responses to spatial pattern of nutrient supply in six herbaceous species. J Ecol 89:972–983

Wisniewska J, Xu J, Seifertová D, Brewer PB, Ruzicka K, Blilou I, Rouquié D, Benková E, Scheres B, Friml J (2006) Polar PIN localization directs auxin flow in plants. Science 312:883

Yao XL, Jenkins EC, Wisniewski HM (1994) Effect of aluminum-chloride on mitogenesis, mitosis, and cell-cycle in human short-term whole blood cultures: lower concentrations enhance mitosis. J Cell Biochem 54:473–477

Zhang WH, Rengel Z, Yan G 1999 Aluminium effects on pollen germination and tube growth of *Chamelaucium uncinatum*. A comparison with other Ca^{2+} antagonists. Ann Bot 84:559–564

Zheng K, Pan JW, Ye L, Fu Y, Peng HZ, Wan BY, Gu Q, Bian HW, Han N, Wang JH, Kang B, Pan JH, Shao HH, Wang WZ, Zhu MY (2007) Programmed cell death-involved aluminum toxicity in yeast alleviated by antiapoptotic members with decreased calcium signals. Plant Physiol 143:38–49

Communication and Signaling in the Plant–Fungus Symbiosis: The Mycorrhiza

Pascale Seddas, Vivienne Gianinazzi-Pearson, Benoit Schoefs, Helge Küster, and Daniel Wipf

Abstract The study of symbiotic mycorrhizal associations is of fundamental and practical interest, raising questions about not only interorganism coevolution but also the ecological significance of the symbiosis in sustainable plant production systems. The partners in these associations belong to the Basidiomycota, Ascomycota or Glomeromycota, and about 95% of extant land plants. Successful colonization of roots by mycorrhizal fungi and subsequent effects on plant processes depend on recognition processes resulting from coordinated genetic programs in both partners and must be driven, at each stage, by reciprocal signaling events. This chapter summarizes current knowledge on communication and signaling in the two most frequent mycorrhizal associations: arbuscular mycorrhiza and ectomycorrhiza.

1 Introduction

The term "mycorrhiza" refers to a symbiosis between plants and soil-borne fungi that colonize the cortical tissues of roots during periods of active plant growth. The partners in this association belong to the Basidiomycota, Ascomycota or Glomeromycota, and about 95% of extant land plants (Smith and Read 2008). Bidirectional movement of nutrients characterizes most types of mycorrhizal symbiosis: carbon (C) flows to the fungus whilst nutrients and water move via the fungus to the plant, thereby providing

P. Seddas, V. Gianinazzi-Pearson, B. Schoefs, and D. Wipf (✉)
Plante-Microbe-Environnement, INRA-CMSE UMR INRA 1088/CNRS 5184/Université Bourgogne BP 86510, 21065 Dijon Cedex, France,
e-mail: Daniel.wipf@dijon.inra.fr

H. Küster
Institute for Plant Genetics, Leibniz Universität Hannover, Herrenhäuser Str. 2, D-30419, Hannover, Germany

F. Baluška (ed.), *Plant-Environment Interactions*, Signaling and Communication in Plants, 45
DOI: 10.1007/978-3-540-89230-4_3, © Springer-Verlag Berlin Heidelberg 2009

a critical linkage between the plant root system and the soil. In depleted soils, nutrient uptake by mycorrhizal fungi can lead to improved plant vigor and reproduction. As a result, mycorrhizal plants are often more competitive and better able to tolerate environmental stresses (pathogen attack, drought…) than nonmycorrhizal plants.

At least seven different types of mycorrhizal associations have been defined: arbuscular mycorrhiza, ectomycorrhiza, orchid mycorrhiza, ericoid mycorrhiza, ectendomycorrhiza, arbutoid and monotropoid mycorrhiza, involving different groups of fungi and host plants and distinct morphology patterns (Brundrett et al. 1996; Smith and Read 2008). Orchids form mycorrhizas with basidiomycetes of various affinities where the fungi produce coils of hyphae within roots or protocorms of the plants. Here the fungi, some of which are saprophytes or parasites of other plants, transfer organic C to protocorms or heterotrophic orchids, and mineral nutrients to photosynthetic orchids. Ericoid mycorrhizas are formed between members of the Ericales and Ascomycota which develop hyphal coils in outer cells of the narrow “hair roots” of plants to which they transport mineral nutrients from the soil. Ectendomycorrhiza, arbutoid and monotropoid mycorrhiza associations are similar to ectomycorrhizal associations (see below), but have specialized anatomical features. In ectendomycorrhizas, formed primarily by *Pinus* and *Larix* species, the fungal mantle on the root surface may be reduced or absent, the Hartig net is usually well developed, but the hyphae penetrate into plant cells. The same species of fungus may form ectomycorrhiza with one plant species and ectendomycorrhiza with another. Some ericaceous plants form arbutoid mycorrhizas, where a mantle and Hartig net are present but, in addition, there is extensive intracellular development of hyphal coils in the outer cell layers of roots. Monotropoid mycorrhizas developed by achlorophyllous monotropes are somewhat similar in structure to arbutoid and ectendomycorrhizas except that they do not have a true haustorium-like structure but rather a short hyphal “peg” which penetrates the epidermal cells.

The most studied mycorrhizal associations are the arbuscular mycorrhiza (AM) and the ectomycorrhiza (ECM). AM associations, which represent the most ancient root symbiosis (estimated to exist since the Silurian/Ordovician period, ~450 Mya), result from interactions between fungi specific to the phylum Glomeromycota and the large majority of land plants (Krings et al. 2007; Redecker et al. 2000; Remy et al. 1994; Taylor et al. 1995). They are now ubiquitous and, in spite of the wide range of plant families involved, the structural and functional characteristics of AM are relatively constant. Here, the fungal symbiont colonizes the internal cortical tissues of roots, where it develops characteristic, ramified intracellular structures called arbuscules which gave their name to this type of mycorrhiza. Ectomycorrhizas subsequently evolved (about 200 Mya) as the organic matter content of soils increased. Today, in forest soils in the northern hemisphere, more than 95% of the root tips of boreal forest trees form ectomycorrhizal symbioses (Fransson et al. 2000). The diagnostic features of EM are the presence at the root surface of a mantle of fungal tissue, which can vary widely in thickness, color or texture, and a network-like structure of intercellular hyphae called the Hartig net, which penetrates between the outer cortical cells.

The study of mycorrhizal symbioses is of fundamental and practical interest, since it raises questions about not only interorganism coevolution but also about the ecological significance of mycorrhizal symbioses in sustainable agriculture and forestry. While early events leading to the appearance of mycorrhizal symbioses may have involved reciprocal genetic changes in ancestral plants and free-living fungi, the available evidence points largely to ongoing parallel evolution of the partners in response to environmental changes (Axelrod 1986; Cairney 2000). Successful colonization of roots by beneficial mycorrhizal fungi and subsequent microbial effects on plant processes depend on recognition processes, which result from coordinated genetical programs in both partners and must be driven, at each stage of the partner interactions, by reciprocal signaling events. Although the benefits of mycorrhizal symbioses to both plant and fungal partners are well described (Smith and Read 2008), our understanding of the molecular cross-talk and genetic programs driving plant–fungal recognition, mycorrhizal development and the maintenance of symbiotic interfaces, is still in its infancy, mainly due to difficulties in synchronizing developmental events in the mycorrhizal symbionts (Gianinazzi-Pearson et al. 2006, 2007; Harrison 1998, 2005; Martin et al. 2001, 2008). This chapter aims to summarize and interpret current knowledge on communication and signaling in AM and ECM associations, and to indicate future research routes in the quest for a more comprehensive picture of the events driving their formation and functioning.

2 Communication and Signaling in Arbuscular Mycorrhiza

2.1 Presymbiotic Events

AM symbiosis is established stepwise and comprises several well-defined stages which begin with the germination of fungal spores, the asymbiotic development of germ tube hyphae and presymbiotic morphomolecular modifications in fungal and plant cell behavior (Gianinazzi-Pearson 1996; Harrison 2005; Hause and Fester 2005). Spore germination occurs spontaneously in the absence of a host plant, but if the fungus does not sense a host root to colonize, the whole germ tube septates, the contents retract and the spore reverts to dormancy. Requena et al. (2002) have suggested that a gene coding a putative hedgehog protein with GTPase activity could be involved in this programmed cell death of hyphae, and this may occur due to a lack of stimulatory host compounds (Buée et al. 2000; Gianinazzi-Pearson et al. 2007) or the release of inhibitory compounds in the presence of a nonhost root (Gadkar et al. 2003; Nagahashi and Douds 2000).

2.1.1 Fungal Perception of Plant Signals Prior to Cell Contact

Plant signals in root exudates alert the AM fungi to the presence of a potential host and hence a vital source of carbon. They are perceived by germinated AM fungal spores prior to cell-to-cell contact, and they trigger a switch from the asymbiotic stage of development to an active presymbiotic growth phase which leads to intense hyphal branching in the vicinity of the root (Giovannetti et al. 1994; Buée et al. 2000). These changes, which convert germ tubes with limited growth potential into mycelium that has the capacity to initiate colonization of roots, are preceded by a rapid increase in mitochondrial activity, respiration rate and fungal gene expression (Tamasloukht et al. 2003; Besserer et al. 2006; Gianinazzi-Pearson et al. 2007; Bücking et al. 2008; Seddas et al.,in press). The vicinity of a host root or incubation with host root exudates stimulates H^+ effluxes in the subapical regions of hyphae, which are probably critical zones for the perception of root signals (Ramos et al. 2008). Such a response, which could generate electrical signals promoting the formation of a sufficiently important stimulus to depolarize the fungal membrane, could reflect the recognition of certain host molecule(s) by the fungal cell (Fromm and Lautner 2007; Ramos et al. 2008). The resulting H^+ ion gradients transmitted along the membrane surface may then drive a cascade of events leading to enhanced hyphal branching and growth.

Genetic screens have identified AM-defective plant mutants affecting spore germination and hyphal growth responses associated with early recognition (*pmi1* and *pmi2* in tomato, David-Schwartz et al. 2001, 2003; *nope1* in maize, Paszkowski et al. 2006). The occurrence of such a phenotype suggests that the mutations could have occurred in genes that are active in the biosynthetic pathway of a plant-derived signal (Paszkowski et al. 2006). The nature of these stimulatory signals has been discussed for a long time because roots release a variety of different compounds into the rhizosphere which could play the roles of stimulators or inhibitors of presymbiotic AM fungal growth (Dakora and Phillips 2002). Whether a single compound or multiple plant signals trigger the different responses in spores during the presymbiotic growth phase is still unknown (Jones et al. 2004). Proposed plant compounds that could be involved in early signaling include flavonoids (Gianinazzi-Pearson et al. 1989; Morandi 1996; Vierheilig et al. 1998; Vierheilig and Piché 2002; Vierheilig 2004; Soares et al. 2005; Scervino et al. 2005), volatiles (Bécard et al. 1992), mannitol (Kuwada et al. 2005) and strigolactone derivatives of the apocarotenoid biosynthetic pathway (Akiyama et al. 2005; Akiyama and Hayashi 2006; Bouwmeester et al. 2007). Recently, Bücking et al. (2008) reported that changes in catabolic metabolism as a response of an AM fungus to root exudates are not associated with significant changes in fungal gene expression and vice versa, indicating that some of the molecular processes are regulated at a post-translational rather than a transcriptional level.

2.1.2 Plant Perception of Fungal Signals Prior to Cell Contact

Some evidence has been provided to support the hypothesis that AM fungal signals (myc factors) are perceived by a host plant before cell contact. Root flavonoid levels increase when mycelium and spores are in the vicinity of a host root (Larose et al. 2002). Furthermore, the expression of plant genes such as *MtENOD11* (Kosuta et al. 2003) or those encoding proteins related to calcium-dependent signal transduction pathways (Weidmann et al. 2004) is activated by fungal molecules diffusing across membranes from germinated spores. More recently, Navazio et al. (2007) and Kosuta et al. (2008) demonstrated that diffusible AM fungal factors activate a rapid calcium response in soybean cell cultures or *Medicago truncatula* root hair cells before direct fungal contact. Calcium oscillations, which are only induced by branched hyphae in root hair cells, are likely to prime host cells for fungal colonization. The nature of these inductive AM fungal signals (myc factors) is unknown, but their perception is dependent on symbiosis-related plant genes and is altered in plant mutants where the fungus is unable to gain entry to epidermal cells (Weidmann et al. 2004; Kosuta et al. 2008).

2.2 AM Fungal Contact with Host Roots

AM fungi differentiate slightly swollen fungal hyphae, called appressoria, upon the first physical contact with a host root. This morphogenetic event, which only occurs on the epidermis of a host root, is a prerequisite for the fungus to penetrate the rhizodermal root cell layer before invading the root cortex (Giovannetti et al. 1993). It is the consequence of presymbiotic recognition between the plant and fungal symbionts, but studies of cell processes related to this developmental stage are still very much in their infancy.

2.2.1 Fungal Perception of Plant Signals During Appressoria Formation

Analysis of transcriptome modifications in germinated sporocarps of *Glomus mosseae*, triggered in synchrony with appressorium formation on parsley roots, identified 27 upregulated genes coding proteins with functions in signaling, transduction, general cell metabolism, defence/stress responses, or those of unknown function (Breuninger and Requena 2004). Among other upregulated fungal genes, two encoded proteins with a potential role in calcium-based signaling pathways and one a putative 14-3-3-like protein for which the same type of protein is induced during appressorium formation in the interaction between the hemibiotrophic pathogen *Magnaporthe grisea* and rice (Takano et al. 2003). It was suggested that a protein regulator could be acting by direct protein–protein interactions and that Ca^{2+} could be involved in the perception of a plant signal leading to appressorium formation (Breuninger and Requena 2004). In a study targeting a subset of *G. intraradices* genes with predicted functions in transcription, protein synthesis, primary/secondary metabolism or which have an unknown function, most of the genes were upregulated during appressorium development in compatible interactions with *M. truncatula* (Seddas et al. in press), whilst several showed no significant activation or downregulation during incompatible interactions with plants mutated for the symbiosis-related (SR) genes *MtDMI1* (coding a putative channel protein; Ané et al. 2004), *MtDMI2/MtSYM2* (coding a receptor-like leucine-rich kinase; Stracke et al. 2002; Endre et al. 2002) or *MtDMI3/MtSYM13* (coding a calcium/calmodulin-dependent protein kinase; Lévy et al. 2004; Mitra et al. 2004). These observations provide a first indication that symbiosis-related (SR) plant genes regulate AM fungal activity through stimulatory pathways and/or by controlling inhibitory factors. In line with this hypothesis, certain transcription factor genes are active in appressoria and upregulated in *G. intraradices* during intercellular root penetration of wild-type *M. truncatula*, but not during interactions with the mycorrhiza-defective *dmi3/Mtsym13* mutant (Gianinazzi-Pearson et al. in press). Inactivation of SR plant genes may modify fungal signaling events which could interfere with plant perception of the fungal symbiont and so impact on its morphological transition from appressorium differentiation to the biotrophic phase of root colonization. Host plants are able to control not only rhizodermal opening for fungal entry, but also fungal passage through the rhizodermis and intracellular passage through cortex cells (Marsh and Schultze 2001; Parniske 2004; Paszkowski et al. 2006).

2.2.2 Plant Perception of Fungal Signals Linked to Appressoria Formation

At the appressoria stage of AM interactions, fungal signals alert the plant, which then prepares the way for intracellular penetration, and this can be detected at both cellular and molecular levels. Before any cell penetration occurs, the epidermal cell nucleus is repositioned immediately under the appressorium (even in symbiosis-defective *M. truncatula* mutants) leading to the specific formation of a special prepenetration apparatus (PPA) (Genre et al. 2005). Fungal entry and growth of penetrating hyphae is then guided through the root epidermal cell by a cytoplasmic tunnel defined by this PPA (Genre et al. 2005). The PPA itself is not formed in symbiosis-defective *M. truncatula* mutants, suggesting that inactivation of symbiosis-related plant genes must affect cell activity involved in the fungal penetration processes (Gianinazzi-Pearson et al. 2007). The way in which putative fungal "myc factors" are perceived and transduced to facilitate hyphal penetration via the PPA is still unknown. Transcriptional activation of plant genes (such as nodulation-related genes, transcription factors, or those implicated in signal transduction or defence) has been demonstrated when appressoria are formed on host roots (Albrecht et al. 1998; Ruiz-Lozano et al. 1999; Chabaud et al. 2002; Weidmann et al. 2004; Sanchez et al. 2005), reflecting the activation of signal perception and transduction pathways in host roots by the fungus. Signal perception/transduction-related genes are not activated when appressoria develop at the root surface of *M. truncatula* mutated for the MtDMI3/MtSYM13 gene, suggesting again that calcium could be an intracellular messenger which plays a role in early root responses to arbuscular mycorrhizal fungi. Another signaling molecule that has been hypothesized to be involved in arbuscular mycorrhiza interactions is nitric oxide (NO) (Vieweg et al. 2005; Weidmann et al. 2004). Transcripts encoding nitrate reductase and nitrite reductase accumulate in *M. truncatula* roots in response to appressorium formation by *G. intraradices*, whereas their transcription is not affected when appressoria develop on roots of the mycorrhiza-deficient *Mtdmi3/Mtsym13* mutant (Gianinazzi-Pearson et al. 2007).

First analyses of Medicago GeneChip (Benedito et al. 2008) transcriptome profiles of *G. intraradices*-inoculated versus uninoculated *M. truncatula* wild-type and symbiosis-defective mutant roots have revealed a high number of differentially regulated genes in each plant genotype (Seddas, Kuester, Becker, Gianinazzi-Pearson, unpublished data). The expression of about 400 genes is significantly modulated (250 upregulated) in wild-type plants, 700 (320 downregulated) in the *Mtdmi1* mutant, 250 (180 downregulated) in the *Mtdmi2/Mtsym2* mutant, and 865 (670 downregulated) in the *Mtdmi3/Mtsym13* mutant. Among these modulated genes, only a few transcription factor genes are modulated in wild type (seven) and *Mtdmi2/Mtsym2* (five) roots, whilst more than 25 and 50 are downregulated in *Mtdmi1* and *Mtdmi3/Mtsym13* mutant roots, respectively, when appressoria are formed. Likewise, fewer genes implicated in cellular signalization are modulated in wild-type and *Mtdmi2/Mtsym2* roots, as compared to *Mtdmi1* and *Mtdmi3/Mtsym13* mutant roots. Among the genes modulated after *G. intraradices* inoculation, 11 that are upregulated in wild-type plants are downregulated in one or two symbiosis-defective mutants, whereas 63 genes that are not modulated in wild-type plants are downregulated (45) or upregulated (18) in one or two of the symbiosis-related mutants. This underlines the very complex molecular mechanisms that must be triggered when an AM fungus comes into contact with roots, and reveals that the mutation of only one symbiosis-related gene can lead to the modulation of several hundred others (Seddas, Kuester, Becker, Gianinazzi-Pearson, unpublished data). Moreover, some genes implicated in primary metabolism, membrane transport or plast metabolism are activated only in wild-type and in *Mtdmi1* roots. This could be due to the fact that *Mtdmi1* is not a tight mutant; under optimum mycorrhizal conditions it allows root penetration and intracellular hyphal development (Morandi et al. 2005). In *Mtdmi2/Mtsym2* and *Mtdmi3/Mtsym13* roots, genes involved in cell wall synthesis or responses to pathogens are activated. These observations reinforce the hypothesize that nonpenetration of an AM fungus in symbiosis-defective mutant roots could be related to the elicitation of defence reactions usually associated with plant responses to pathogens (Gollotte et al. 1993; Ruiz-Lozano et al. 1999; Gianinazzi-Pearson et al. 2007; Garcia-Garrido and Ocampo 2002), the suppression of which during initial interactions between AM symbionts would favor establishment of the symbiosis (Pozo and Azcon-Aguilar 2007). The mechanisms underlying the control of defence responses during mycorrhizal interactions and the role of symbiosis-related plant genes in such a process remain to be elucidated.

2.3 Arbuscule and Symbiotic Interface Development

Once inside the root tissues, an AM fungus proliferates inter- and/or intracellularly throughout the parenchymal cortex and differentiates highly branched haustoria (arbuscules) within cortical cells, which create an extended symbiotic interface consisting of the fungal plasma membrane and cell wall separated from a plant-derived periarbuscular membrane by an interfacial matrix layer. This symbiotic interface is assumed to be the primary site of bidirectional nutrient transfer between the symbionts (Gianinazzi-Pearson 1996). Arbuscules are ephemeral structures that remain active for only a few days and then senesce and collapse. Although there is a relatively large volume of literature describing the structural characteristics of symbiotic interfaces in arbuscule-containing plant cells (Smith and Read 2008), nothing is known about the molecular mechanisms controlling their development or function.

2.3.1 Fungal Perception of Plant Signals Within the Symbiosis

Determining molecular events linked to the symbiotic stages of AM fungal development is a difficult task because fungal tissues are imbricated with the root tissues. However, transcript profiling during the establishment of a functional AM does suggest that the host plant may exert some control over fungal gene expression in symbiotic tissues. A limited number of AM fungal genes, mainly related to membrane transport and nutrient exchange processes with host cells, have been reported to be differentially modulated within the established symbiosis (Balestrini and Lanfranco 2006). More recent monitoring of a subset of *G. intraradices* genes implicated in transcription, protein synthesis, primary/secondary metabolism or which have an unknown function revealed a clear enhancement of fungal gene expression when arbuscules are developed within *M. truncatula* roots (Seddas et al., in press). Expression of the same set of genes was downregulated when *G. intraradices* developed in an arbuscule-defective pea mutant (*Pssym36*; Duc et al. 1989), whereas it was upregulated in a mutant characterized by more rapid arbuscule turnover (*Pssym40*; Jacobi et al. 2003; Kuznetsova et al., unpublished). These observations suggest that the plant does indeed control arbuscule formation and/or functioning, and that the fungal symbiont perceives plant signals that modulate its development and activity inside the root. They are in agreement with conclusions, based on mutants such as *Pram1* of maize (Paszkowski et al. 2006), *nts1007* of soybean (Meixner et al. 2005) or *Pssym33* and *Pssym40* of pea (Jacobi et al. 2003), that the timing and progress of AM fungal development in the symbiosis can be accelerated or slowed down by factor(s) encoded by the host (Paskowski et al. 2006). The recent development of microdissection and in situ RT–PCR techniques to localize fungal transcripts within mycorrhizal tissues (Siciliano et al. 2007; Seddas et al. 2008) provides the possibility of obtaining more information about molecular responses of fungal structures during arbuscule ontogenesis and, consequently, a better understanding of the processes driving the function of these structures in symbiotic interactions with host roots.

2.3.2 Plant Perception of Fungal Signals Within the Symbiosis

It is possible that AM fungal signals prepare cortical cells of host roots for colonization and for the differentiation of the periarbuscular membrane linked to the development of arbuscules. In this context, a recent in vivo cellular investigation of host cell responses during AM fungal colonization of carrot and *M. truncatula* roots has shown that nuclear repositioning and the assembly of a PPA-like intracellular structure precedes not only epidermal cell colonization (see Sect. 2.2) but also arbuscule formation in the inner cortex. Furthermore, PPAs can be induced in adjacent cortical cells ahead of the advancing fungus, which argues in favor of sequential cell-to-cell signaling. Such cellular reorganization, together with changes in plant gene expression and (re)localization of membrane and matrix proteins that facilitate nutrient transfer between the symbionts (Harrison 2005), is probably linked to the perception of fungal signal(s) by the plant, but the molecular nature of these has not yet been identified. Likewise, fungal–plant communication must be involved in the localized activation of defence-related responses in host cells accommodating arbuscule development (Dumas-Gaudot et al. 2000). Whilst such host reactions may somehow regulate AM fungal development within root tissues (Catford et al. 2006; Larose et al. 2002; Vierheilig 2004), their expression must be compatible with symbiosis establishment and activity. For example, Pozo and Azcon-Aguilar (2007) have proposed that the partial suppression of salicylic acid-dependent plant defense responses associated with the initial stages of AM development is compensated for by the enhancement of jasmonic acid-regulated responses during arbuscule formation (see 2.4.2). In addition, fungal proliferation in host cortical cells could be facilitated by the induction of a reactive oxygen species-inactivating system in signal transduction between the symbionts (Lanfranco et al. 2005) and of a hemoglobin-encoding gene in the suppression of NO-based defense processes (Vieweg et al. 2005).

2.4 *Role of Plastids in Communication in AM*

Plastids represent a plant cell compartment which plays a crucial role in plants because most of the cellular anabolic reactions take place there, both under normal conditions and in the case of stress. Apart from their capacity to produce carbohydrates through photosynthesis, plastids are involved in many biochemical pathways that are used to synthesize other "elementary" molecules and in the production of compounds somehow involved in cell–cell and/or plant–plant communications (Bick and Lange 2003; Walter et al. 2002; Dudareva et al. 2005; Okada et al. 2007) (Fig. 1). There is an increasing amount of recent data in favor of the involvement of plastids in plant–fungal communication in the AM symbiosis.

2.4.1 Signal Reception and Transduction

The identification of DMI1 protein as part of the signaling pathway associated with appressoria formation has been already mentioned (Sect. 2.1). DMI1, which is involved in the generation of calcium oscillations, is localized in the nucleus envelope

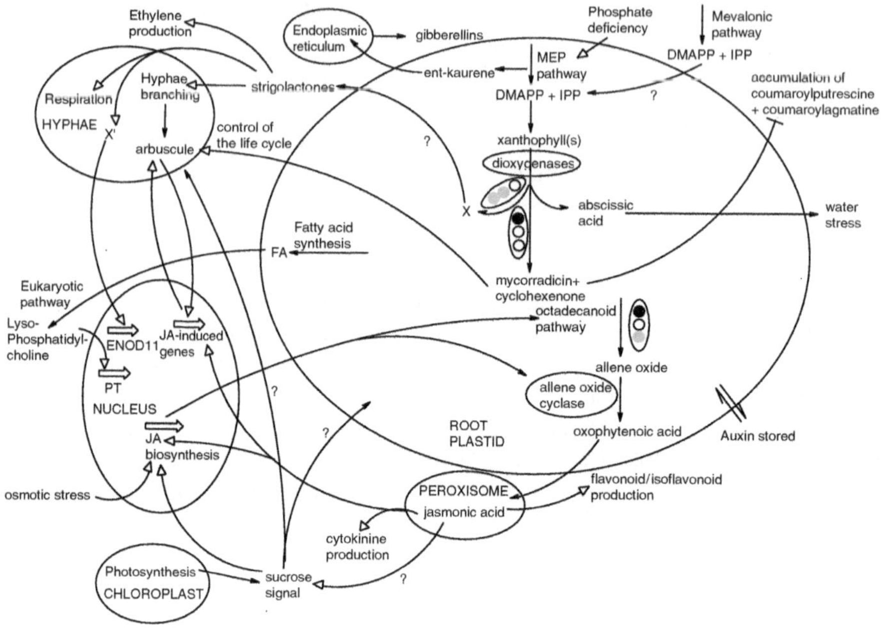

Fig. 1 Root plastids: important partners in plant–fungus communication. Plastids are involved in many biochemical pathways that are used to synthesize various "elementary" molecules and also other molecules such as hormones that are involved in private (cell–cell) and/or public (plant–plant) communications. The biochemical pathways are indicated by *arrows with closed heads*. The positive and negative actions of compounds are indicated by *arrows with open heads* and *T*, respectively. The *spots* indicate when the biosynthetic pathway is active (*top*, prearbuscule; *middle*, arbuscule; *bottom*, senescent arbuscule; *black*, inactive; *white*, active; *gray*, no data)

(Riely et al. 2007), and the protein homologs CASTOR and POLLUX of *Lotus japonicus* carry a plastid transit peptide (Imaizumi-Anraku et al. 2005), suggesting the involvement of plastids in signaling in AM. Recently, it was suggested that the proteins NUP133 and NUP96 of the nuclear pore (Paullilo and Fahrenkrog 2008) are also involved in the calcium oscillations that occur during the early steps of symbiotic root colonization (Kanamori et al. 2006). How plastids and nucleus cooperate in the signal transduction remains to be understood.

Root colonization by AM fungi is accompanied by a tremendous increase in mitochondria and plastid numbers in arbuscule-containing cells (Fester 2008). Metabolic profiling of roots of *M. truncatula* has shown that upon root colonization by an AM fungus, root plastid metabolism is reoriented to the synthesis of several types of compounds, including amino acids, fatty acids and secondary carotenoids (Lohse et al. 2005; Schliemann et al. 2008) (see below). Other plant taxa have to be tested in order to determine whether the modifications that occur in root plastids in response to arbuscule formation constitute a general feature. Aside from the possibility of the direct involvement of root plastids in the signaling between fungus and plants, plastids could partially or completely synthesize molecules such as the phytohormones that may participate in the communication network between the two partners.

2.4.2 *Lipid and Lipid Derivatives as Signaling Molecules*

Lysophosphatidylcholine

Phosphate is probably the most important nutrient transferred from fungal to plant cells in AM symbiosis. Phosphate transporters (PT) are necessary for this transfer, and several mycorrhiza-inducible PT have been identified (Javot et al. 2007; Karandashov and Bucher 2005; Karandashov et al. 2004). In potato and tomato, the signaling molecule that induces the transcription of *PT3* and *PT4* genes is the lysolipid lysophosphatidylcholine (Drissner et al. 2007) (Fig. 1). Synthesis of this signal may require cooperation between plastid and cytosol compartments in the host cell. In plant cells, the acyl carrier protein (ACP)-dependent de novo fatty acid synthesis is restricted to organelles (Ohlrogge et al. 1979), and essentially all acyl chains are produced in plastids (Ohlrogge and Browse 1995; Schwender and Ohlrogge 2002). The pathway of incorporation for the initial products of fatty acid synthesis esterified to ACP that predominates in root cells involves the hydrolysis of the acyl–ACP thioester bond during the export of acyl from the plastids prior to fatty acid reactivation and then its incorporation into glycerolipids by acyltransferases in the cytosol (Roughan and Slack 1982; Somerville and Browse 1991) (Fig. 1).

Secondary Apocarotenoids

Isoprenoid lipid derivatives constitute one of the largest groups of plant metabolites (Withers and Keasling 2007). Many of these are produced through primary metabolism and are essential for growth and development, while others are synthesized through the secondary metabolism and are involved in responses to modifications in the environment (Fester 2008; Lemoine et al. 2008). A continuous supply of the precursors isopenthenyl diphosphate (IPP) and dimethylallyl diphosphate (DMAPP) is ensured in plants through two pathways that take place in separate cell compartments. The first one is located in the cytosol, whereas the second one is localized in plastids and is generally known as the nonmevalonate or methylerythritol phosphate (MEP) pathway (e.g., Lemoine et al. 2008; Fester 2008) (Fig. 1). The cytoplasmic pathway serves to synthesize sesquiterpenes, triterpenes and polyterpenes, whereas the plastid pathway serves to synthesize monoterpenes, diterpenes, tetraterpenes, platoquinones and tocopherols and also sesquiterpenes (Dudareva et al. 2005; for a review, see Chappell 2002). Several terpernoids are thought to be involved in the communication between plant and fungal cells in mycorrhizal interactions: strigolactones, mycorradicins, and cyclohexenones.

Strigolactones in root exsudates are able to induce hyphal branching (Akiyama et al. 2005; Akiyama and Hayashi 2006) and to activate respiration (Besserer et al. 2006) of AM fungi. They are widely occurring molecules (Bouwmeester et al. 2007; Yoneyama et al. 2008) which belong to a group of sesquiterpenoid lactones derived from the cleavage of a *cis*-epoxycarotenoids (Matusova et al. 2005; Humphrey and Beale 2006; Bouwmeester et al. 2007). Strigolactone production and exudation by sorghum roots is promoted by nitrogen and/or phosphorus deficiency (Yoneyama et al. 2007). Besides their effects on hyphal branching, a chemoattractive role of root exsudates in guiding hyphae to the host root has also been suggested (Sbrana and Giovannetti 2005). Plastids and cytosol may cooperate to produce the IPP molecules necessary for strigolactone production (Humphrey and Beale 2006) (Fig. 1), with the cleaved carotenoid fragment being exported to the cytosol to be transformed to strigol (Humphrey and Beale 2006).

Mycorrhizal development in some plant species results in a yellowing of root tissues (Jones 1924; Klingner et al. 1995a, b; Walter et al. 2000; Fester et al. 2002a, b), reflecting the reorientation of plastid metabolic activity towards the synthesis of secondary carotenoids and apocarotenoids (Fester et al. 2002a). In the case of AM, these have been named mycorradicins and identified as acyclic C14 apocarotenoid polyenes (Bothe et al. 1994; Klinger et al. 1995a; Schliemann et al. 2006). Although Fester et al. (2002a) demonstrated that some highly mycorrhizal roots may completely lack mycorradicin, apocarotenoid synthesis seems to be important for AM establishment because mutants deficient of (Fester et al. 2002a) or with reduced carotenoid biosynthesis capacity (Fester 2008) show a reduced development of functional symbiotic structures (Floß et al. 2008). In addition to mycorradicins, esterified mycorradicins and glycosylated C13 cyclohexenone apocarotenoid derivatives can accumulate in mycorrhizal tissues (Maier et al. 1995; Strack and Fester 2006; Schliemann et al. 2008). However, the application of derivatives of the cyclohexenone blumenin to AM roots strongly inhibits fungal colonization and triggers a reduction in arbuscule formation during the early stages of mycorrhizal development. The fact that increases in secondary apocarotenoid levels occur after the onset of mycorrhizal formation strongly suggests that they participate in the network regulating plant cell and/or fungal hyphal development and not in the early recognition phase of the symbiosis. The time course of mycorradicin accumulation coincides with the accumulation of ROS, which is known to be abundant in the vicinity of arbuscules (Salzer et al. 1999; Fester and Hause 2005). The apocarotenoids which accumulate with mycorrhization are derived from xanthophyll molecules that have still to be identified.

Phytohormone Signaling Pathways

Several plant hormones are apocarotenoids (abscisic acid), terpenoid derivatives (gibberellins) or lipid-derived molecules (jasmonic acid), and their production can be modulated during mycorrhizal interactions.

A strong increase in abscisic acid levels has been reported in mycorrhizal roots of *Zea mays* and *Glycine max*, but not in those of *Boutelia gracilis* (Allen et al. 1982; Hause et al. 2007). Abscisic acid is an apocarotenoid that is derived from the xanthophyll *cis*-neoxanthin through the catalytic action of 9-*cis*-epoxy carotenoid dioxygenase in plastids (Fig. 1). Allen et al. (1982) found, on the other hand, that gibberellin concentrations decrease in mycorrhizal roots. Gibberellins are derived from tetracyclic terpenoids and are therefore made from isoprenoid units. The first steps of their biosynthetic pathway, up to ent-kaurene production, are catalyzed by plastid enzymes (Hedden and Phillips 2000; Helliwell et al. 2001) (Fig. 1).

The possible involvement of jasmonic acid (JA) in the process of mycorrhization was first inferred from leaf application experiments (Regvar et al. 1996; Ludwig-Muller et al. 2002). The amount of JA and its conjugates increases concomitantly in cells containing arbuscules through a cell-specific expression of genes coding for JA biosynthetic enzymes and of jasmonate-induced genes (Hause et al. 2002; Strassner et al. 2002; Hause et al. 2007). The first biosynthetic steps of JA, up to the formation of oxophytodienoic acid, are localized in the plastids (Hause et al. 2007), and the last steps of the biosynthetic pathway occur in peroxisomes (Strassner et al. 2002) (Fig. 1). Increases in jasmonate levels only occur after the onset of mycorrhization, so these molecules must somehow be associated with late plant–fungal interactions and not the early recognition phase of the symbiosis (Hause et al. 2007; Vierheilig 2004). The JA or derivatives synthesized in colonized cells may regulate the metabolism of other cells, because jasmonates have been shown to act as mobile signals (Schilmiller and Howe 2005). On the other hand, reductions in the level of allene oxide cyclase (AOC1), the last enzyme in the plastid pathway, reduce JA levels in roots, which in turn leads to an overall reduction in arbuscule frequency and alterations in their development or in the root colonization program as a whole (Isayenkov et al. 2005) (Fig. 1). Jasmonates could enhance the carbon sink strength of mycorrhizal tissues, therefore increasing carbohydrate biosynthesis in chloroplasts and their transportation to the root. This view is supported by the fact that the genes involved in coding for enzymes that function in sink/source relationships, such as an extracellular invertase, are jasmonic acid responsive (Thoma et al. 2003) and expressed in cells that require a high carbohydrate supply (Godt and Roitsch 1997), like those containing arbuscules.

3 Communication and Signaling in Ectomycorrhiza (ECM)

In contrast to some other plant–microbe interactions (Dénarié et al. 1996; Moller and Chua 1999), and even to AM (see Sect. 3), much less is known about the nature of the signaling molecules and the molecular basis of signal perception and transduction in ECM. Host plants release critical metabolites into the rhizosphere that are able to trigger basidiospore germination (Fries et al. 1987), growth of hyphae towards the root (Horan and Chilvers 1990), and the early developmental steps of ECM formation (Béguiristain and Lapeyrie 1997; Salzer et al. 1997; Ditengou and

Lapeyrie 2000). According to Martin et al. (2001), molecular control of interactions between symbionts can be classified as follows:

- Tropism of hyphae towards host tissues via rhizospheric signals
- Hyphal attachment and invasion of host tissues by hyphae via adhesins and hydrolases
- Induction of organogenetic programs in both fungal and root cells via hormones and secondary signals
- Facilitating survival of the mycobiont despite plant defense responses
- Coordinating strategies for exchanging carbon and other metabolites (e.g., vitamins) for in planta colonization and for growth and activity of the soil fungal web in mineral transfer from the soil

3.1 Possible Signals in the ECM

Early morphological changes during ectomycorrhizal development have been identified (Kottke and Oberwinkler 1987; Horan et al. 1988). Based on current knowledge of the molecules released in other plant–microbe interactions, the early plant host signals secreted into the rhizosphere can include flavonoids, diterpenes, hormones and various nutrients (Martin et al. 2001). Several plant metabolites have been shown to induce striking modifications in hyphal morphology. Rutin, a phenol compound found in eucalyptus root exudates, may be a signal in ectomycorrhizal symbiosis, as it stimulates the hyphal growth of certain *Pisolithus tinctorius* strains at picomolar concentrations (Lagrange et al. 2001). On the other hand, the tryptophan derivative hypaphorine is secreted by *P. tinctorius* and can arrest root hair elongation and stimulate the formation of short roots in the plant host, possibly acting as an antagonist of the plant hormone auxin (Martin et al. 2001). On the fungal side, root exudates have been shown to stimulate an enhanced accumulation of fungal molecules such as hypaphorine, the betaine of tryptophan (Béguiristain and Lapeyrie 1997), that can induce morphological changes in root hairs of seedlings. Hypaphorine is produced in larger amounts by *P. tinctorius* during mycorrhizal development (Béguiristain and Lapeyrie 1997). Ditengou and Lapeyrie (2000) report an antagonistic effect of hypaphorine on indole-3-acetic acid (IAA).

The production of hormones, including auxins, cytokinins, abscisic acid and ethylene, by ectomycorrhizal fungi was first reported in the early 1990s (Gogala 1991). Many studies indicate that changes in auxin balance are a prerequisite for mycorrhiza organogenesis (Rupp et al. 1989; Gay et al. 1994; Karabaghli-Degron et al. 1998; Kaska et al. 1999) (e.g., short root development). The presence of plant-derived molecules in the rhizosphere could be sufficient to enhance the biosynthesis of hormones by ectomycorrhizal fungi (Rupp et al. 1989), which induce morphological changes leading to symbiosis development. For example, an IAA-upregulated cDNA, referred to as Pp-C61, was isolated by the differential screening of a cDNA library constructed from auxin-treated roots of the ECM host tree *Pinus pinaster*

(Reddy et al. 2003). Pp-C61 is present as a single copy in the *P. pinaster* genome, and homologous genes were detected in other gymnosperm and angiosperm trees. The fact that Pp-C61 is transcriptionally regulated by auxin suggests that Pp-C61 activation corresponds to a reaction in response to fungal colonization.

Hydrophobins, a class of fungal cell wall proteins involved in establishing cell–cell or cell–surface contact, are also probably involved in fungus–plant communication in ECM. A class I hydrophobin (HYD1) was purified from the culture supernatant of *Tricholoma terreum* (Mankel et al. 2002). The coding gene (hyd1) expression pattern suggests that hydrophobins might be involved in host recognition and in the host tree specificity of the fungus.

Mitogen-activated protein kinase (MAPK) signal transduction cascades are used by fungi to modulate their cellular responses to environmental conditions, in mating, and for cell-wall integrity. The yeast extracellular signal-regulated kinase (YERK1) is the most thoroughly investigated MAPK subfamily involved in mating response (Fus3) and nitrogen starvation (Kss1). The first MAPK from an ectomycorrhizal fungus was cloned from *Tuber borchii* (TBMK) (Menotta et al. 2006). It belongs to the YERK1 (yeast extracellular regulated kinase subfamily). TBMK is present as a single copy in the genome, and the codified protein was phosphorylated during the interaction with the host plant, *Tilia americana*. TBMK partially restores the invasive growth of *Fusarium oxysporum* that lack the *fmk1* gene. This suggests that protein kinase activity may play an important role during the interaction of *T. borchii* with its host plant by modulating the genes needed to establish symbiosis, leading to the synthesis of functional ectomycorrhizae.

3.2 Cytoskeleton and Signal Transduction

The integration of signals received by a cell and their transduction to targets are essential actions for all cellular responses. The cytoskeleton has been identified as being a major target of signaling cascades in animal, plant and yeast cells (Alberts et al. 2002). The cytoskeleton, which is unique to eukaryotic cells, is a dynamic three-dimensional (3D) structure that fills the cytoplasm with an extensive system of protein filaments enabling eukaryotic cells to organize their interiors and to perform various directed functions. The most abundant components of the cytoskeleton in mammalian cells are microfilaments (MFs), microtubules (MTs), and intermediate filaments. The latter structures, however, are yet to be identified in plant and fungal cells. The structures of MTs and MFs and the expression of their structural subunits, actin and tubulins, have been investigated in ectomycorrhizal fungi, non-mycorrhizal roots of Scots pine, and symbiotic ectomycorrhizal roots (Salo et al. 1989; Niini and Raudaskoski 1993; Timonen et al. 1993; Niini et al. 1996; Raudaskoski et al. 2001, 2004). Scots pine roots have three α- and three β-tubulin isoforms, whilst two additional isoforms of α-tubulin are detected in ectomycorrhizal roots, which suggests there are cellular level changes involving MTs when the two partners come into contact (Niini et al. 1996). Three α- and two β-tubulins that remain unchanged, even during the symbiosis, have been similarly identified in the ectomycorrhizal fungus *S. bovinus*. The presence of two and four actin isoforms in *P. sylvestris* lateral root tips and short roots, respectively, and two actin isoforms in *S. bovinus* has also been reported (Niini et al. 1996). The fungal tubulins (Niini and Raudaskoski 1998) and actins (Tarkka et al. 2000) are constitutively expressed at the mRNA and protein levels, suggesting that the reorganization of the cytoskeleton during ectomycorrhizal formation of *S. bovinus* with the *P. sylvestris* short roots is not mediated via differential expression of these genes. Ectomycorrhizal association, however, leads to major changes in the growth patterns of both plant and fungal partners (Niini 1998; Barlow and Baluska 2000; Raudaskoski et al. 2001). On the basis of the visualization of the MTs and MFs in vegetative hyphae of *S. bovinus* and in ectomycorrhiza (Timonen et al. 1993; Raudaskoski et al. 2001, 2004), it has been deduced that the cytoskeleton plays a role in fungal morphogenesis during the formation of ectomycorrhiza.

The small GTPases Cdc42 and Rac1, the regulators of the actin cytoskeleton in eukaryotes, have been isolated from the ectomycorrhizal fungus *Suillus bovinus* (Hanif 2004). IIF microscopic analysis suggests that the small GTPases Cdc42 may play a significant role in the polarized growth of *S. bovinus* hyphae and may regulate fungal morphogenesis during ectomycorrhizal formation by reorganizing the actin cytoskeleton. A small GTPase (TbRhoGDI) was more recently isolated from the ectomycorrhizal fungus *T. borchii* (Menotta et al. 2008). The specificity of the actions of TbRhoGDI was underscored by its inability to elicit a growth defect in *Saccharomyces cerevisiae* or to compensate for the loss of a *Dictyostelium discoideum* RhoGDI.

3.3 Impact of Nutrient Levels and Transport in Plant–Fungus Communication

Nutrient transport, namely hyphal absorption from soil solution, nutrient transfer from the fungus to plant, and carbon movement from plant to fungus, is a key feature of mycorrhizal symbioses. Phosphorus, nitrogen and carbohydrate are considered to be the main nutrients transferred by the mycorrhizal symbiosis, although a supply of water and trace elements can also be very important under certain conditions (Smith and Read 2008). Due to the importance of trophic exchanges in the mycorrhiza, it is likely that the levels/concentrations of the nutrients play an important role in the communication between both partners. Previous studies have clearly demonstrated that changes in the diversity of microbial consortia associated with plant roots dramatically alter the physiological status of host plants, including nitrogen, phosphorus and carbon allocation and sequestration (Liu et al. 2007). On the other hand, changes in the host physiology induce striking changes in symbiotic and pathogenic microbial populations through modifications of the carbon flux. It has been suggested that when a fungus colonizes a plant, the nature and dynamics of nutrient exchange determine the outcome of their interaction (Divon and Fluhr 2001). Plant and fungal cells must be "reprogrammed" to fulfil the task of massive nutrient transfer.

Nutrient-dependent regulation of gene expression in ectomycorrhiza has been investigated for sugar (Nehls et al. 1998; Nehls 2004) and nitrogen (Benjdia et al. 2006; Müller et al. 2007) using hexose importer genes and di- and tripeptide importer genes respectively. In *Hebeloma cylindrosporum* cultures, the expression of di- and tripeptide importer genes was under the control of both the external concentration (and nature) of nitrogen and the internal concentration of amino acids. Further studies have shown that the expression of several transporter genes from this mycorrhizal model fungus is under the control of the external C/N ratio (Avolio et al., unpublished).

In an axenic *Ammanita muscaria* culture, the expression of sugar importer genes is regulated by a threshold response mechanism that is dependent on the extracellular monosaccharide concentration (Nehls et al. 2001a). In functional ectomycorrhizas, elevated hexose transporter gene expression was exclusively observed in hyphae of the Hartig net (Nehls et al. 2001a). Differences in the apoplastic hexose concentration at the Hartig net versus the fungal sheath could be a signal that regulates fungal physiological heterogeneity in ectomycorrhizas (Nehls et al. 2001b; Nehls 2004). A microarray hybridization (800 tentative genes) assay indicates that (for *A. muscaria*) sugar-dependent regulation of fungal gene expression caused by differences in the apoplastic hexose concentration at the plant–fungus interface versus the fungal sheath may explain some of the local adaptations of fungal physiology in functional ectomycorrhizas. Results obtained for the same gene families in *Laccaria bicolor* show that the extent of fine-tuning of EM fungal physiology by sugar regulation might be species dependent, and this issue must be further addressed in the future (Nehls 2008).

3.4 How Do ECM Fungi Bypass Plant Defense Reactions?

To develop the symbiosis, prior to establishing functional exchanges, mycorrhizal fungi must be able to bypass host defense mechanisms. Some nonspecific, broad-spectrum responses (e.g., those involving metallothioneins, chitinases, glutathione S-transferases and peroxidases) are clearly activated in host plants when ectomycorrhizal fungi approach the roots (Frettinger et al. 2007), and as they penetrate the root and progress through the apoplastic space (Duplessis et al. 2005; Le Quéré et al. 2005). A challenge for future research is to identify the mechanisms that allow the fungus to escape plant defense mechanisms. An effect of fungal volatile compounds (VOC) on the growth and oxidative metabolism of a nonhost plants has recently been described (Splivallo et al. 2007). In a closed chamber bioassay, the volatiles produced by *Tuber melanosporum*, *Tuber indicum* and *T. borchii* fruiting bodies inhibited *A. thaliana* growth in terms of root length and cotyledon leaf size, and in some cases induced a bleaching of the seedlings, thus indicating toxicity.

Even though limited to laboratory observations, these results highlight a hitherto unknown function of fungal VOC: as molecules that mediate fungal–plant interactions in ECM.

3.5 Toward the Identification of Ectomycorrhiza-Specific Genes?

Variation in gene expression reflects modifications in the development/formation of the ectomycorrhiza. In the last decade several transcriptomic studies have shown variations in gene expression patterns related to changes in the morphology during symbiosis development (e.g. Duplessis et al. 2005; Herrmann and Buscot 2007; Kruger et al. 2004; Le Quéré et al. 2004, 2005, 2006; Menotta et al. 2004; Wright et al. 2005). So far, no ectomycorrhiza-specific gene has been identified. Nevertheless, the recent release of genome sequences from the host tree *Populus trichocarpa* (Tuskan et al. 2004) and the ectomycorrhizal fungus *Laccaria bicolor* (Martin et al. 2008) offer new perspectives. For example, analysis of the *L. bicolor* genome revealed that this ECM basidiomycete must have both saprotrophic and mutualistic abilities (Martin et al. 2007, 2008). By comparing the *L. bicolor* genome with closely related saprophytic fungi such as *Coprinus cinerea*, it should be possible to catalog the genetic differences that might underlie their different life habits and thus the interactions with the plant partner.

4 Conclusion and Future Prospects

As shown by the present review, we are only now beginning to identify and analyze the nature of signals exchanged in mycorrhizal symbiosis as well as their transduction between plant and fungal partners. A combination of ecological, biochemical and molecular approaches (e.g., availability of new genome sequences) may help us to identify signals, pathways, etc., and to get a clearer picture of the functioning of the mycorrhiza, which will enable better use of mycorrhiza in sustainable agriculture and forest management.

References

Akiyama K, Hayashi H (2006) Strigolactones: chemical signals for fungal symbionts and parasitic weeds in plant roots. Ann Bot 97:925–931

Akiyama K, Matsuzaki K, Hayashi H (2005) Plant sesquiterpenes induce hyphal branching in arbuscular mycorrhizal fungi. Nature 435:824–827

Alberts, B., Johnson, A., Lewis, J., Raff, M., Roberts, K., and Walter, P. 2002. The Cytoskeleton, p. 907-982, Molecular Biology of The Cell, fourth ed. Garland Science, a member of the Taylor and Francis group, 29 west 35th street, New York, NY 10001–2299.

Albrecht C, Geurts R, Lapeyrie F, Bisseling T (1998) Endomycorrhizae and rhizobial Nod factors both require *SYM8* to induce the expression of the early nodulin genes *PsENOD5* and *PsENOD12A*. Plant J 15:605–614

Allen MF, Moore TS, Christensen M (1982) Phytohormone changes in *Bouteloua gracilis* infected by vesicular-arbuscular mycorrhizae. II. Altered levels of gibberellin-like substances and abscisic acid in the host plant. Can J Bot 60:468–471

Ané J-M, Kiss GB, Riely BK, Penmetsa RV, Oldroyd GED, Ayax C, Lévy J, Debellé F, Baek J-M, Kalo P, Rosenberg C, Roe BA, Long SR, Dénarié J, Cook DR (2004) *Medicago truncatula DMI1* required for bacterial and fungal symbioses in legumes. Science 303:1364–1367

Axelrod DI (1986) Ceoconic history of some western American pines. Ann Bot Gard 73:565–641

Balestrini R, Lanfranco L (2006) Fungal and plant gene expression in arbuscular mycorrhizal symbiosis. Mycorrhiza 16:509–524

Barlow PW, Baluska F (2000) Cytoskeletal perspectives on root growth and morphogenesis. Ann Rev Plant Physiol Plant Mol Biol 51:289–322

Bécard G, Douds DD, Pfeffer PE (1992) Extensive hyphal growth of vesicular-arbuscular mycorrhizal fungi in the presence of CO_2 and flavonols. Appl Environ Microbiol 58:821–825

Béguiristain T, Lapeyrie F (1997) Host plant stimulates hypaphorine accumulation in *Pisolithus tinctorius* hyphae during ectomycorrhizal infection while excreted fungal hypaphorine controls root hair development. New Phytol 136:525–532

Benedito V, Torez-Jerez I, Murray J, Andriankaja A, Allen S, Kakar K, Wandrey M, Verdier J, Zuber H, Ott T, et al. (2008) A gene expression atlas of the model legume *Medicago truncatula*. Plant J 55:504–513

Benjdia M, Rikirsch E, Müller T, Morel M, Zimmermann S, Chalot M, Frommer WB, Wipf D (2006) Peptide uptake in the ectomycorrhizal fungus *Hebeloma cylindrosporum*: characterization of two di- and tripeptide transporters (HcPTR2A and B). New Phytol 170:401–410

Besserer A, Puech-Pagès V, Kiefer P, Gomez-Roldan V, Jauneau A, Roy S, Portais JC, Roux C, Bécard G, Séjalon-Delmas N (2006) Strigolactones stimulate arbuscular mycorrhizal fungi by activating mitochondria. PLoS Biol 4:e226

Bick JA, Lange BM (2003) Metabolic cross talk between cytosolic and plastidial pathways of isoprenoid biosynthesis: unidirectional transport of intermediates across the chloroplast envelope membrane. Arch Biochem Biophys 415:146–154

Bothe H, Klingner A, Kaldorf M, Schmitz O, Esch H, Hundeshagen B, Kernebeck H (1994) Biochemical approaches to the study of plant–fungal interactions in arbuscular mycorrhiza. Experientia 50:919–925

Bouwmeester HJ, Roux C, Lopez-Raez JA, Bécard G (2007) Rhizosphere communication of plants, parasitic plants and AM fungi. Trends Plant Sci 12:224–230

Breuninger M, Requena N (2004) Recognition events in AM symbiosis: analysis of fungal gene expression at the early appressorium stage. Fungal Gen Biol 41:794–804

Brundrett M, Bougher N, Dell B, Grove T, Malajczuk N (1996) Working with mycorrhizas in forestry and agriculture. Australian Centre for International Agricultural Research, Canberra

Buée M, Rossignol M, Jauneau A, Ranjeva R, Bécard G (2000) The presymbiotic growth of arbuscular mycorrhizal fungi is induced by a branching factor partially purified from plant root exudates. Mol Plant Microbe Interact 13:693–698

Bucking H, Abubaker J, Govindarajan M, Tala M, Pfeffer PE, Hegahashi G, Lammers P, Shachar-Hill Y(2008)Root exudates stimulate the uptake and metabolism of organic carbon in germinating spores of glomus intraradices. New Phytol(inpress)

Cairney JW (2000) Evolution of mycorrhiza systems. Naturwissenschaften 87:467–475

Catford JG, Staehelin C, Larose G, Piche Y, Vierheilig H (2006) Systemically suppressed isoflavonoids and their stimulating effects on nodulation and mycorrhization in alfalfa split-root systems. Plant Soil 285:257–266

Chabaud M, Venard C, Defaux-Petras A, Bécard G, Barker DG (2002) Targeted inoculation of *Medicago truncatula in vitro* root cultures reveals *MtENOD11* expression during early stages of infection by arbuscular mycorrhizal fungi. New Phytol 156:265–273

Chappell J (2002) The genetics and molecular genetics of terpene and sterol origami. Curr Opin Plant Biol 5:151–157

Dakora FD, Phillips DA (2002) Root exudates as mediators of mineral acquisition in low-nutrient environments. Plant Soil 245:35–47

David-Schwartz R, Badani H, Smadar W, Levy AA, Galili G, Kapulnik Y (2001) Identification of a novel genetically controlled step in mycorrhizal colonization: plant resistance to infection by fungal spores but not extra-radical hyphae. Plant J 27:561–569

David-Schwartz R, Gadkar V, Wininger S, Bendov R, Galili G, Levy AA, Kapulnik Y (2003) Isolation of a premycorrhizal infection (*pmi2*) mutant of tomato, resistant to arbuscular mycorrhizal fungal colonization. Mol Plant Microbe Interact 16:382–388

Dénarié J, Debellé F, Promé JC (1996) *Rhizobium* lipochitooligosaccharide nodulation factors: signaling molecules mediating recognition and morphogenesis. Ann Rev Biochem 65:503–535

Ditengou FA, Lapeyrie F (2000) Hypaphorine from the ectomycorrhizal fungus *Pisolithus tinctorius* counteracts activities of indole-3-acetic acid and ethylene but not synthetic auxins in eucalypt seedlings. Mol. Plant Microbe Interact 13:151–158

Divon HH, Fluhr R (2001) Nutrition acquisition strategies during fungal infection of plants. FEMS Microbiol Lett 266:65–74

Drissner D, Kunze G, Callewaert N, Gehrig P, Tamasloukht M, Boller T, Felix G, Amrhein N, Bucher M (2007) Lyso-Phosphatidylcholine Is a Signal in the Arbuscular Mycorrhizal Symbiosis. Science 318: 265–268

Duc G, Trouvelot A, Gianinazzi-Pearson V, Gianinazzi S (1989) First report of non-mycorrhizal plant mutants (Myc–) obtained in pea (*Pisum sativum* L.) and fababean (*Vicia faba* L.). Plant Sci 60:215–222

Dudareva N, Negre F (2005) Practical applications of research into the regulation of plant volatile emission. Curr Opin Plant Biol 8:113–118

Dumas-Gaudot E, Gollotte A, Cordier C, Gianinazzi S, Gianinazzi-Pearson V (2000) Modulation of host defense systems. In: Douds D, Kapulnik Y (eds) Arbuscular mycorrhizas: molecular biology and physiology. Kluwer, Boston, MA, pp 173–200

Duplessis S, Courty PE, Tagu D, Martin F (2005) Transcript patterns associated with ectomycorrhiza development in *Eucalyptus globulus* and *Pisolithus microcarpus*. New Phytol 165:599–611

Endre G, Kereszt A, Kevei Z, Mihacea S, Kalo P, Kiss GB (2002) A receptor kinase gene regulating symbiotic nodule development. Nature 417:962–966

Fester T (2008) Plastid reorganization in arbuscular mycorrhizal roots. In: Schoefs B (ed) Plant cell compartments. Selected Topics Research Signpost Research Signpost. Trivandrum 338–345.

Fester T, Hause B (2005) Accumulation of reactive oxygen species in arbuscular mycorrhizal roots. Mycorrhiza 15:373–379

Fester T, Hause B, Schmidt D, Halfmann K, Schmidt J, Wray V, Hause G, Strack D (2002a) Occurrence and localization of apocarotenoids in arbuscular mycorrhizal plant roots. Plant Cell Physiol 43:256–265

Fester T, Schmidt D, Lohse S, Walter MH, Giuliano G, Bramley PM, Fraser PD, Hause B, Strack D (2002b) Stimulation of carotenoid metabolism in arbuscular mycorrhizal roots. Planta 216:148–154

Floß DS, Hause B, Lange PR, Küster H, Strack D, Walter MH (2008) Knock-down of the MEP pathway isogene 1-deoxy-D-xylulose 5-phosphate synthase 2 inhibits formation of arbuscular mycorrhiza-induced apocarotenoids, and abolishes normal expression of mycorrhiza-specific plant marker genes. Plant J 56:86–100

Fransson PM, Taylor AF, Finlay RD (2000) Effects of continuous optimal fertilization on belowground ectomycorrhizal community structure in a Norway spruce forest. Tree Physiol 20:599–606

Frettinger P, Derory J, Herrmann S, Plomion C, Lapeyrie F, Oelmu ller R, Martin F, Buscot F (2007) Transcriptional changes in two types of pre-mycorrhizal roots and in ectomycorrhizas of oak microcuttings inoculated with *Piloderma croceum*. Planta 225:331–340

Fries N, Serck-Hanssen K, Häll Dimberg L, Theander O (1987) Abietic acid, an activator of basidiospore germination in ectomycorrhizal species of the genus *Suillus* (*Boletaceae*). Exp Mycol 11:360–363

Fromm J, Lautner S (2007) Electrical signals and their physiological significance in plants. Plant Cell Environ 30:249–257

Gadkar V, David-Schwartz R, Nagahashi G, Douds DD, Wininger S, Kapulnik Y (2003) Root exudate of *pmi* tomato mutant M161 reduces AM fungal proliferation *in vitro*. FEMS Microbiol Lett 223:193–198

Garcia-Garrido JM, Ocampo JA (2002) Regulation of the plant defence response in arbuscular mycorrhzal symbiosis. J Exp Bot 53:1377–1386

Gay G, Normand L, Marmeisse R, Sotta B, Debaud JC (1994) Auxin overproducer mutants of *Hebeloma cylindrosporum* Romagnési have increased mycorrhizal activity. New Phytol 128:645–657

Genre A, Chabaud M, Timmers T, Bonfante P, Barker DG (2005) Arbuscular mycorrhizal fungi elicit a novel intracellular apparatus in *Medicago truncatula* root epidermal cells before infection. Plant Cell 17:3489–3499

Gianinazzi-Pearson V (1996) Plant cell responses to arbuscular mycorrhizal fungi: getting to the roots of the symbiosis. Plant Cell 8:1871–1883

Gianinazzi-Pearson V, Branzanti B, Gianinazzi S (1989) In vitro enhancement of spore germination and hyphal growth of a vesicular-arbuscular mycorrhizal fungus by host root exudates and plant flavonoids. Symbiosis 7:243–255

Gianinazzi-Pearson V., Tollot M., Seddas P. Dissection of genetic cell programmes driving early arbuscular mycorrhiza interactions. In: Mycorrhizas: functional processes and ecological impacts, Springer Verlag, Heidelberg (in press)

Gianinazzi-Pearson V, Weidmann S, Seddas P, Massoumou M, van Tuinen D, Gianinazzi S (2006) Signal-related gene responses in beneficial root interactions with non-rhizobial microorganisms. In: Sanchez F, Quinto C, Geiger O (eds) Biology of plant–microbe interactions, vol 5. IS-MPMI, St. Paul, MN, pp 516–524

Gianinazzi-Pearson V, Séjalon-Delmas N, Genre A, Jeandroz S, Bonfante P (2007) Plants and arbuscular mycorrhizal fungi: cues and communication in the early steps of symbiotic interactions. Adv Bot Res 46:181–219

Giovannetti M, Sbrana C, Avio L, Citernesi AS (1993) Factors affecting appressorium development in the vesicular-arbuscular mycorrhizal fungus *Glomus mosseae*. New Phytol 123:114–122

Giovannetti M, Sbrana C, Logi C (1994) Early processes involved in host recognition by arbuscular mycorrhizal fungi. New Phytol 127:703–709

Godt DE, Roitsch T (1997) Regulation and tissue-specific distribution of mRNAs for three extracellular invertase isoenzymes of tomato suggests an important function in establishing and maintaining sink metabolism. Plant Physiol 115:273–282

Gogala N (1991) Regulation of mycorrhizal infection by hormonal factors produced by hosts and fungi. Experimentia 47:331–339

Gollotte A, Gianinazzi-Pearson V, Giovannetti M, Sbrana C, Avio L, Gianinazzi S (1993) Cellular localization and cytochemical probing of resistance reactions to arbuscular mycorrhizal fungi in a "locus A" myc– mutant of *Pisum sativum* (L.). Planta 191:112–122

Hanif M (2004) Characterization of small GTPase Cdc42 from the ectomycorrhizal fungus *Suillus bovinus* and *Agrobacterium tumefaciens*-mediated transformation of fungi. Dissertation, University of Helsinki, Finland

Harrison MJ (1998) Development of the arbuscular mycorrhizal symbiosis. Curr Opin Plant Biol 1:360–365

Harrison MJ (2005) Signaling in the arbuscular mycorrhizal symbiosis. Annu Rev Microbiol 59:19–42

Hause B, Fester T (2005) Molecular and cell biology of arbuscular mycorrhizal symbiosis. Planta 221:184–196

Hause B, Maier W, Miersch O, Kramell R, Strack C (2002) Induction of jasmonate biosynthesis in arbuscular mycorrhizal barley roots. Plant Physiol 130:1213–1220

Hause B, Mrosk C, Isayenkov S, Strack D (2007) Jasmonates in arbuscular mycorrhizal interactions. Phytochemistry 68(1):101–110

Hedden P, Phillips AI (2000) Gibberellin metabolism: new insights revealed by the genes. Trends Plant Sci 5:523–530

Helliwell CA, Sullivan JA, Mould RM, Gray JC, Peacock WJ, Dennis ES (2001) A plastid envelope location of *Arabidopsis ent*-kaurene oxidase links the plastids to the endoplasmic reticulum steps of the gibberellin biosynthesis pathway. Plant J 28:201–208

Herrmann S, Buscot F (2007) Cross talks at the morphogenetic, physiological and gene regulation levels between the mycobiont *Piloderma croceum* and oak microcuttings (*Quercus robur*) during formation of ectomycorrhizas. Phytochemistry 68:52–67

Horan DP, Chilvers GA (1990) Chemotropism; the key to ectomycorrhizal formation? New Phytol 116:297–301

Horan DP, Chilvers GA, Lapeyrie FF (1988) Time sequence of the infection process in eucalypt ectomycorrhizas. New Phytol 109:451–458

Humphrey AJ, Beale MH (2006) Strigol: biosynthesis and physiological activity. Phytochemistry 67:636–640

Imaizumi-Anraku H, Takeda N, Charpentier M, Perry J, Miwa H, Umehara Y, Kouchi H, Murakami Y, Mulder L, Vickers K, Pike J, Downie JA, Wang T, Sato S, Asamizu E, Tabata S, Yoshikawa M, Murooka Y, Wu GJ, Kawaguchi M, Kawasaki S, Parniske M, Hayashi M (2005) Plastid proteins crucial for symbiotic fungal and bacterial entry into plant roots. Nature 433:527–531

Isayenkov S, Mrosk C, Stenzel I, Strack D, Hause B (2005) Suppression of allele oxide cyclase in hairy roots of *Medicago truncatula* reduces jasmonate levels and the degree of mycorrhization with *Glomus intraradices*. Plant Physiol 139:1401–1410

Jacobi LM, Zubkova LA, Barmicheva EM, Tsyganov VE, Borisov AY, Tikhonovich IA (2003) Effect of mutations in the pea genes Sym33 and Sym40 II. Dynamics of arbuscule development and turnover. Mycorrhiza 13:9–16

Javot H, Pemetsa RV, Terzaghi N, Cook DR, Harrisson MJ (2007) A *Medicago truncatula* phosphate transporter indispensable for the arbuscular mycorrhizal symbiosis. Proc Natl Acad Sci USA 104:1720–1725

Jones DL, Hodge A, Kuzyakov Y (2004) Plant and mycorrhizal regulation of rhizodeposition. New Phytol 163:459–480

Kanamori N, Madsen LH, Radutoiu S, Frantescu M, Quistgaard EMH, Miwa H, Downie JA, James EK, Felle HH, Haaning LL, Jensen TH, Sato S, Nakamura Y, Tabata S, Sandal N, Stougaard J (2006) A nucleoporin is required for induction of Ca^{2+} spiking in legume nodule development and essential for rhizobial and fungal symbiosis. Proc Natl Acad Sci USA 103:359–364

Karabaghli-Degron C, Sotta B, Bonnet M, Gay G, Le Tacon F (1998) The auxin transport inhibitor 2,3,5-triiodobenzoic acid (TIBA) inhibits the stimulation of in vitro lateral root formation and the colonization of the tap-root cortex of Norway spruce (*Picea abies*) seedlings by the ectomycorrhizal fungus *Laccaria bicolor*. New Phytol 140:723–733

Karandashov V, Bucher M (2005) Symbiotic phosphate transport in arbuscular mycorrhizas. Trends Plant Sci 10:22–29

Karandashov V, Nagy R, Wegmuller S, Amrhein N, Bucher M (2004) Evolutionary conservation of a phosphate transporter in the arbuscular mycorrhizal symbiosis. Proc Natl Acad Sci USA 101:6285–6290

Kaska DD, Myllylä R, Cooper JB (1999) Auxin transport inhibitors act through ethylene to regulate dichotomous branching of lateral root meristems in pine. New Phytol 142:49–58

Klingner A, Bothe H, Wray V, Marner FJ (1995a) Identification of a yellow pigment formed in maize roots upon mycorrhizal colonization. Phytochemistry 38:53–55

Klingner A, Hundeshagen B, Kernebeck H, Bothe H (1995b) Localization of the yellow pigment formed in roots of gramineous plants colonized by arbuscular fungi. Protoplasma 185:50–57

Kosuta S, Chabaud M, Lougnon G, Gough C, Dénarie J, Barker DG, Bécard G (2003) A diffusible factor from arbuscular mycorrhizal fungi induces symbiosis-specific *MtENOD11* expression in roots of *Medicago truncatula*. Plant Physiol 131:952–962

Kosuta S, Hazledine S, Sun J, Miwa H, Morris RJ, Downie JA, Oldroyd GED (2008) Differential and chaotic calcium signatures in the symbiosis signalling pathway of legumes. Proc Nat Acad Sci USA 105:9823–9828

Krings M, Taylor TN, Hass H, Kerp H, Dozler N, Hermsen EJ (2007) Fungal endophytes in a 400-million-yr-old land plant: infection pathways, spatial distribution, and host responses. New Phytol 174:648–657

Kruger A, Peskan-Berghofer T, Frettinger P, Herrmann S, Buscot F, Oelmuller R (2004) Identification of premycorrhiza-related plant genes in the association between *Quercus robur* and *Piloderma croceum*. New Phytol 163:149–157

Kuwada K, Kuramoto M, Utamura M, Matsushita I, Shibata Y, Ishii T (2005) Effect of mannitol from *Laminaria japonica*, other sugar alcohols, and marine alga polysaccharides on in vitro hyphal growth of *Gigaspora margarita* and root colonization of trifoliate orange. Plant Soil 276:279–286

Lagrange H, Jay-Allemand C, Lapeyrie F (2001) Rutin, the phenolglycoside from Eucalyptus root exudates, stimulates *Pisolithus* hyphal growth at picomolar concentrations. New Phytol 150:349–355

Lanfranco L, Novero M, Bonfante P (2005) The mycorrhizal fungus *Gigaspora margarita* possesses a Cu/Zn superoxide dismutase that is up-regulated during symbiosis with legume hosts. Plant Physiol 137:1319–1330

Larose G, Chênevert R, Moutoglis P, Gagne S, Piché Y, Vierheilig H (2002) Flavonoid levels in roots of *Medicago sativa* are modulated by the development stage of the symbiosis and the root colonizing arbuscular-mycorrhizal fungus. J Plant Physiol 159:1329–1339

Lemoine Y, Rmiki N-E, Créach A, Rachidi J, Schoefs B (2008) Cytoplasmic accumulation of astaxanthin by the green alga *Haematococcus pluvialis* (flotow). In: Schoefs B (ed) Plant cell compartments. Selected Topics Research Signpost, Research Signpost. Trivandrum 251–284

Le Quéré A, Schutzendubel A, Rajashekar B, Canback B, Hedh J, Erland S, Johansson T, Tunlid A (2004) Divergence in gene expression related to variation in host specificity of an ectomycorrhizal fungus. Mol Ecol 13:3809–3819

Le Quéré A, Wright D, Söderström B, Tunlid A, Johansson T (2005) Global patterns of gene regulation associated with the development of ectomycorrhiza between birch (*Betula pendula* Roth.) and *Paxillus involutus* (Batsch) Fr. Mol Plant Microbe Interact 18:659–673

Le Quéré A, Eriksen KA, Rajashekar B, Schutzenbubel A, Canback B, Johansson T, Tunlid A (2006) Screening for rapidly evolving genes in the ectomycorrhizal fungus *Paxillus involutus* using cDNA microarrays. Mol Ecol 15:535–550

Lévy J, Bres C, Geurts R, Chalhoub B, Kulikova O, Duc G, Journet E-P, Ané J-M, Lauber E, Bisseling T, Dénarié J, Rosenberg C, Debellé F (2004) A putative Ca^{2+} and calmodulin-dependent protein kinase required for bacterial and fungal symbioses. Science 303:1361–1364

Liu J, Maldonado-Mendoza I, Lopez-Meyer M, Cheung F, Town CD, Harrison MJ (2007) Arbuscular mycorrhizal symbiosis is accompanied by local and systemic alterations in gene expression and an increase in disease resistance in the shoots. Plant J 50:529–544

Lohse S, Schliemann W, Ammer C, Kopka J, Strack D, Fester T (2005) Organization and metabolism of plastids and mitochondria in arbuscular mycorrhizal roots of *Medicago truncatula*. Plant Physiol 139:329–340

Ludwig-Müller J, Bennett, RN, garcia-Garrido JM, Piché Y, Vierheilig H (2002) Reduced arbuscular mycorrhizal root colonization in *Tropaeolum majus* and *Carica papaya* after jasmonic acid application can not be attributed to increased glucosinolate levels. J Plant Physiol 517-523

Maier W, Peipp H, Schmidt J, Wray V, Strack D (1995) Levels of a terpenoid glycoside (blumenin) and cell wall-bound phenolics in some cereal mycorrhizas. Plant Physiol 109:465–470

Mankel A, Krause K, Kothe E (2002) Identification of a Hydrophobin Gene That is Developmentally Regulated in the Ectomycorrhizal Fungus Tricholoma terreum. Appl Environ Microbiol 68(3):1408–13

Marsh J, Schultze M (2001) Analysis of arbuscular mycorrhizas using symbiosis-defective plant mutants. New Phytol 150:525–532

Martin F, Duplessis S, Ditengou F, Lagrange H, Voiblet C, Lapeyrie F (2001) Developmental cross talking in the ectomycorrhizal symbiosis: signals and communication genes. New Phytol 151:145–154

Martin F, Kohler A, Duplessis S (2007) Living in harmony in the wood underground: ectomycorrhizal genomics. Curr Opin Plant Biol 10:204–210

Martin F, Aerts A, Ahren D, Brun A, Danchin EGJ, Duchaussoy F, Gibon, Kohler A, Lindquist E, Pereda V, et al. (2008) The genome of *Laccaria bicolor* provides insights into mycorrhizal symbiosis. Nature 452:88–97

Matusova R, Rani K, Verstappen FWA, Franssen MCR, Beale MH, Bouwmeester HJ (2005) The strigolactone germination stimulants of the plant-parasitic *Striga* and *Orobanche* spp. are derived from the carotenoid pathway. Plant Physiol 139:920–934

Meixner C, Ludwig-Muller J, Miersch O, Gresshoff P, Staehelin C, Vierheilig H (2005) Lack of mycorrhizal autoregulation and phytohormonal changes in the supernodulating soybean mutant nts1007. Planta 222:709–715

Menotta M, Amicucci A, Sisti D, Gioacchini AM, Stocchi V (2004) Differential gene expression during pre-symbiotic interaction between *Tuber borchii* Vittad. and *Tilia americana* L. Curr Genet 46:158–165

Menotta M, Pierleoni R, Amicucci A, Sisti D, Cerasi A, Millo E, Chiarantini L, Stocchi V (2006) Characterization and complementation of a Fus3/Kss1 type MAPK from *Tuber borchii*, TBMK. Mol Genet Genom 276:126–134

Menotta M, Amicucci A, Basili G, Polidori E, Stocchi V, Rivero F (2008) Molecular and functional characterization of a Rho GDP dissociation inhibitor in the filamentous fungus *Tuber borchii*. BMC Microbiol 8:57

Mitra RM, Gleason CA, Edwards A, Hadfield J, Downie JA, Oldroyd GED, Long SR (2004) A Ca^{2+}/calmodulin-dependent protein kinase required for symbiotic nodule development: gene identification by transcript-based cloning. Proc Natl Acad Sci USA 101:4701–4705

Moller SG, Chua NH (1999) Interactions and intersections of plant signaling pathays. J Mol Biol 293:219–234

Morandi D 1996 Occurrence of phytoalexins and phenolic compounds in endomycorrhizal interactions, and their potential role in biological control. Plant and Soil 185, 223–232

Morandi D, Prado E, Sagan M, Duc G (2005) Characterisation of new symbiotic *Medicago truncatula* (Gaertn.) mutants, and phenotypic or genotypic complementary information on previously described mutants. Mycorrhiza 15:283–289

Müller T, Avolio M, Olivi M, Benjdia M, Rikirsch E, Kasaras A, Fitz M, Chalot M, Wipf D (2007) Organic nitrogen transport in the ectomycorrhiza on the basis of the *Hebeloma cylindrosporum-Pinus pinaster* association. Phytochemistry 68:41–51

Nagahashi G, Douds DD (2000) Partial separation of root exudate components and their effects upon the growth of germinated spores of AM fungi. Mycol Res 104:1453–1464

Navazio N, Moscatiello R, Genre A, Novero M, Baldan B, Bonfante P, Mariani P (2007) A diffusible signal from arbuscular mycorrhizal fungi elicits a transient cytosolic calcium elevation in host plant cells. Plant Physiol 144:673–681

Nehls U (2004) Carbohydrates and nitrogen: nutrients and signals in ectomycorrhizas. In: Varma A, Abbott L, Werner D, Hampp R (eds) Plant surface microbiology. Springer, Berlin, pp 373–392

Nehls U (2008) Mastering ectomycorrhizal symbiosis: the impact of carbohydrates. J Exp Bot 59:1097–1108

Nehls U, Wiese J, Guttenberger M, Hampp R (1998) Carbon allocation in ectomycorrhizas: identification and expression analysis of an *Amanita muscaria* monosaccharide transporter. Mol Plant Microbe Interact 11:167–176

Nehls U, Bock A, Ecke M, Hampp R (2001a) Differential expression of hexose-regulated fungal genes within *Amanita muscaria/Populus tremula* × *tremuloides* ectomycorrhizas. New Phytol 150:583–589

Nehls U, Mikolajewski S, Magel E, Hampp R (2001b) The role of carbohydrates in ectomycorrhizal functioning: gene expression and metabolic control. New Phytol 150:533–541

Niini SS, Raudaskoski M (1993) Response of ectomycorrhizal fungi to benomyl and nocodazole, growth inhibition and microtubule depolymerization. Mycorrhiza 3:83–91

Niini SS, Tarkka MT, Raudaskoski M (1996) Tubulin and actin protein patterns in Scots pine (*Pinus sylvestris*) roots and developing ectomycorrhiza with *Suillus bovinus*. Physiol Plant 96:186–192

Niini SS, (1998) Growth pattern and cytoskeleton in *Suillius bovinus* hyphae, *Pinus Syslvestris* roots and in *Pinus sylvestris- Suillus bovinus* ectomycorrhiza. PhD Thesis, University of Helsinki. 16: ISBN 951-45-8258–6.

Niini S, Raudaskoski M, (1998) Growth patterns in non-mycorrhizal and mycorrhizal short roots of *Pinus sylvestris*. Symbiosis 25:101–114.

Okada A, Shimizu T, Okada K, Kuzuyama T, Koga J, Shibuya N, Nojiri H, Yamane H (2007) Elicitor induced activation of the methylerythritol phosphate pathway toward phytoalexins biosynthesis in rice. Plant Mol Biol 65:177–187

Ohlrogge JB, Browse J (1995) Lipid biosynthesis. Plant Cell 7:957–970

Ohlrogge JB, Kuhn DN, Stumpf PK (1979) Subcellular localization of acyl carrier protein in leaf protoplasts of *Spinacia oleracea*. Proc Natl Acad Sci USA 76:1194–1198

Parniske M (2004) Molecular genetics of the arbuscular mycorrhizal symbiosis. Curr Opin Plant Biol 7:414–421

Paszkowski U, Jakovleva L, Boller T (2006) Maize mutants affected at distinct stages of the arbuscular mycorrizal symbiosis. Plant J 47:165–173

Paullilo SM, Fahrenkrog B (2008) The nuclear pore complex: from higher eukaryotes to plants. In: Schoefs B (ed) Plant cell compartments. Selected Topics Research Signpost Research Signpost. trivandrum 1–18

Pozo MJ, Azcon-Aguilar C (2007) Unraveling mycorrhiza-induced resistance. Curr Opin Plant Biol 10:393–398

Ramos AC, Façanha AR, Feijo JA (2008) Proton (H^+) flux signature for the presymbiotic development of the arbuscular mycorrhizal fungi. New Phytol 178:177–188

Raudaskoski M, Pardo AG, Tarkka MT, Gorfer M, Hanif M, Laitiainen E (2001) Small GTPases, cytoskeleton and signal transduction in filamentous homobasidiomycetes. In: Geitmann A, Cresti M, Heath IB (eds) Cell biology of plant and fungal tip growth (NATO Science Series 328). IOS, Amsterdam, pp 123–136

Raudaskoski M, Tarkka MT, Niini S (2004) Mycorrhizal development and cytoskeleton. In: Varma A, Abott L, Werner D, Hampp R (eds) Plant surface microbiology. Springer, Berlin, pp 293–329

Reddy SM, Pandey AK, Melayah D, Marmeisse R, Gay G (2003) The auxin responsive gene Pp-C61 is up-regulated in *Pinus pinaster* roots following inoculation with ectomycorrhizal fungi. Plant Cell Environ 26:681–691

Redecker D, Kodner R, Graham LE (2000) Glomalean fungi from the Ordovician. Science 289:1920–1921

Regvar M, Gogala N, Zalar P (1996) Effects of jasmonic acid on mycorrhizal *Allium sativum*. New Phytol 134:703–707

Remy W, Taylor TN, Hass H, Kerp H (1994) Four-hundred-million-year-old vesicular arbuscular mycorrhizae. Proc Natl Acad Sci USA 91:11841–11843

Requena N, Mann P, Hampp R, Franken P (2002) Early developmentally regulated genes in the arbuscular mycorrhizal fungus *Glomus mosseae*: identification of *GmGIN1*, a novel gene with homology to the C-terminus of metazoan hedgehog proteins. Plant Soil 244:129–139

Riely BK, Lougnon G, Ane JM, Cook DR (2007) The symbiotic ion channel homolog DMI1 is localized in the nuclear membrane of *Medicago truncatula* roots. Plant J 49:208–216

Roughan PG, Slack CR (1982) Cellular organization of glycerolipid metabolism. Annu Rev Plant Physiol Plant Mol Biol 33:97–132

Ruiz-Lozano JM, Roussel H, Gianinazzi S, Gianinazzi-Pearson V (1999) Defence genes are differentially induced by a mycorrhizal fungus and *Rhizobium* in wild type and symbiosis-defective pea genotypes. Mol Plant Microbe Interact 12:976–984

Rupp LA, Mudge KW, Negm FB (1989) Involvement of ethylene in ectomycorrhiza formation and dichotomous branching of roots of mugo pine seedlings. Can J Bot 67:477–482

Salo V, Niini SS, Virtanen I, Raudaskoski M (1989) Comparative immunocytochemistry of the cytoskeleton in filamentous fungi with dikaryotic and multinucleate hyphae. J Cell Sci 94:11–24

Salzer P, Hebe G, Hager A (1997) Cleavage of chitinous elicitors from the ectomycorrhizal fungus *Hebeloma crustuliniforme* by host chitinases prevents induction of K^+ and Cl^{l-} release, extracellular alkalinization and H_2O_2 synthesis of *Picea abies* cells. Planta 203:470–479

Salzer P, Corbiere H, Boller T (1999) Hydrogen peroxide accumulation in *Medicago truncatula* roots colonized by the arbuscular mycorrhiza-forming fungus *Glomus intraradices*. Planta 208:319–325

Sanchez L, Weidmann S, Arnould C, Bernard AR, Gianinazzi S, Gianinazzi-Pearson V (2005) *P. fluorescens* and *G. mosseae* trigger *DMI3*-dependent activation of genes related to a signal transduction pathway in roots of *Medicago truncatula*. Plant Physiol 139:1065–1077

Sbrana C, Giovannetti M (2005) Chemotropism in the arbuscular mycorrhizal fungus *Glomus mossae*. Mycorrhiza 15:539–545

Scervino JM, Ponce MA, Erra-Bassells R, Bompadre MJ, Vierheilig H, Ocampo JA, Godeas A (2005) Glycosidation of apigenin results in a loss of its activity on different growth parameters of arbuscular mycorrhizal fungi from the genus *Glomus* and *Gigaspora*. Soil Biol Biochem 38:2919–2922

Schilmiller AL, Howe GA (2005) Systemic signaling in the wound response. Curr Opin Plant Biol 8:369–377

Schliemann W, Schmidt J, Nimtz M, Wray V, Thomas F, Strack D (2006) Accumulation of apocarotenoids in mycorrhizal roots of *Ornithogalum umbellatum*. Phytochemistry 67:1196–1205

Schliemann W, Ammer C, Strack D (2008) Metabolite profiling of mycorhizal root of *Medicago truncatula*. Phytochemistry 69:112–146

Schwender J, Ohlrogge JB (2002) Probing *in vivo* metabolism by stable isotope labelling of storage lipids and proteins in developing *Brassica napus* embryos. Plant Physiol 130:347–361

Seddas PMA, Arias CM, Arnould C., van Tuinen D., Godfroy O., Aït Benhassou H., Gouzy J., Morandi D., Dessaint F., Gianinazzi-Pearson V. 2008. Symbiosis-related plant genes modulate molecular responses in an arbuscular mycorrhizal fungus during early root interactions. Molecular Plant-Microbe Interactions (in press)

Seddas PM, Arnould C, Tollot M, Arias CM, Gianinazzi-Pearson V (2008) Spatial monitoring of gene activity in extraradical and intraradical developmental stages of arbuscular mycorrhizal fungi by direct fluorescent in situ RT-PCR. Fungal Genet Biol 45:1155–1165

Siciliano V, Genre A, Balestrini R, Cappellazzo G, deWit PJGM, Bonfante P (2007) Transcriptome analysis of arbuscular mycorrhizal roots during development of the prepenetration apparatus. Plant Physiol 144:1455–1466

Smith SE, Read DJ (2008) Mycorrhiza symbiosis, 3rd edn. Academic, San Diego

Soares ACF, Martins MA, Mathias L, Freitas MSM (2005) Arbuscular mycorrhizal fungi and the occurence of flavonoids in roots of passion fruit seedlings. Sci Agric 62:331–336

Somerville C, Browse J (1991) Plant lipids: metabolism, mutants, and membranes. Science 252:80–87

Splivallo R, Novero M, Bertea CM, Bossi S, Bonfante P (2007) Truffle volatiles inhibit growth and induce an oxidative burst in *Arabidopsis thaliana*. New Phytol 175:417–424

Strack D, Fester T (2006) Isoprenoid metabolism and plastid reorganization in arbuscular mycorrhizal roots. New Phytol 172:22–34

Stracke S, Kistner C, Yoshida S, Mulder L, Sato S, Kaneto T, Tabata S, Sandal N, Stougaard J, Szczyglowski K, Parniske M (2002) A plant receptor-like kinase required for both bacterial and fungal symbiosis. Nature 417:959–962

Strassner J, Schaller F, Frick U, Howe G, Weiler E, Amrhein N, Macheroux P, Schaller A (2002) Characterization and cDNA-microarray expression analysis of 12-oxophytodienoate reductase reveals differential roles for octadecanoid biosynthesis in the local versus of the systemic wound response. Plant J 32:585–601

Takano Y, Choi W, Mitchell T, Okuno T, Dean RA (2003) Large scale parallel analysis of gene expression during infection-related morphogenesis of *Magnaporthe grisea*. Mol Plant Pathol 4:337–346

Tamasloukht MB, Séjalon-Delmas N, Kluever A, Jauneau A, Roux C, Bécard G, Franken P (2003) Root factors induce mitochondrial-related gene expression and fungal respiration during the developmental switch from asymbiosis to presymbiosis in the arbuscular mycorrhizal fungus *Gigaspora rosea*. Plant Physiol 131:1468–1478

Tarkka, MT, Vasara R, Gorfer M, Raudaskoski M (2000) Molecular characterization of actin genes from homobasidiomycetes: two different actin genes from *Schizophyllum commune* and *Suillus bovinus*. Gene 251:27–35

Taylor TN, Remy W, Hass H, Kerp H (1995) Fossil arbuscular mycorrhizae from the early Devonian. Mycologia 87:560–573

Thoma I, Loeffler C, Sinha A, Gupta M, Krischke M, Steffan B, Roistch T, Mueller M (2003) Cyclopentenone isoprostanes induced by reactive oxygen species trigger defense gene activation and phytoalexin accumulation in plants. Plant J 34:363–375

Timonen S, Finlay RD, Soderstrom B, Raudaskoski M (1993) Identification of cytoskeletal components in pine ectomycorrhizas. New Phytol 124:83–92

Tuskan G, DiFazio S, Jansson S, Bohlmann J, Grigoriev I, Hellsten U, Putnam N, Ralph S, Vierheilig H (2004) Regulatory mechanisms during the plant-arbuscular mycorrhizal fungus interaction. Can J Bot 82:1166–1176

Vierheilig H 2004 Regulatory mechanisms during the plant –arbuscular mycorrhizal fungus interaction. Can. J. Bot. 82: 1166–1176

Vierheilig H, Piché Y (2002) Signalling in arbuscular mycorrhiza: facts and hypotheses. In: Buslig B, Manthey J (eds) Flavonoids in cell function. Kluwer, New York, pp 23–29

Vierheilig H, Bago B, Albrecht C, Poulin M-P, Piché Y (1998) Flavonoids and arbuscular-mycorrhizal fungi. In: Manthey JA, Buslig BS (eds) Flavonoids in the living system. Plenum, New York, pp 9–33

Vieweg MF, Hohnjec N, Küster H (2005) Two genes encoding truncated hemoglobins are regulated during root nodule and arbuscular mycorrhiza symbioses of *Medicago truncatula*. Planta 220:757–766

Walter MH, Fester T, Strack D (2000) Arbuscular mycorrhizal fungi induce the non-mevalonate methylerythritol phosphate pathway of isoprenoid biosynthesis correlated with accumulation of the "yellow pigment" and other apocarotenoids. Plant J 21:571–578

Walter MH, Hans J, Strack D (2002) Two distantly related genes encoding 1-deoxy-D-xylulose 5-phosphate synthases: differential regulation in shoots and apocarotenoid-accumulating mycorrhizal roots. Plant J 31:243–254

Weidmann S, Sanchez L, Descombin J, Chatagnier O, Gianinazzi S, Gianinazzi-Pearson V (2004) Fungal elicitation of signal transduction-related plant genes precedes mycorrhiza establishment and requires the *dmi*3 gene in *Medicago truncatula*. Mol Plant Microb Interact 17:1385–1393

Withers ST, Keasling JD (2007) Biosynthesis and engineering of isoprenoid small molecules. Appl Microbiol Biotechnol 73:980–990

Wright DP, Johansson T, Le Quere A, Soderstrom B, Tunlid A (2005) Spatial patterns of gene expression in the extramatrical mycelium and mycorrhizal root tips formed by the ectomycorrhizal fungus *Paxillus involutus* in association with birch (*Betula pendula*) seedlings in soil microcosms. New Phytol 167:579–596

Yoneyama K, Xie XN, Kasumoto D, Sekimoto H, Sugimoto Y, Takeuchi Y, Yoneyama K (2007) Nitrogen deficiency as well as phosphorus deficiency in sorghum promotes the production and exudation of 5-deoxystrigol, the host recognition signal for arbuscular mycorrhizal fungi and root parasites. Planta 227:125–132

Yoneyama K, Xie X, Sekimoto H, Takeuchi Y, Ogasawara S, Akiyama K, Hayashi H, Yoneyama K (2008) Strigolactones, host recognition signals for root parasitic plants and arbuscular mycorrhizal fungi, from Fabaceae plants. New Phytol 179:484–494

Role of γ-Aminobutyrate and γ-Hydroxybutyrate in Plant Communication

Barry J. Shelp, Wendy L. Allan, and Denis Faure

Abstract The neurotransmitters gamma-aminobutyrate (GABA) and gamma-hydroxybutyrate (GHB) are found in virtually all prokaryotic and eukaryotic organisms. The physiological roles of these metabolites in plants are not yet clear, but both readily accumulate in response to stress through a combination of biochemical and transcriptional processes. GABA accumulation has been associated with the appearance of extracellular GABA, and evidence is available for a role of extracellular GABA in communications between plants and animals, fungi, bacteria or other plants, although the mechanisms by which GABA functions in communication appear to be diverse. As yet there is no evidence from plants of GHB receptors, GHB signaling or extracellular GHB, although the level of the quorum-sensing signal in *Agrobacterium* is known to be modulated by GHB.

1 Introduction

γ-Aminobutyrate (GABA), a nonprotein amino acid, and γ-hydroxybutyrate (GHB), a short-chain fatty acid that closely resembles GABA (Fig. 1), are found in virtually all prokaryotic and eukaryotic organisms. They are endogenous constituents of the mammalian nervous system, wherein GABA plays a role in neural transmission and development, and functions through interactions with specialized receptors ($GABA_A$, $GABA_B$, $GABA_C$) and transporters, and GHB serves as a neurotransmitter or neuromodulator postulated to act via a $GABA_B$ receptor or an independent GHB-specific receptor (see review by Fait et al. 2006). When administered, GABA does not cross

B.J. Shelp (✉) and W.L. Allen
Department of Plant Agriculture, Bovey Bldg., Rm 4237, University of Guelph, Guelph, ON, Canada N1G 2W1
e-mail: bshelp@uoguelph.ca, wallan@uoguelph.ca

D. Faure
Institut des Sciences du Végétal, Centre National de la Recherche Scientifique, Gif-sur-Yvette, 91 198, France
e-mail: Denis.Faure@isv.cnrs-gif.fr

F. Baluška (ed.), *Plant-Environment Interactions*, Signaling and Communication in Plants,
DOI: 10.1007/978-3-540-89230-4_4, © Springer-Verlag Berlin Heidelberg 2009

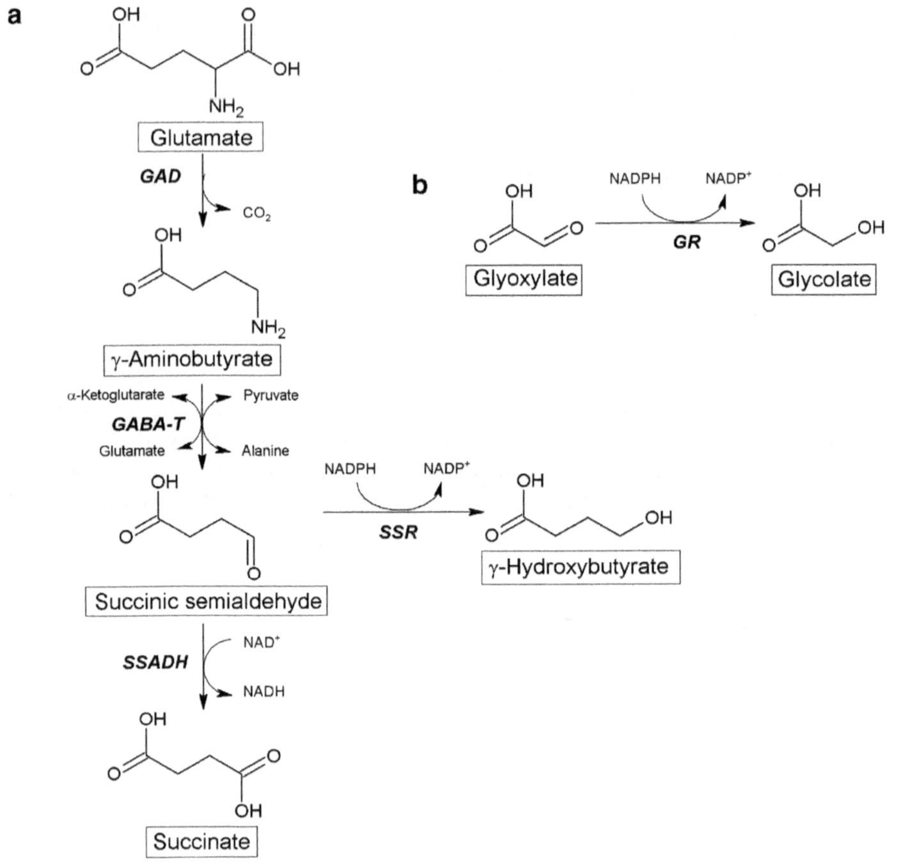

Fig. 1 **a** Alternative pathways for GABA metabolism via succinic semialdehyde. **b** Glyoxylate reductase reaction. *Enzymes* are in *italics*. *GAD*, glutamate decarboxylase; *GABA-T*, GABA transaminase; *GR*, glyoxylate reductase; *SSADH*, succinic semialdehyde dehydrogenase; *SSR*, succinic semialdehyde reductase. (Adapted from Hoover et al. 2007a)

the blood–brain barrier, whereas GHB does so with ease, penetrating the brain and producing diverse neuropharmacological and neurophysiological effects. For further details on the roles of GABA and GHB in animals, refer to reviews by Mamelak (1989) and Fait et al. (2006).

Evidence for the existence of GABA receptors in plants and the notion that GABA serves as a signaling molecule is emerging: (1) the growth of *Stellaria longipes* and duckweed is sensitive to GABA, GABA isomers, and GABA antagonists or agonists (Kathiresan et al. 1998; Kinnersley and Lin 2000); (2) the N-terminal regions of the superfamily of ionotropic glutamate receptors are highly homologous to members of the $GABA_B$ receptors (Lacombe et al. 2001; Bouché et al. 2003a, b); (3) a GABA gradient is required for the guidance of the pollen tube through the apoplastic spaces within the *Arabidopsis* pistil to the female gametophyte (Palanivelu et al. 2003); (4) proteins capable of transporting GABA are present in the plasma membrane of *Arabidopsis* (Meyer et al. 2006); (5) GABA binding sites are found on the protoplast membrane of both pollen and somatic cells of tobacco, and these sites are involved in the regulation of endogenous Ca^{2+} level (Yu et al. 2006); (6) *Arabidopsis* 14-3-3 expression is regulated by GABA in a calcium-dependent manner (Lancien and Roberts 2006); (7) *E*-2-hexanal responses in *Arabidopsis* are mediated by GABA (Mirabella et al. 2008); (8) GABA is translocated in phloem, and changes in phloem GABA are positively correlated with nitrate influx during nitrogen deprivation and over the growth cycle of rape (Bown and Shelp 1989; Beuvé et al. 2004), and; (9) extracellular GABA induces expression of a plasma membrane-located nitrate transporter and stimulates $^{15}NO_3$ influx by the root system (Beuvé et al. 2004). To date, there is no direct evidence for GHB receptorsor GHB signaling in plants.

While the physiological roles of GABA and GHB in plants are not yet clear, evidence indicates that both metabolites readily accumulate in response to stress (Shelp et al. 1999; Allan et al. 2008). GABA accumulation has been associated with the appearance of extracellular GABA, either in the apoplast or external medium (Secor and Schrader 1985; Chung et al. 1992; Crawford et al. 1994; Solomon and Oliver 2001; Bown et al. 2006). Herein, the evidence for and the mechanisms involved in the accumulation of GABA and GHB are reviewed. This is followed by a description of evidence for their role in communication between plants and other organisms.

2 GABA and GHB Metabolism

In plants, GABA is derived primarily via the H^+-consuming α-decarboxylation of glutamate in an irreversible reaction catalyzed by cytosolic-localized glutamate decarboxylase (GAD) that proceeds optimally at acidic pH (Fig.1; Shelp et al. 1999). While increasing cytosolic H^+ concentration can result in GABA accumulation, there is abundant evidence for a mechanism involving Ca^{2+}-dependent binding of calmodulin to GAD proteins at neutral pH, thereby relieving the enzyme from autoinhibition and stimulating enzymatic activity (reviewed by Shelp et al. 1999). Thus, calmodulin links GABA accumulation with increasing cytosolic Ca^{2+}, which typically accompanies stress. Research has identified multiple GAD genes from *Petunia*, tomato, tobacco, *Arabidopsis* and rice, and differential organ localization of two isoforms in both *Arabidopsis* and tobacco (Shelp et al. 1999; Yevtushenko et al. 2003; Akama and Takaiwa 2007; Miyashita and Good 2008), implying that they may have specific functions. For example, GAD1 is predominantly expressed in roots, while GAD2 expression is evident in all organs; expression of the other three GAD genes is weak (Miyashita and Good 2007). Phenotypic analysis of loss-of-function *gad1* mutants revealed that GABA levels in roots are dramatically lower than in wild-type roots, and that heat-induced GABA accumulation is prevented in *gad1* mutants (Bouché et al. 2004). Moreover, antisense suppression of GAD results in the accumulation of glutamate in transgenic tomato fruit (Kisaka et al. 2006). Transcriptional induction of one or more *GAD* forms is often observed in response to low oxygen, water deficit, salinity or *Agrobacterium* infection (Klok et al. 2002; Deeken et al. 2006; Cramer et al. 2007; Miyashita and Good 2007; Pasentsis et al. 2007).

GABA is then transaminated to succinic semialdehyde (SSA) via a mitochondrial-localized GABA transaminase (GABA-T) that is probably reversible (Fig.1; Van Cauwenberghe and Shelp 1999; Van Cauwenberghe et al. 2002). Both pyruvate- and 2-oxoglutarate-dependent activities are found in crude tobacco plant extracts; however, only the gene for pyruvate-dependent activity (*GABA-T1*) in *Arabidopsis* has been identified to date (Van Cauwenberghe et al. 2002). Research has identified highly homologous proteins in pepper, tomato and rice (Ansari et al. 2005; Wu et al. 2006), although protein function has not been examined. The expression of *GABA-T1* is detected in all *Arabidopsis* organs and the vegetative phenotype appears normal, but a *gaba-t1* mutant lacks a GABA gradient from the stigma to the embryo sac and pollen tube growth is misdirected, thereby causing a reduced-seed phenotype, while GABA-T activity is decreased to negligible levels in both shoots and roots and GABA accumulates in roots (Palanivelu et al. 2003; Miyashita and Good 2007). Significant transcriptional change typically occurs in *GABA-T1* under low oxygen, water deficit and salinity (Klok et al. 2002; Cramer et al. 2007), although not always (Miyashita and Good 2007).

SSA dehydrogenase (SSADH) catalyzes the irreversible, NAD-dependent oxidation of SSA to succinate in the mitochondrion (Fig. 1). The enzyme is competitively inhibited by NADH and AMP, noncompetitively inhibited by ATP, and inhibited by ADP via both competitive and noncompetitive means (Busch and Fromm 1999). SSADH occurs as a single-copy gene in *Arabidopsis*, and *ssadh* mutants contain elevated levels of reactive oxygen species, are hypersensitive to heat and light stress, and have a stunted and necrotic phenotype (Bouché et al. 2003a).

Other research suggests an additional coping mechanism for the detoxification of SSA, which involves its reduction to GHB (Fig. 1). A number of strategies, including complementation of a SSADH-deficient yeast mutant with an *Arabidopsis* cDNA library, recombinant expression in *Escherichia coli*, and transient expression in tobacco BY-2 cells, were used to identify two highly homologous proteins (designated as *Arabidopsis* glyoxylate reductases 1 and 2, or *At*GR1 and *At*GR2) that catalyze the conversion of both SSA to GHB and glyoxylate to glycolate via an essentially irreversible, NADPH-based ordered mechanism, although they are located in different cellular compartments (cytosol, plastid) (Breitkreuz et al. 2003; Hoover et al. 2007a, b; Simpson et al. 2008). $NADP^+$ is an effective competitive inhibitor with respect to NADPH, suggesting that the ratio of $NADPH/NADP^+$ regulates the activities of both isoforms in planta. Time course experiments revealed that GHB accumulates in leaves of both *Arabidopsis* and tobacco plants subjected to stress, and that this accumulation is associated with higher GABA levels, higher $NADPH/NADP^+$ ratios, and lower glutamate levels (Fig. 2; see also Allan et al. 2008). Expression analysis of *Arabidopsis* leaves revealed that the relative abundance of the *AtGR1* and *AtGR2* transcripts is enhanced by stress (Allan et al. 2008). Thus, it was proposed that the *At*GR isoforms are involved in redox homeostasis, and that they represent alternative means for the detoxification of SSA as well as glyoxylate, but further in planta work is required to substantiate this hypothesis.

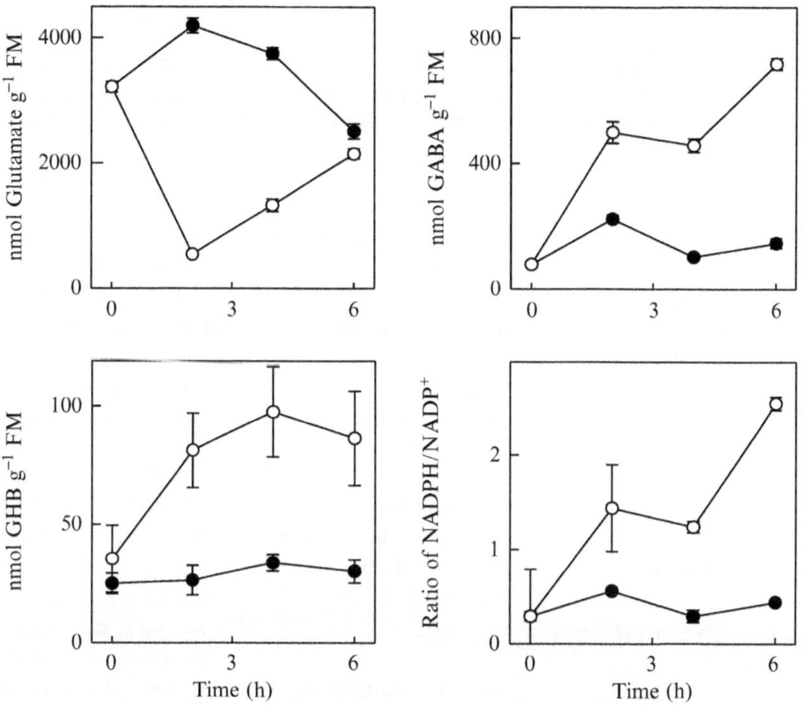

Fig. 2 Response of glutamate, GABA, GHB and NADPH/NADP+ ratio in mature rosette leaves of *Arabidopsis* plants subjected to submergence. Control plants were maintained in the dark at the same temperature. *Closed* and *open symbols* represent control and experimental plants, respectively. Data represent the mean ± SE; where the bar is not shown, it is within the symbol

3 Accumulation of GABA and GHB is a General Response to Stress

A large number of studies have reported the accumulation of GABA in plant tissues and transport fluids in response to many biotic and abiotic stresses (Table 1). These include temperature shock, oxygen deficiency, cytosolic acidification, water stress and UV stress, as well as mechanical stimulation and damage, which are commonly associated with the activities of invertebrate pests during foraging and feeding. In some cases, the response is rapid, often within seconds, suggesting that the biochemical control, rather than transcriptional control, is involved, although there is some evidence for the induction of GAD and GABA-T in the longer term (see Sect. 2).

The first evidence for the occurrence of GHB in plants and its accumulation was presented in 2003 (Table 1). For example, oxygen deficiency increases GHB concentrations from about 10 to 155 nmol g^{-1} fresh mass in soybean sprouts, and from 273 to 739 nmol g^{-1} dry mass in green tea leaves (Allan et al. 2003). Furthermore, the concentrations of GHB and GABA increase in *Arabidopsis* plants under various

Table 1 Biotic and abiotic stresses stimulating GABA and GHB accumulation

Metabolite	Treatment	Tissue/fluid	Reference
GABA	Mechanical stimulation	Soybean leaves and hypocotyl tissue	Wallace et al. (1984); Bown and Zhang (2000)
	Mechanical damage	Soybean and tobacco leaves	Ramputh and Bown (1996); Bown et al. (2002); Hall et al. (2004)
		Alfalfa and tomato phloem exudate	Girousse et al. (1996); Valle et al. (1998)
	Fungal infection	Tomato cell apoplast	Solomon and Oliver (2001)
	Agrobacterium infection	*Arabidopsis* tumors	Deeken et al. (2006)
	Rhizobium infection	Legume nodule	Vance and Heichel (1991)
	Cold stress	Soybean and *Arabidopsis* leaves	Wallace et al. (1984); Kaplan et al. (2007); Allan et al. (2008)
		Asparagus mesophyll cells	Cholewa et al. (1997)
		Barley and wheat seedlings	Mazzucotelli et al. (2006)
	Heat stress	Cowpea cell cultures	Mayer et al. (1990)
		Arabidopsis leaves	Allan et al. (2008)
	Oxygen deficiency	Rice roots	Reggiani et al. (1988); Aurisano et al. (1995)
		Tea leaves, soybean sprouts, tobacco and *Arabidopsis* leaves	Tsishida and Murai (1987); Allan et al. (2003); Breitkreuz et al. (2003); Allan et al. (2008)
		Medicago seedlings	Ricoult et al. (2005)
		Rice cotyledons	Kato-Noguchi and Ohashi (2006)
		Broccoli florets	Hansen et al. (2001)
	Cytosolic acidification	Asparagus mesophyll cells	Crawford et al. (1994)
		Carrot cell suspensions	Carroll et al. (1994)
	Water stress	Tomato roots and leaves	Bolarin et al. (1995)
		Soybean nodules and xylem sap	Serraj et al. (1998)
		Wheat seedlings	Bartyzel et al., (2003–2004)
		Arabidopsis leaves	Allan et al. (2008)
	Phytohormones	*Datura* root cultures	Ford et al. (1996)
	Carbon dioxide enrichment	Cherimoya fruit	Merodio et al. (1998)
		Broccoli florets	Hansen et al. (2001)
	UV stress	*Arabidopsis* plants	Fait et al. (2005)
GHB	Oxygen deficiency	Tea leaves, soybean sprouts, tobacco and *Arabidopsis* leaves	Allan et al. (2003, 2008); Breitkreuz et al. (2003)
	Cold stress	*Arabidopsis* leaves	Kaplan et al. (2007); Allan et al. (2008)
	Heat or water stress	*Arabidopsis* leaves	Allan et al. (2008)
	UV stress	*Arabidopsis* plants	Fait et al. (2005)

stress conditions that should increase the cellular $NADH:NAD^+$ ratio and decrease the adenylate energy charge, thereby inhibiting SSADH activity and diverting carbon from succinate (Shelp et al. 1995, 1999; Busch et al. 1999; Breitkreuz et al. 2003; Allan et al. 2008). Other work revealed that: (1) *ssadh* mutant *Arabidopsis* plants grown under high UV light have five times the normal level of GHB and high levels of ROS (Fait et al. 2005), and; (2) the pattern of GHB in cold-acclimated *Arabidopsis* plants is consistent with the rise and fall of GABA (Kaplan et al. 2007). Together, these data indicate that the accumulation of GHB in plants, as well as GABA, is a general response to abiotic stress.

4 GABA and GHB Signaling Between Plants and Other Organisms

Several papers demonstrate that plant-derived extracellular GABA and possibly GHB mediate communications between plants and animals, fungi, bacteria and other plants. (1) Chemosensory recognition of GABA-mimetic molecules uniquely associated with the surface of crustose red algae induces the motile planktonic larvae of the large red abalone of the eastern Pacific to settle, attach to substrata and metamorphose into benthic juveniles, which feed nondestructively on the algal surface (Morse et al. 1979; Morse and Morse 1984; Trapido-Rosenthal and Morse 1986). (2) The ingestion of elevated GABA concentrations, either in synthetic diets or in transgenic tobacco plants overexpressing GAD, interferes with physiological and developmental processes of several invertebrate pests (Ramputh and Bown 1996; MacGregor et al. 2003; McLean et al. 2003), a result attributed to activation by excess GABA of chloride channels at neuromuscular junctions (Bown et al. 2006). (3) Elevated GABA concentrations in the apoplast of tomato cells infected with the fungus *Cladosporium fulvum* are associated with the induction of the fungal GABA-T and SSADH, indicating that the GABA is being utilized as a nutrient source (Solomon and Oliver 2001, 2002; Oliver and Solomon 2004). (4) High GABA levels are present in *Rhizobium*-induced nodules (see review by Vance and Heichel 1991), as well as in *Rhizobium* bacteroids (Miller et al. 1991) which exhibit active GABA metabolism during symbiosis (Prell et al. 2002). (5) Elevated GABA concentrations in tobacco *GAD* overexpression mutants, wounded tomato stems or culture solution enters *Agrobacterium tumefaciens* cells via the GABA transporter Bra and controls the level of the quorum-sensing signal, thereby resulting in a decline in *Agrobacterium* virulence (Chevrot et al. 2006). It is noteworthy that the level of the quorum-sensing signal is also modulated by GHB (Carlier et al. 2004; Chai et al. 2007), suggesting that plant-derived extracellular GHB might be effective in controlling *Agrobacterium* virulence; however, further research is required to test this hypothesis. (6) A GABA-T-mediated gradient of GABA through apoplastic spaces within the *Arabidopsis* pistil to the female gametophyte is required to guide the pollen tube (Palanivelu et al. 2003), providing evidence for the role of GABA in cell-to-cell communication within plants (Bouché et al. 2003b; Palanivelu et al.

2003) and between plants (Shelp et al. 2006). For further discussion of these papers, refer to a recent review by Shelp et al. (2006).

5 Conclusions and Future Prospects

The neurotransmitters GABA and GHB are found in virtually all prokaryotic and eukaryotic organisms. Recent studies suggest that GABA receptors exist in plants and that GABA serves as a signaling molecule within plants. The physiological roles of GABA and GHB in plants are not yet clear, but both metabolites readily accumulate in response to stress by a combination of biochemical and transcriptional processes. GABA accumulation has been associated with the appearance of extracellular GABA, and evidence is available for a role of extracellular GABA in communications between plants and animals, fungi, bacteria or other plants, although the mechanisms by which GABA functions in communication appear to be diverse. There is no evidence from plants of GHB receptors, GHB signaling or extracellular GHB yet, although the level of the quorum-sensing signalin *Agrobacterium* is known to be modulated by GHB. Future studies should attempt to address these issues and to uncover further examples and the mechanisms by which extracellular GABA is employed to mediate plant communication with other organisms.

Acknowledgments The authors acknowledge research support from the Natural Science and Engineering Research Council of Canada and the Ontario Ministry of Agriculture and Food to B.J.S., and the Centre National de la Recherche Scientifique to D.F.

References

Akama K, Takaiwa F (2007) C-terminal extension of rice glutamate decarboxylase (OsGAD2) functions as an autoinhibitory domain and overexpression of a truncated mutant results in the accumulation of extremely high levels of GABA in plant cells. J Exp Bot 58:2699–2707

Allan WL, Peiris C, Bown AW, Shelp BJ (2003) Gamma-hydroxybutyrate accumulates in green tea leaves and soybean sprouts in response to oxygen deficiency. Can J Plant Sci 83:951–953

Allan WL, Simpson JP, Clark SM, Shelp BJ (2008) γ-Hydroxybutyrate accumulation in *Arabidopsis* and tobacco plants is a general response to abiotic stress: putative regulation by redox balance and glyoxylate reductase isoforms. J Exp Bot 59(9):2545–2554

Ansari MI, Lee RH, Chen RG (2005) A novel senescence-associated gene encoding γ-aminobutyric acid (GABA): pyruvate transaminase is upregulated during rice leaf senescence. Physiol Plant 123:1–8

Aurisano N, Bertani A, Regianni R (1995) Anaerobic accumulation of 4-aminobutyrate in rice seedlings: causes and significance. Phytochemistry 38:1147–1150

Bartyzel I, Pelczar K, Paskowski A (2003/4) Functioning of the γ-aminobutyrate pathway in wheat seedlings affected by osmotic stress. Biol Plant 47:221–225

Beuvé N, Rispail N, Laine P, Cliquet J-B, Ourry A, Le Deunff E (2004) Putative role of γ-aminobutyric acid as a long-distance signal in up-regulation of nitrate uptake in *Brassica napus* L. Plant Cell Environ 27:1035–1046

Bolarin MC, Santa-Cruz A, Cayuela E, Perez-Alfocea F (1995) Short-term solute changes in leaves and roots of cultivated and wil tomato seedlings under salinity. J Plant Physiol 147:463–468

Bouché N, Fait A, Bouchez D, Moller SG, Fromm H (2003a) Mitochondrial succinic-semialdehyde dehydrogenase of the gamma-aminobutyrate shunt is required to restrict levels of reactive oxygen intermediates in plants. Proc Natl Acad Sci USA 100:6843–6848

Bouché N, Lacomb B, Fromm H (2003b) GABA signalling: a conserved and ubiquitous mechanism. Trends Cell Biol 13:607–610

Bown AW, Shelp BJ (1989) The metabolism and physiological roles of 4-aminobutyric acid. Biochemistry (Life Sci Adv) 8:21–25

Bown AW, Zhang G (2000) Mechanical stimulation, 4-aminobutyric acid (GABA) synthesis, and growth inhibition in soybean hypocotyl tissue. Can J Bot 78:119–1123

Bown AW, Hall DE, MacGregor KB (2002) Insect footsteps on leaves stimulate the accumulation of 4-aminobutyrate and can be visualized through increased chlorophyll fluorescence and superoxide production. Plant Physiol 129:1430–1434

Bown AW, MacGregor KB, Shelp BJ (2006) Gamma-aminobutyrate: defense against invertebrate pests?. Trends Plant Sci 11:424–427

Breitkreuz KE, Allan WL, Van Cauwenberghe OR, Jakobs C, Talibi D, Andre B, Shelp BJ (2003) A novel gamma-hydroxybutyrate dehydrogenase: identification and expression of an *Arabidopsis* cDNA and potential role under oxygen deficiency. J Biol Chem 278:41552–41556

Busch KB, Fromm H (1999) Plant sucinic semialdehyde dehydrogenase. Cloning, purification, localization in mitochondria, and regulation by adenine nucleotides. Plant Physiol 121:589–597

Carlier A, Chevrot R, Dessaux Y, Faure D (2004) The assimilation of γ-butyrolactone in *Agrobacterium tumfaciens* C58 interferes with the accumulation of the *N*-acyl-homoserine lactone signal. Mol Plant Microbe Interact 17:951–957

Carroll AD, Fox GC, Laurie S, Phillips R, Ratcliffe RG, Stewart GR (1994) Ammonium assimilation and the role of γ-aminobutyric acid in pH homeostasis in carrot cell suspensions. Plant Physiol 106:513–520

Chai Y, Tsai CS, Cho H, Winans SC (2007) Reconstitution of the biochemical activities of the AttJ repressor and the AttK, AttL, and AttM catabolic enzymes of *Agrobacterium tumefaciens*. J Bacteriol 189:3674–3679

Chevrot R, Rosen R, Haudecoeur E, Cirou A, Shelp BJ, Ron E, Faure D (2006) GABA controls the level of quorum-sensing signal in *Agrobacterium tumefaciens*. Proc Natl Acad Sci USA 103:7460–7464

Cholewa E, Cholewinski A J, Shelp BJ, Snedden WA, Bown AW (1997) Cold shock-stimulated γ-aminobutyric acid synthesis is mediated by an increase in cytosolic Ca^{2+}, not by an increase in cytosolic H^+. Can J Bot 75:375–382

Chung I, Bown AW, Shelp BJ (1992) The production and efflux of 4-aminobutyrate in isolated mesophyll cells. Plant Physiol 99:659–664

Cramer GR, Ergül A, Grimplet J, Tillett RL, Tattersall EAR, Bohlman MC, Vincent D, Sonderegger J, Evans J, Osborne C, Quilici D, Schlauch KA, Schooley DA, Cushman JC (2007) Water and salinity stress in grapevines: early and late changes in transcript and metabolite profiles. Funct Integr Genomics 7:111–134

Crawford LA, Bown AW, Breitkreuz KE, Guinel F (1994) The synthesis of γ-aminobutyric acid in response to treatments reducing cytosolic pH. Plant Physiol 104:865–871

Deeken R, Engelmann JC, Efetova M, Czirjak T, Müller T, Kaiser WM, Tietz O, Krischke M, Mueller MJ, Palme K, Dandekar T, Hedrich R (2006) An integrated view of gene expression and solute profiles of Arabidopsis tumors: a genome-wide approach. Plant Cell 18:3617–3634

Fait A, Yellin A, Fromm H (2005) GABA shunt deficiencies and accumulation of reactive oxygen intermediates: insight from *Arabidopsis* mutants. FEBS Lett 579:415–420

Ford Y-Y, Ratcliffe RG, Robins RJ (1996) Phytohormone-induced GABA production in transformed root cultures of *Datura strammonium*: an in vivo ^{15}N NMR study. J Exp Bot 47:865–871

Girousse C, Bournoville R, Bonnemain J-L (1996) Water deficit-induced changes in concentrations of proline and some other amino acids in the phloem sap of alfalfa. Plant Physiol 111:109–113

Hall DE, MacGregor KB, Nijisse J, Bown AW (2004) Footsteps from insect larvae damage leaf surfaces and initiate rapid responses. Eur J Plant Pathol 110:441–447

Hansen ME, Sørensen H, Cantwell M (2001) Changes in acetaldehyde, ethanol and amino acids in broccoli florets during aire and controlled atmosphere storage. Postharv Biol Technol 22:227–237

Hoover GJ, Van Cauwenberghe OR, Breitkreuz KE, Clark SM, Merrill AR, Shelp BJ (2007a) Characteristics of an *Arabidopsis* glyoxylate reductase: general biochemical properties and substrate specificity for the recombinant protein, and developmental expression and implications for glyoxylate and succinic semialdehyde metabolism in planta. Can J Bot 85:883–895

Hoover GJ, Prentice GA, Merrill AR, Shelp BJ (2007b) Kinetic mechanism of an *Arabidopsis* glyoxylate reductase: studies of initial velocity, dead-end inhibition and product inhibition. Can J Bot 95:896–902

Kaplan F, Kopka J, Sung DY, Zhao W, Popp W, Porat R, Guy CL (2007) Transcript and metabolite profiling during cold acclimation of *Arabidopsis* reveals an intricate relationship of cold-regulated gene expression with modifications in metabolite content. Plant J 50:967–981

Kathiresan A, Miranda J, Chinnapa CC, Reid DD (1998) γ-Aminobutyric acid promotes elongation in *Stellaria longipes*: the role of ethylene. Plant Growth Regul 26:131–137

Kato-Noguchi H, Ohashi C (2006) Effects of anoxia on amino acid levels in rice cotyledons. Plant Prod Sci 9:383–387

Kinnersley AM, Lin F (2000) Receptor modifiers indicate that 4-aminobutyric acid (GABA) is a potential modulator of ion transport in plants. Plant Growth Regul 32:65–76

Kisaka H, Kida T, Miwa T (2006) Antisense suppression of glutamate decarboxylase in tomato (*Lycopersicon esculentum* L.) results in accumulation of glutamate in transgenic tomato fruits. Plant Biotech 23:267–274

Klok EJ, Wilson IW, Wilson D, Chapman, SC, Ewing RM, Somerville SC, Peacock WJ, Dolferus R, Dennis ES (2002) Expression profile analysis of the low-oxygen response in *Arabidopsis* root cultures. Plant Cell 14:2481–2494

Lacombe B, Becker D, Hedrich R, DeSalle R, Hollmann M, Kwak JM, Schroeder JI, Le Novere N, Nam HG, Spalding EP, Tester M, Turano FJ, Chiu J, Coruzzi (2001) The identity of plant glutamate receptors. Science 292:1486–1487

Lancien M, Roberts MR (2006) Regulation of *Arabidopsis thaliana* 14-3-3 gene expression by γ-aminobutyric acid. Plant Cell Environ 29:1430–1436

MacGregor KB, Shelp BJ, Peiris SE, Bown AW (2003) Overexpression of glutamate decarboxylase in transgenic tobacco deters feeding by phytophagous insect larvae. J Chem Ecol 29:2177–2182

Mamelak M (1989) Gammahydroxybutyrate: an endogenous regulator of energy metabolism. Neurosci Biobehav Rev 13:187–197

Mayer R, Cheery J, Rhodes D (1990) Effect of heat shock on amino acid metabolism. Plant Physiol 94:796–810

Mazzucotelli E, Tartari A, Cattivelli L, Forlani G (2006) Metabolism of γ-aminobutyric acid during cold acclimation and freezing and its relationship to frost tolerance in barley and wheat. J Exp Bot 57:3755–3766

McLean MD, Yevtushenko DP, Deschene D, Van Cauwenberghe OR, Makhmoudova A, Potter JW, Bown AW, Shelp BJ (2003) Overexpression of glutamate decarboxylase in transgenic tobacco plants confers resistance to the northern root-knot nematode. Mol Breed 11:277–285

Merodio C, Muñoz MT, Del Cure B, Buitrango D, Escribano MI (1998) Effect of high CO_2 on the titres of γ-aminobutyric acid, total polyamines and some pathogenesis-related proteins in cherimoya fruit stored at low temperature. J Exp Bot 49:1339–1347

Meyer A, Eskandari S, Grallath S, Rentsch D (2006) AtGAT1, a high affinity transporter for γ-aminobutyric acid in *Arabidopsis thaliana*. J Biol Chem 281:7197–7204

Miller RW, McRae DG, Joy K (1991) Glutamate and γ-aminobutyrate metabolism in isolated *Rhizobium meliloti* bacteroids. Mol Plant Microbes Interact 4:37–45

Mirabella R, Rauwerda H, Struys EA, Jakobs C, Triantaphylidès C, Haring MA, Schuurink RC (2008) The *Arabidopsis her1* mutant implicates GABA in *E*-2-hexanal responsiveness. Plant J 53:197–213

Miyashita Y, Good AG (2008) Contribution of the GABA shunt to hypoxia-induced alanine accumulation in roots of *Arabidopsis thaliana*. Plant Cell Physiol 49:92–102

Morse DE, Hooker N, Duncan H, Jensen L (1979) γ-aminobutyric acid, a neurotransmitter, induces planktonic abalone larvae to settle and begin metamorphosis. Science 204:407–410

Morse ANC, Morse DE (1984) Recruitment and metamorphosis of *Haliotis* larvae induced by molecules uniquely available at the surface of crustose red algae. J Exp Mar Biol Ecol 75:191–215

Oliver RP, Solomon PS (2004) Does the oxidative stress used by plants for defence provide a source of nutrients for pathogenic fungi?. Trends Plant Sci 9:472–473

Palanivelu R, Brass L, Edlund AF, Preuss D (2003) Pollen tube growth and guidance is regulated by *POP2*, an *Arabidopsis* gene that controls GABA levels. Cell 114:47–59

Pasentsis K, Falara V, Pateraki I, Gerasopoulos D, Kanellis AK (2007) Identification and expression profiling of low oxygen regulated genes from *Citrus flavedo* tissues using RT-PCR differential display. J Exp Bot 58:2203–2216

Prell J, Boesten B, Poole P, Priefer UB (2002) The *Rhizobium leguminosarum* bv. *viciae* VF39 γ-aminobutyrate (GABA) aminotransferase gene (*gabT*) is induced by GABA and highly expressed in bacteroids. Microbiology 148:615–623

Ramputh A-I, Bown AW (1996) Rapid γ-aminobutyric acid synthesis and the inhibition of the growth and development of oblique-banded leaf-roller larvae. Plant Physiol 111:1349–1352

Reggiani R, Cantu CA, Brimballa I, Britani A (1988) Accumulation and interconversion of amino acids in rice roots under anoxia. Plant Cell Physiol 29:981–987

Ricoult C, Cliquet J-B, Limami AM (2005) Stimulation of alanine amino transferase (*AlaAT*) gene expression and alanine accumulation in embryo axis of the model legume *Medicago truncatula* contribute to anoxia stress tolerance. Physiol Plant 123:30–39

Secor J, Schrader LE (1985) Amino acid efflux from cells and leaf discs. In: Shibles R. (ed) World Soybean Conference III: Proceedings. Westview Co. pp 749-758

Serraj R, Shelp BJ, Sinclair TR (1998) Accumulation of γ-aminobutyric acid in nodulated soybean in response to drought stress. Physiol Plant 102:79–86

Shelp BJ, Walton CS, Snedden WA, Tuin LG, Oresnik IJ, Layzell DB (1995) GABA shunt in developing soybean seeds is associated with hypoxia. Physiol Plant 94:219–228

Shelp BJ, Bown AW, McLean MD (1999) Metabolism and functions of gamma-aminobutyric acid. Trends Plant Sci 4:446–452

Shelp BJ, Bown AW, Faure D (2006) Extracellular γ-aminobutyrate mediates communication between plants and other organisms. Plant Physiol 142:1350–1352

Simpson JP, Di Leo R, Dhanoa PK, Allan WL, Makhmoudova A, Clark SM, Hoover GJ, Mullen RT, Shelp BJ (2008) Identification and characterization of a plastid-localized *Arabidopsis* glyoxylate reductase isoform: comparison with a cytosolic isoform and implications for cellular redox homeostasis and aldehyde detoxification. J Exp Bot 59(9):2545–2554

Solomon PS, Oliver RP (2001) The nitrogen content of the tomato leaf apoplast increases during infection by *Cladosporium fulvum*. Planta 213:241–249

Solomon PS, Oliver RP (2002) Evidence that γ-aminobutyric acid is a major nitrogen source during *Cladosporium fulvum* infection of tomato. Planta 214:414–420

Trapido-Rosenthal HG, Morse DE (1986) Availability of chemosensory receptors is down-regulated by habituation of larvae to a morphogenetic signal. Proc Natl Acad Sci USA 83:7658–7662

Tsishida T, Murai T (1987) Conversion of glutamic acid to γ-aminobutyric acid in tea leaves under anaerobic conditions. Agric Biol Chem 51:2805–2871

Valle EM, Boggio SB, Heldt HW (1998) Free amino acid composition of phloem sap and growing fruit of *Lycopersicon esculentum*. Plant Cell Physiol 39:458–461

Van Cauwenberghe OR, Shelp BJ (1999) Biochemical characterization of partially purified GABA: pyruvate transaminase from *Nicotiana tabacum*. Phytochemistry 52:575–581

Van Cauwenberghe OR, Makhmoudova A, McLean MD, Clark SM, Shelp BJ (2002) Plant pyruvate-dependent gamma-aminobutyrate transaminase: identification of an *Arabidopsis* cDNA and its expression in *Escherichia coli*. Can J Bot 80:933–941

Vance CP, Heichel GH (1991) Carbon in N_2 fixation: limitation or exquisite adaptation. Annu Rev Plant Physiol Plant Mol Biol 42:373–392

Wallace W, Secor J, Schrader LE (1984) Rapid accumulation of gamma-aminobutyric acid and alanine in soybean leaves in response to an abrupt transfer to lower temperature, darkness, or mechanical manipulation. Plant Physiol 75:170–175

Wu C, Zhou S, Zhang Q, Zhao W, Peng Y (2006) Molecular cloning and differential expression of an γ-aminobutyrate transaminase gene, *OsGABA-T*, in rice (*Oryza sativa*) leaves infected with blast fungus. J Plant Res 119:663–669

Yevtushenko D, McLean MD, Peiris SE, Van Cauwenberghe OR, Shelp BJ (2003) Calcium/calmodulin activation of two divergent glutamate decarboxylases from tobacco. J Exp Bot 54:2001–2002

Yu G, Liang J, He Z, Sun M (2006) Quantum dot-mediated detection of γ-aminobutyric acid binding sites on the surface of living pollen protoplasts in tobacco. Chem Biol 13:723–731

Hemiparasitic Plants: Exploiting Their Host's Inherent Nature to Talk

John I. Yoder, Pradeepa C. Gunathilake, and Denneal Jamison-McClung

Abstract Parasitic plants invade and rob host plants of water, minerals and carbohydrates. Host attachment, invasion and resource acquisition is mediated through a parasite-encoded organ called the haustorium. Since the vast majority of plants don't develop haustoria, it is of interest to understand the genetic mechanisms that provide parasites with this novel organ. Host–parasite signaling has been most extensively investigated in the Orobanchaceae, a family of root parasites that includes some of the world's worst agricultural weeds. The need for host resources varies widely among different Orobanchaceae species. Facultative hemiparasites, essentially autotrophic plants that are able to make haustoria, grow fine without ever attacking a host. In contrast, obligate holoparasites are incapable of photosynthesis and require host attachment soon after germination to survive. While morphologically quite different, all parasitic Orobanchaceae develop haustoria in response to chemical and tactile cues provided by their host plants. This review will focus on host signal recognition by hemiparasites, since they represent the earliest stage in the evolutionary transition from autotrophy to heterotrophy. Parasitic plant–host plant interactions provide an excellent illustration of how plants respond to signals in their environments, and how they in turn alter the environment in which they live.

1 Introduction

Introductory biology courses teach that plants are free-living autotrophic organisms capable of independently satisfying their water and mineral needs by absorption and their carbohydrate needs through photosynthesis. In reality, however, plants are

J.I. Yoder (✉) and P.C. Gunathilake
Department of Plant Sciences, University of California–Davis, Davis, CA 95616 USA

D. Jamison-McClung
UC Davis Biotechnology Program, University of California–Davis, Davis, CA 95616 USA

F. Baluška (ed.), *Plant-Environment Interactions*, Signaling and Communication in Plants, DOI: 10.1007/978-3-540-89230-4_5,

of course continually engaged in numerous symbioses with a huge variety of organisms. Some symbioses are considered mutually beneficial to both partners, such as the colonization of plant roots by nitrogen-fixing bacteria or phosphate-acquiring mycorrhizae (Harrison 1999; Jones et al. 2007). Others, such as associations with plant pathogens, are detrimental to the plant host (Jones and Dangl 2006). Most plant symbioses, both mutualistic and parasitic, are realized through a series of developmental processes regulated by biotic and abiotic signals that specify the interaction (Pieterse and Dicke 2007). Identifying the molecular mechanisms associated with symbiosis is a critical step in being able to modify interorganism interactions for enhanced agricultural plant performance. The overall theme of this review is to address the question of how plants interpret and respond to signals from other plants in order to optimize their symbiotic potential.

Parasitic plants invade the tissues of other plants in order to rob them of water and essential nutrients (Kuijt 1969; Press and Graves 1995). Invasion of host plant tissue occurs through a parasite-encoded organ called the haustorium (Visser and Dorr 1987; Riopel and Timko 1995). Haustoria facilitate the attachment of parasitic plants to their hosts, the invasion of host tissues, and the establishment of a vascular continuity between the vascular reserves of the host and those of the parasite. The multiple functions generally attributed to haustoria in parasitic plants fulfill the functions of appressoria and haustoria in fungal plant pathogens (Mendgen and Deising 1993). Haustorium development distinguishes parasitic plants from non-parasitic mycoheterotrophs, such as *Monotropa,* which obtains its carbohydrates via fungal intermediates with other plants (Leake 1994), and epiphytes (such as orchids and Spanish moss), that attach to other plants for physical support but do not directly invade their hosts (Garth 1964). The haustorium is the defining feature of parasitic plants: “It is the organ which...embodies the very idea of parasitism” (Kuijt 1969).

It is estimated that over 4,500 species from 19 angiosperm families are parasitic on other plants (Nickrent 2007). Parasitic plants vary in terms of the degree to which they require host resources. About 10% of parasitic plants are obligate parasites that are incapable of photosynthesis and completely dependent on their hosts for fixed carbohydrates; these species are considered holoparasitic. Holoparasitism, the most evolutionarily advanced incarnation of plant parasitism, will be discussed in the chapter by Mark Mescher. Most parasitic plants are hemiparasites that are able to provide at least some of their own carbon resources autotrophically. While all hemiparasites are capable of photosynthesis, they differ dramatically in their photosynthetic efficiency and thus in their need for host resources. Facultative hemiparasites, like *Triphysaria* (formerly *Orthocarpus* (Chuang and Heckard 1991), can survive to maturity in the absence of host plants, taking up water and inorganic nutrients through their roots and fixing carbon dioxide in their green leaves. In contrast, *Striga hermonthica,* a closely related hemiparasite, has a subterranean growth phase that is heterotrophic and an aboveground phase which is photosynthetic (Parker and Riches 1993). However the aboveground, photosynthetic stage of S. *hermonthica* does not provide sufficient carbon for *S. hermonthica* to survive independently (Press 1995). Because *Striga* fixes carbon but still requires host resources for survival, it is considered an obligate hemiparasite. The most evolutionarily advanced stage of parasitism is represented by the achlorotic holoparasites, such as *Orobanche, Rafflesia* and *Hydnora* (Barkman et al. 2004; Tennakoon et al. 2007). These are incapable of photosynthesis and hence reliant on host carbohydrates at all stages of their lives.

2 Purpose of Review

Intergeneric communications between plants are wonderfully demonstrated by the symbioses between host and parasitic plants. This chapter will overview how parasitic plants perceive and respond to host-derived cues in ways that promote their success as parasites, and how the parasites, in turn, alter the biota in their immediate environments as well as their larger ecology. In this review, we accept Theodosius Dobzhansky's declaration that, "nothing makes sense in biology except in the light of evolution" (Dobzhansky 1964). Phylogenic placement of parasitic plants using morphological characters has been historically problematic, because evolutionarily rapid changes in plant morphology are associated with the acquisition of heterotrophy in plants (Young et al. 1999). Recent studies using DNA sequence polymorphisms as characters has dramatically improved our understanding of the relatedness of parasitic plant lineages and their nearest nonparasitic relatives (Nickrent et al. 1998). These results will be briefly reviewed with respect to what they tell us about the origin of parasitism in plants.

3 Evolution of Parasitism

Parasitic organisms have evolved from free-living ancestors in most major clades of prokaryotic and eukaryotic organisms (Combes 2001). Correspondingly, the fundamental hypothesis of parasitic plant evolution is that parasitism evolved from nonparasitic plants (Kuijt 1969). The phylogenic placement of parasitic lineages has repeatedly concluded that haustorium development originated multiple times in angiosperm evolution; estimates ranges between eight and thirteen independent evolutions of parasitism (Kuijt 1969; Nickrent et al. 1998; Barkman et al. 2007). Interestingly, haustoria only evolved in dicotyledonous lineages: there are no known parasitic monocots and only one species of parasitic gymnosperm, the rare, red-wine-colored *Parasitaxus usta,* whose infection lifecycle combines haustorium invasion for water access and mycorrhizae fungi for carbon acquisition (Feild and Brodribb 2005).

The current models of parasitic plant evolution can be discussed in three phases: (1) the transition from autotrophic to facultative hemiparasite via acquisition of haustoria; (2) the increasing reliance on host resources and consequent refinement of host selection; and, (3) the loss of genes encoding autotroph-specific events.

3.1 Transition from Autotroph to Facultative Hemiparasite: The Origin of Haustoria

There are two general, nonexclusive, hypotheses regarding the evolutionary origins of genes encoding haustorium development; (1) haustorial genes evolved following the duplication and neofunctionalization of genes that exist in nonparasitic plants, or (2) haustorial genes were introduced into the first parasites from nonplant organisms by endosymbiosis and/or horizontal gene transfer. Gene duplications, which can occur in either the whole genome or at a more localized, gene level, are common in plants, and redundant genes can provide novel functions or subfunctions to the plant (Roth et al. 2007; Hegarty and Hiscock 2008). This appears to be the case for many of the genes involved in flower development. By comparing the genomes of flowering plants, gymnosperms, the moss *Physcomitrella* and the lycophyte *Selaginella*, it was clear that nonflowering plants have genes homologous to those regulating flower development (Floyd and Bowman 2007). A second example is DM13, a Ca^{2+}/calmodulin-dependent protein kinase that is required for symbiotic nodule development in legumes; the gene has high homology to genes in tobacco, rice and other non-nodulating plants, indicating that it has alternative functions in nonleguminous plants (Raka et al. 2004). Similarly, the LATD gene of *Medicago truncatula* is required for both nodule and root development, suggesting that both developmental pathways have a common, endogenous origin (Bright et al. 2005).

The second hypothesis for the origin of haustorial genes is that they are of exogenous origin and were introduced into parasitic plants by endosymbiosis or horizontal gene transfer. The superficial resemblance of parasitic haustoria to crown galls, nodules and other microbe-induced modifications led Atsatt to hypothesize that haustoria originated from the endophytic establishment of a plant pathogen, probably a bacterium (Atsatt 1973). Kuijt also hypothesized an exogenous origin for haustoria, where it evolved from a mycoheterotrophic interaction in which the plant first became parasitic on a mycorrhizal fungus which itself was acquiring carbon from another plant host (Kuijt 1969).

There are clear cases of horizontal gene transfer between microbes, microbes and plants and between distinct plants (Zaneveld et al. 2008). In both natural and research settings, horizontal gene transfer occurs during *Agrobacterium* infection of plant tissue via transfer or Ti or Ri plasmids to host plant cells (Nester et al. 2005). Horizontal gene transfer from host to parasitic plant has been inferred from the uniquely discordant phylogenetic placement of the mitochondrial gene *nad1B-C* of *Rafflesia* sp. into a group closely related to its host *Tetrastigma* (Davis and Wurdack 2004). In another example, three species of *Plantago* contain a duplicate pseudogene of the mitochondrial gene *atp1* that phylogenetically clusters with the *atp1* homolog found in *Cuscutta* sp., a distantly related parasite of *Plantago* (Mower et al. 2004). In this latter case, the nucleic acid moved from the parasite to the host plant.

Genome projects to identify genes responsible for haustorium development are underway (Matvienko et al. 2001a; Torres et al. 2005). In one study of about 10,000 transcripts sequenced from roots of *Triphysaria* undergoing haustorium development, all of the transcripts appeared to originate from plants, and none had significant sequence homologies to sequences in the microbial or fungal databases (Torres et al. 2005). While it is not possible to discount the exogenous origin hypothesis until the entire pathway of haustorium development genes is identified, at this point there is no evidence that horizontal gene transfer accounts for the origination of haustorial genes in Orobanchaceae. This is consistent with recent analyses suggesting that similarities between fungal and plant parasitism are largely superficial, and most of the fundamental mechanisms controlling successful parasitism are dissimilar (Mayer 2006).

Regardless of the mechanism of origin, the first parasitic plants were facultative hemiparasites whose newly evolved haustoria could supplement the parasites' water, nitrogen and mineral requirements.

3.2 Facultative Hemiparasite to Obligate Hemiparasite: Increased Host Specificity

Facultative hemiparasites tend to be generalist feeders with a broad range of potential hosts (Atsatt and Strong 1970; Gibson and Watkinson 1989). In field studies, *Triphysaria* was observed growing in association with at least 27 families of plants (Thurman 1966). *Rhinanthus minor*, a related hemiparasite, can parasitize at least 50 species in 18 different families (Gibson and Watkinson 1989). In general, hemiparasites grow better after attachment to a host, but not all host plants are equally as effective at supporting parasite growth (Govier et al. 1967; Gibson and Watkinson 1989). In some cases, attachment to a particular host may be detrimental. It was observed, for example, that *Orthocarpus purpurascens* performed better by several measures when grown in pots autotrophically compared to when it was grown with *Trifolium repens* (Atsatt and Strong 1970). The generalist nature of these plants allows attachment to more than one host species simultaneously, increasing the likelihood of obtaining beneficial compounds and ameliorating potential costs of associating with a poor host (Atsatt and Strong 1970; Marvier 1998).

It is reasonable that the increase in fecundity resulting from the selection of a good host, and the avoidance of a bad one, will result in increased host specificity over time. This seems to be the case; obligate hemiparasites tend to be more specialized than facultative hemiparasites. While facultative hemiparasites like *Triphysaria* can parasitize a wide range of monocot and dicot host families, *Striga* species are much more host specific. The highest degree of host specificity to date has been observed in the obligate hemiparasite *S. gesnerioides*, where seven different races have been identified based on their differential ability to parasitize a tester panel of cowpea lines (Botanga and Timko 2006). The distinction between host races demonstrates a very high degree of host specificity. Similarly, host specialization of plant pathogens generally increases as the organism becomes more dependent on host resources (Kohmoto et al. 1995).

There is therefore a delicate balance between relying on specific beneficial hosts at the expense of environmental adaptability, and indiscriminately parasitizing hosts that may reduce, rather than increase, relative fitness. Factors directly influencing this balance include the level of fluctuation in yearly host populations, combined with the level of genetic variation for both autotrophic and heterotrophic abilities maintained in parasite populations.

3.3 Obligate Hemiparasite to Holoparasite: Loss of Autotrophic Functions

The increasing use of and dependence on host carbohydrates allows a relaxation in parasite photosynthesis. This is accompanied in many cases by rearrangements and deletions to the parasite chloroplast genome (Morden et al. 1991; Bommer et al. 1993; Delavault et al. 1996). Of course, once the chloroplast genome has undergone extensive deletions, the parasite is fixed as a heterotrophic holoparasite.

Haustorium formation differs between facultative and obligate parasites (Goldwasser et al. 2002). Obligate parasites develop primary haustoria that need to successfully invade host roots before further development occurs (Riopel and Baird 1987). Once the primary haustorium is established, secondary roots form on which secondary haustoria develop (Baird and Riopel 1984). Facultative parasites, on the other hand, start their lifecycles without a host, and haustoria are originated near the root apical meristem (Heide-Jørgensen and Kuijt 1995). In these plants, haustoria development is the starting point of the parasitic lifecycle.

4 Hemiparasite Families

Of the eleven independent clades of parasitic plants, five contain hemiparasitic plants. I will very briefly describe these hemiparasitic families. For more detailed information, the reader is directed towards the Parasitic Plant Connection website, which was the starting point for much of the following information (Nickrent 2007).

4.1 Orobanchaceae

Plants in the Orobanchaceae parasitize host plants through their roots (Musselman 1980). The family, which has been recently revised to include hemiparasites previously in the Scrophulariaceae, contains 89 genera (about two thousand species), and all but one are parasitic (Tank et al. 2006). The entire trophic range of parasitism is represented in the Orobanchaceae from the single autotroph *Lindendbergia philippinensis*, the nonparasitic sister to the parasite clade, to facultative and obligate hemiparasites like *Triphysaria* and *Striga*, to achlorophyllous holoparasites like *Orobanche*. Despite the broad range of trophic levels in Orobanchaceae, it is a monophyletic group, which indicates that parasitism has a common origin within the Orobanchaceae (dePamphilis et al. 1997; Nickrent et al. 1998).

Striga and *Orobanche* are both notorious agricultural weeds that are particularly devastating in the poorly nourished soils common in underdeveloped countries (Parker and Riches 1993; Scholes and Press 2008). *Striga* infests about 60% of the agricultural regions in sub-Saharan Africa; when established in a field, it can cause complete yield losses by reducing host resources as well as non-resource-dependent pathogenesis (Musselman 1980; Rank et al. 2004). The genus *Striga* has a broad host range, including monocots and dicots, but individual species are much more specialized. S. *hermonthica* and *S. asiatica* are monocot specific and their hosts include all the major tropical cereals (maize, sorghum, rice and millet). In contrast, *S. gesnerioides* parasitizes dicotyledonous hosts, most notably Leguminosae (Parker and Riches 1993).

The success of Orobanchaceae as plant pests is related to their ability to integrate their lifestyles into that of their host through chemical communications. Haustoria develop on the roots of Orobanchaceae in response to chemical and tactile signals from their hosts (Riopel and Timko 1995). Some Orobanchaceae also require host plant signals in order to germinate (Bouwmeester et al. 2007). Because the Orobanchaceae alter their growth and development in response to host plant signals in ways that are easily visualized, they provide excellent models of plant–plant communications.

4.2 Santalales

Santalales is a large group of approximately 160 plant genera, including nonparasites, root parasites, and aerial, stem parasites (Der and Nickrent 2008). The order comprises five families: Loranthaceae, Misodendraceae, Olacaceae, Opiliaceae, and Santalaceae (which now includes Viscaceae). Phylogenetic analyses suggest that aerial parasitism arose five times and root parasitism at least once in this order (Malécot and Nickrent 2008). Aerial stem parasites in Santalales go by the common name of mistletoes. Mistletoe species grow on a wide range of host trees, commonly reducing their growth or even killing them with heavy infestation (Parker and Riches 1993). The genus *Arceuthobium* (dwarf mistletoe) is particularly harmful and is considered the most damaging pathogen in North American coniferous forests, where it causes timber losses estimated at about 3 billion board feet of lumber per year (Hawksworth and Wiens 1996).

Mistletoes are hemiparasitic, although the amount of carbon fixed by different genera is quite variable. Leaves of the leafy mistletoe *Phoradendron* have a chlorophyll content and a photosynthetic capacity comparable to that of the host (Hull and Leonard 1964). When labeled C is applied to *Phoradendron* shoots, the authors observed translocation of photosynthate from mistletoe shoots to the endophytic system of the parasite. In contrast, aerial shoots of *Arceuthobium* also contain chlorophyll but at concentrations that are only about 10% of those found in host foliage. Carbon fixed by dwarf mistletoe shoots was never observed to translocate into the endophytic system or into the host tissue (Hull and Leonard 1964).

4.3 *Convolvulaceae*

Convolvulaceae, or the Morning Glory family, contains about 60 genera, only one of which (*Cuscutta*) is parasitic. The genus *Cuscutta* contains about 170 hemiparasitic and holoparasitic species. Molecular studies of the chloroplast genome and physiological studies of photosynthetic enzymes show that *Cuscuta reflexa* retains a deleted yet functional plastid genome (Haberhausen et al. 1992). Conversely, the plastid genome of *C. europaea* has sustained greater losses and shows no RUBISCO activity (Machado and Zetsche 1990).

4.4 *Lauraceae*

The laurel family contains a single parasitic genus, *Cassytha*, that grows as a yellowish-brown vine, similar in appearance (but not origin) to *Cuscutta*. The 17 *Cassytha* species are distributed principally in Australia, but some species are found in southern Asia, Africa, northern South America, Central America, southern Florida and Japan (Nickrent 2007).

4.5 *Krameriaceae*

This family, commonly known as Rhatany, comprises a single genus, *Krameria*, with 17 species. *Krameria* are root parasites that grow as perennial shrubs in South and Central America as well as the southwestern region of North America (Nickrent 2007).

5 The Parasitism Process with Specific Reference to Host Determination

The mechanisms associated with plant parasitism have been most thoroughly investigated in Orobanchaceae. This is due in large part to their agricultural significance and in part to the tractability of the family to in vitro studies. Based on phenotypes of host resistances, there are likely to be many stages at which chemical or physical signals are exchanged between the host and parasitic plants. Two stages in the parasite life cycle are known to be influenced by chemical factors; germination and haustorium development.

5.1 *Germination*

Orobanchaceae seeds are generally small and fall directly into the soil, so the movement of plants to new host environments generally involves hitching a ride with animal

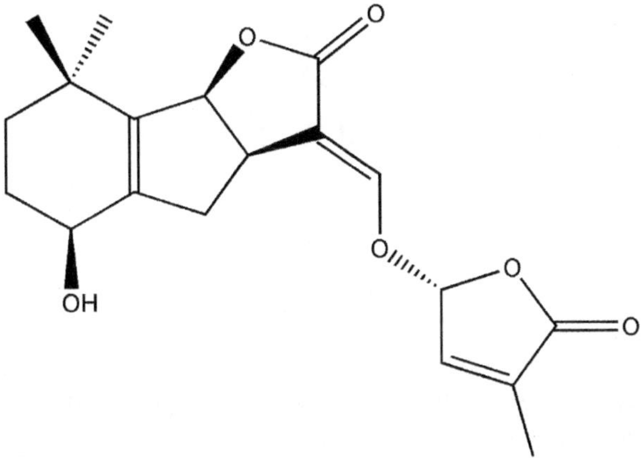

Fig. 1 Strigolactone

vectors (Berner et al. 1994). Most hemiparasitic Orobanchaceae germinate under the appropriate conditions of humidity, temperature and light. Others require host factors in order to germinate. The first molecule identified as a germination stimulant for *Striga* was strigolactone (Fig. 1; Cook et al. 1966); a molecule originally described as a sesquiterpene lactone but which has since been shown to be synthesized in the carotenoid biosynthesis pathway (Matusova et al. 2005). Strigolactone is active at very low concentrations, and its ability to induce hyphal branching in arbuscular mycorrhizal fungi indicates that strigolactone plays additional roles in the rhizosphere (Akiyama et al. 2005; Humphrey and Beale 2006). Strigolactone is not, however, a determinant of host specificity, because even nonhost plants produce strigolactone and germinate *Striga* seed.

5.2 Early Haustorium Development

Haustoria develop on roots of Orobanchaceae in response to host factors, both chemical and tactile (Atsatt et al. 1978; Riopel and Timko 1995). The first haustorium-inducing factor (HIF) to be identified was 2,6-dimethoxy-1,4-benzoquinone (DMBQ) (Chang and Lynn 1986) (Fig. 2). DMBQ is a common component of plant cell walls and has been observed in at least 48 genera belonging to 29 plant families (Handa et al. 1983). Due to its electrophilic, oxidant nature, DMBQ has allelopathic, mutagenic, carcinogenic and cytotoxic characteristics (Brambilla et al. 1988). Cellular damage results from the redox cycling between quinone and semiquinone states, giving rise to reactive oxygen species (Testa 1995).

In fact, it is the redox cycling of quinones to their semiquinone forms that has been hypothesized to induce haustorium development. In DMBQ induction of *Striga* seedlings, addition of spin-trap chemicals, such as cyclopropyl benzoquinone (CPBQ) and tetrafluorobenzo-1,4-quinone (TFBQ), has been shown to inhibit haustorium development (Smith et al. 1996; Zeng et al. 1996). Further, the HIFs

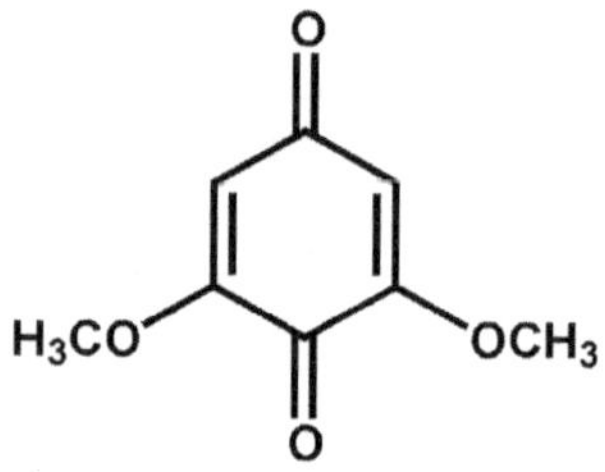

Fig. 2 2,6-Dimethoxybenzoquinone

active in inducing *Striga* seedlings had a narrow range of redox potentials (Smith et al. 1996), and it was shown that phenolics must be converted to quinone forms before they become active HIFs (Kim et al. 1998). The hypothesis put forward by this work on *Striga* is that the free radical associated with redox cycling between the oxidized and reduced forms is the signal that initiates haustorium development.

Genes upregulated in *Triphysaria* roots soon after DMBQ exposure included two NAD(P)H-dependent quinone oxidoreductases, TvQR1 and TvQR2 (Matvienko et al. 2001a, b). In *Triphysaria* there is rapid transcriptional induction of both TvQR1 and TvQR2 as a primary response to DMBQ treatment. TvQR1 exhibits homology to a family of zeta-crystallins and catalyzes a one-electron reduction of quinone to semiquinone, providing a free radical consistent with the redox signaling hypothesis (Fillapova, Petite, Yoder, unpublished). TvQR2 is related to a class of detoxifying enzymes, such as human liver DT-diaphorase, and catalyzes a two-electron reduction of quinones (Wrobel et al. 2002). We hypothesize that TvQR1 and TvQR2 act antagonistically in that TvQR1 generates free radicals and TvQR2 detoxifies them. We propose that if the activity of TvQR1 is greater than TvQR2, haustorium development proceeds; if the activity of TvQR2 is greater, no haustoria form. Haustorium development in this model is proposed to be regulated by the relative activities of two counteracting enzymes.

5.3 Post-Attachment Physiology

Vascular connections made though haustoria provide the route for the molecular trafficking of sugars, water, amino acids, organic acids, and ions between host and parasitic plants (Okonkwo 1966; Hibberd and Jeschke 2001). In addition to nutritional molecules, informational macromolecules can also translocate, including RNAs (Roney et al. 2007), silencing RNAi molecules (Tomilov et al. 2008), proteins (Haupt et al. 2001; Birschwilks et al. 2006) and DNA (Davis and Wurdack 2004; Richardson and Palmer 2006).

Hemiparasites are usually xylem feeders (Hibberd and Jeschke 2001) that depend on their host for xylem-dissolved minerals and some organic compounds such as reduced N in the form of amino acids (Jiang et al. 2008). In *Triphysaria* haustoria, one can generally visualize 1–5 xylem strands with vessel elements connecting host and parasite vasculature (Heide-Jørgensen and Kuijt 1993, 1995). These make bridges with host xylem directly or occasionally host parenchyma of the plate xylem adjacent to the stele of the parasite root. Based on microscopic observations and physiological facts, Kuijt proposed that minerals and organic compounds are transferred directly to the xylem apoplast of the host and then to the xylem of the haustoria at the xylem bridge (Kuijt 1991). In the transfer of organic compounds such as soluble sugars and carbohydrates, parenchyma cells of the plate xylem may act as a sink, and then those compounds could reach the sieve tubes of the parasite's root by symplastic transport.

The haustoria interface in the root hemiparasitic *Olax phyllanthi* consists almost entirely of xylem parenchyma cells that function as transfer cells. Even with a few tracheids present at the host–parasite interface, direct lumen-to-lumen continuity between tracheary elements of the two plants was not observed (Pate et al. 1990). Light, transmission electron and scanning electron microscopy studies on the haustorial interface of *S. hermonthica* and *S. asiatica* have recognized the presence of very specific clustered intrusions and their growth into the host's xylem, mainly into the large vessel elements (Dorr 1997). Later, these intrusions and the haustorial cells lose their protoplasts and transform into structures called "oscula" that are used for water and nutrient uptake, making direct lumen connection with the host xylem (Dorr 1997).

There is no evidence to show that direct phloem tapping by hemiparasites to withdraw phloem-borne photosynthates occurs. However, studies have shown that about 30% of the total carbon in leaves of mature *S. hermonthica* is synthesized in the host (Press et al. 1987; Shah et al. 1987). Moreover, mistletoes tapping host xylem can withdraw between 5 and 63% of their carbon requirement from the host (Marshall et al. 1994). When *Olax* parasitize *Acacia*, 40% of the total carbon is host derived, and this value is about 10% when *Hordeum* is parasitized by Rhinanthus.

Most hemiparasites maintain transpiration rates that are double (or more) those of the hosts, especially under low soil water potential (Shen et al. 2006). These high rates may generate strong mass flow, making water, carbon and mineral nutrients flow from the host to the parasitic plants (Press and Whittaker 1993). Another mechanism is that hemiparasites maintain a lower water potential than their hosts (Tennakoon et al. 1997). This may generate a high osmotic gradient between the parasite and the attached host, establishing a driving force for transpiration and nutrient transfer (Ehleringer and Marshall 1995). High concentrations of soluble carbohydrates such as glucose, fructose and mannitol and inorganic ions such as potassium and magnesium in the hemiparasite vascular system are directly related to the high osmotic potential gradient between host and parasite (Press et al. 1990). For example, *Striga* seems to absorb resources from the host by maintaining a strong osmotic pull achieved through the biosynthesis of polyhydric alcohols such as mannitol (Press 1995; Robert et al. 1999). Both organic and inorganic forms of nitrogen compounds found in the transpiration stream contribute to the high osmotic potential in hemiparasites. Since the nitrogen concentration in the transpiration stream can affect the leaf stomatal conductance of hemiparasites, it is hypothesized that nitrogen may be a key factor in regulating the solute transfer from host to parasite (Press 1995).

There is evidence for the bidirectional movement of molecules from hemiparasites into host. For example, dwarf mistletoes alter the growth of host trees by stimulating the production of host growth regulators or by transferring hormones directly into hosts (Knutson 1979; Livingston et al. 1984). Similarly, *Striga* infection has a pathological effect on host plants, which leads to a reduction in host growth that is more than can be accounted for by loss of nutrients alone (Musselman 1980; Rank et al. 2004). A recent study has shown that RNAi targeting a transgene can traffic from a lettuce host to *T. versicolor* roots cultured in vitro (Tomilov et al. 2008). Transgenic *T. versicolor* root cultures containing GUS were attached to lettuce generating a double-stranded RNA for GUS (dsGUS). Histochemical staining and semi-quantitative RT-PCR showed the silencing of GUS in parasite root tips after haustoria connection with dsGUS-expressing lettuce. Interestingly, when a nontransgenic *Triphysaria* seedling was allowed to infect two lettuce roots, one transgenic for GUS and the second transgenic for dsGUS, a clearing of GUS activity was observed near the haustorial infection site. This indicates that the dsGUS molecule is picked up by the parasite from one plant and transferred to a second plant where it functions (Tomilov et al. 2008)

6 Conclusions

Hemiparasites represent the first evolutionary manifestation of parasitism in plants, the ability to develop haustoria. In some cases haustorium development is induced by chemical and tactile signals from the host. Facultative hemiparasites tend to have broad host ranges and take up a range of molecules, nutritional and informational. The degree of benefit to the parasite is a function of the host species; attachment to some hosts increases parasite performance, while attachment to other host species can be worse than independent growth alone. Host specificity increases over evolutionary time, presumably to allow the parasite to identify the most beneficial host. Specificity increases over evolutionary time until some parasites become obliged to invade a single host species for survival. Once the parasites have identified and adapted to certain species they no longer need to make their own carbohydrates, and the loss of photosynthetic genes from the chloroplast genome is a repeated fate. Because obligate hemiparasites and holoparasites have undergone numerous secondary mutations as a result of their host dependence, facultative hemiparasites offer the prospect of studying one of the earliest events in parasitic plant evolution: haustorium development.

References

Akiyama K, Matsuzaki K, Hayashi H (2005) Plant sesquiterpenes induce hyphal branching in arbuscular mycorrhizal fungi. Nature 435:824–827

Atsatt PR (1973) Parasitic flowering plants: how did they evolve? Am Nat 107:502–510

Atsatt P, Strong D (1970) The population biology of annual grassland hemiparasites: I The host environment. Evol 24:278–291

Atsatt PR, Hearn TF, Nelson RL, Heineman RT (1978) Chemical induction and repression of haustoria in *Orthocarpus purpurascens* (Scophulariaceae). Ann Bot 42:1177–1184

Baird WV, Riopel JL (1984) Experimental studies of haustorium initiation and early development in *Agalinis purpurea* (L.) Raf. (Scrophulariaceae). Am J Bot 71:803–814

Barkman TJ, Lim S-H, Salleh KM, Nais J (2004) Mitochondrial DNA sequences reveal the photosynthetic relatives of *Rafflesia*, the world's largest flower. Proc Natl Acad Sci USA 101:787–792

Barkman T, McNeal J, Lim S-H, Coat G, Croom H, Young N, dePamphilis C (2007) Mitochondrial DNA suggests at least 11 origins of parasitism in angiosperms and reveals genomic chimerism in parasitic plants. BMC Evol Biol 7:248

Berner DK, Cardwell KF, Faturoti BO, Ikie FO, Williams OA (1994) Relative roles of wind, crop seeds, and cattle in dispersal of *Striga* ssp. Plant Dis 78:402–406

Birschwilks M, Haupt S, Hofius D, Neumann S (2006) Transfer of phloem-mobile substances from the host plants to the holoparasite *Cuscuta* sp. J Exp Bot 57:911–921

Bommer D, Haberhausen G, Zetsche K (1993) A large deletion in the plastid DNA of the holoparasitic flowering plant *Cuscuta reflexa* concerning 2 ribosomal proteins (RPL2, RPL23), one transfer RNA (TRNI) and an ORF-2280 homolog. Curr Genet 24:171–176

Botanga CJ, Timko MP (2006) Phenetic relationships among different races of *Striga gesnerioides* (Willd.) Vatke from West Africa. Genome 49:1351–1365

Bouwmeester HJ, Roux C, Lopez-Raez JA, Becard G (2007) Rhizosphere communication of plants, parasitic plants and AM fungi. Trends Plant Sci 12:224–230

Brambilla G, Robbiano L, Cajelli E, Martelli A, Turmolini F, Mazzei M (1988) DNA-damaging and mutagenic properties of 2,6-dimethoxy-1,4-benzoquinone, formed by dimethoprine-nitrite interaction. Int J Clin Pharmacol 244:1011–1015

Bright LJ, Liang Y, Mitchell DM, Harris JM (2005) The LATD gene of *Medicago truncatula* is required for both nodule and root development. Mol Plant Microbe Interact 18:521–532

Chang M, Lynn DG (1986) The haustorium and the chemistry of host recognition in parasitic angiosperms. J Chem Ecol 12:561–579

Chuang TI, Heckard LR (1991) Generic realignment and synopsis of subtribe Castillejinae (Scrophulariaceae—tribe Pedicularеae). Syst Bot 16:644–666

Combes C (2001) Parasitism: the ecology and evolution of intimate interactions. University of Chicago Press, Chicago

Cook CE, Whichard LP, Wall ME, Egely GH (1966) Germination of witchweed (*Striga lutea* Lour.): isolation and properties of a potent stimulant. Science 196:1189–1190

Davis CC, Wurdack KJ (2004) Host-to-parasite gene transfer in flowering plants: phylogenetic evidence from Malpighiales. Science 305:676–678

Delavault PM, Russo NM, Lusson NA, Thalouarn PA (1996) Organization of the reduced plastid genome of *Lathraea clandestina*, an achlorophyllous parasitic plant. Physiol Plant V96:674–682

dePamphilis CW, Young NN, Wolfe AD (1997) Evolution of plastid gene *rps2* in a lineage of hemiparasitic and holoparasitic plants: many losses of photosynthesis and complex patterns of rate variation. Proc Natl Acad Sci USA 93:7367–7372

Der JP, Nickrent DL (2008) A molecular phylogeny of Santalaceae (Santalales). Syst Bot 33:107–116

Dobzhansky T (1964) Biology, molecular and organismic. Am Zool 4:443–452

Dorr I (1997) How *Striga* parasitizes its host: a TEM and SEM study. Ann Bot 79:463–472

Ehleringer JR, Marshall JD (1995) Water relations. In: Press MC, Graves JD (eds) Parasitic plants. Chapman and Hall, London, pp. 124–140

Feild TS, Brodribb TJ (2005) A unique mode of parasitism in the conifer coral tree *Parasitaxus ustus* (Podocarpaceae). Plant Cell Environ 28:1316–1325

Floyd SK, Bowman JL (2007) The ancestral developmental tool kit of land plants. Int J Plant Sci 168:1–35

Garth RE (1964) The ecology of Spanish moss (*Tillandsia usneoides*): its growth and distribution. Ecol 45:470–481

Gibson CC, Watkinson AR (1989) The host range and selectivity of a parasitic plant: *Rhinanthus minor* L. Oecologia 78:401–406

Goldwasser Y, Westwood JH, Yoder JI (2002) The use of *Arabidopsis* to study interactions between parasitic angiosperms and their plant hosts. In: Somerville C, Meyerowitz E (eds) The *Arabidopsis* book. American Society Plant Biologists, Rockville, MD

Govier RN, Nelson MD, Pate JS (1967) Hemiparasitic nutrition in angiosperms: the transfer of organic compounds from host to *Odontites verna* (Bell.) Dum. (Scrophulariaceae). New Phytol 66:285–297

Haberhausen G, Valentin K, Zetsche K (1992) Organization and sequence of photosynthetic genes from the plastid genome of the holoparasitic flowering plant *Cuscuta reflexa*. Mol Gen Genet 232:154–161

Handa S, Kinghorn A, Cordell G, Farnsorth N (1983) Plant anticancer agents XXVI. Constituents of *Peddiea fischeri*. J Nat Prod 46:248–250

Harrison MJ (1999) Molecular and cellular aspects of the arbuscular mycorrhizal symbiosis. Ann Rev Plant Phys Plant Mol Biol 50:361–389

Haupt S, Oparka K, Sauer N, Neumann S (2001) Macromolecular trafficking between *Nicotiana tabacum* and the holoparasite *Cuscuta reflexa*. J Exp Bot 52:173–177

Hawksworth FG, Wiens D (1996) Dwarf mistletoes: biology, pathology, and systematics. United States Department of Agriculture Forest Service, Washington, DC

Hegarty MJ, Hiscock SJ (2008) Genomic clues to the evolutionary success of review polyploid plants. Curr Biol 18:R435–R444

Heide-Jørgensen HS, Kuijt J (1993) Epidermal derivatives as xylem elements and transfer cells: a study of the host–parasite interface in two species of *Triphysaria* (Scrophulariaceae). Protoplasma 174:173–183

Heide-Jørgensen HS, Kuijt J (1995) The haustorium of the root parasite *Triphysaria* (Scrophulariaceae), with special reference to xylem bridge ultrastructure. Am J Bot 82:782–797

Hibberd JM, Jeschke WD (2001) Solute flux into parasitic plants. J Exp Bot 52:2043–2049

Hull RJ, Leonard OA (1964) Physiological aspects of parasitism in mistletoes (*Arceuthobium* and *Phoradendron*). II. The photosynthetic capacity of mistletoe. Plant Physiol 39:1008–1017

Humphrey AJ, Beale MH (2006) Strigol: biogenesis and physiological activity. Phytochemistry 67:636–640

Jiang F, Jeschke WD, Hartung W, Cameron DD (2008) Does legume nitrogen fixation underpin host quality for the hemiparasitic plant *Rhinanthus minor*? J Exp Bot 59:917–925

Jones JDG, Dangl JL (2006) The plant immune system. Nature 444:323–329

Jones KM, Kobayashi H, Davies BW, Taga ME, Walker GC (2007) How rhizobial symbionts invade plants: the *Sinorhizobium-Medicago* model. Nat Rev Microbiol 5:619–633

Kim D, Kocz R, Boone L, Keyes WJ, Lynn DG (1998) On becoming a parasite: evaluating the role of wall oxidases in parasitic plant development. Chem Biol 5:103–117

Knutson DM (1979) How parasitic seed plants induce disease in other plants. In: Horsfall JG, Cowling EB (eds) Plant disease: an advanced treatise. Academic Press, San Diego, pp 293–312

Kohmoto K, Singh US, Singh RP (1995) Pathogenesis and host specificity in plant pathogenic fungi and nematodes. In: Kohmoto K, Singh US, Singh RP (eds) Pathogenesis and host specificity in plant diseases: histopathological, biochemical, genetic and molecular bases, vol. 2. Elsevier, Tarrytown, pp. xxi–xxvii

Kuijt J (1969) The biology of parasitic flowering plants. University of California Press, Berkeley

Kuijt J (1991) The haustorial interface: what does it tell us? In: Ranson J, Musselman LJ, Worsham AD, Parker C (eds) Proceedings of the Fifth International Symposium of Parasitic Weeds. CIM-MYT, Narobi

Leake JR (1994) The biology of myco-heterotrophic ("saprophytic") plants. New Phytol 127:171–216

Livingston WH, Brenner ML, Blanchette RA (1984) Altered concentrations of abscisic acid, indole-3-acetic acid zeatin riboside associated with eastern dwarf mistletoe infections on black spruce. In: Hawksworth FG, Scharpf RF (eds) Biology of Dwarf Mistletoes. USDA Forest Service, Fort Collins, pp 53–61

Machado MA, Zetsche K (1990) A structural, functional and molecular analysis of plastids of the holoparasites *Cuscusta reflexa* and *Cuscuta europaea*. Planta 181:91–96

Malécot V, Nickrent DL (2008) Molecular phylogenetic relationships of Olacaceae and related Santalales. Syst Bot 33:97–106

Marshall JD, Ehleringer JR, Schulze ED, Farquhar G (1994) Carbon-isotope composition, gas-exchange and heterotrophy in Australian mistletoes. Funct Ecol 8:237–241

Marvier MA (1998) A mixed diet improves performance and herbivore resistance of a parasitic plant. Ecol 79:1272–1280

Matusova R, Rani K, Verstappen FWA, Franssen MCR, Beale MH, Bouwmeester HJ (2005) The strigolactone germination stimulants of the plant-parasitic *Striga* and *Orobanche* spp. are derived from the carotenoid pathway. Plant Physiol 139:920–934

Matvienko M, Torres MJ, Yoder JI (2001a) Transcriptional responses in the hemiparasitic plant *Triphysaria versicolor* to host plant signals. Plant Physiol 127:272–282

Matvienko M, Wojtowicz A, Wrobel R, Jamison D, Goldwasser Y, Yoder JI (2001b) Quinone oxidoreductase message levels are differentially regulated in parasitic and non-parasitic plants exposed to allelopathic quinones. Plant J 25:375–387

Mayer AM (2006) Pathogenesis by fungi and by parasitic plants: similarities and differences. Phytoparasitica 34:3–16

Mendgen K, Deising H (1993) Infection structures of fungal plant pathogens: a cytological and physiological evaluation. New Phytol 124:193–213

Morden CW, Wolfe KH, dePamphilis CW, Palmer JD (1991) Plastid translation and transcription genes in a non-photosynthetic plant: intact, missing and pseudo genes. EMBO J 10:3281–3288

Mower JP, Stefanovic S, Young GJ, Palmer JD (2004) Plant genetics: gene transfer from parasitic to host plants. Nature 432:165–166

Musselman LJ (1980) The biology of *Striga*, *Orobanche*, and other root parasitic weeds. Ann Rev Phytopath 18:463–489

Nester EW, Gordon MP, Kerr A (2005) *Agrobacterium tumefaciens*: from plant pathology to biotechnology. APS, St. Paul

Nickrent DL, Duff RJ, Colwell AE, Wolfe AD, Young ND, Steiner KE, dePamphilis CW (1998) Molecular phylogenetic and evolutionary studies of parasitic plants. In: Soltis DE, Soltis PS, Doyle JJ (eds) Molecular systematics of plants II: DNA sequencing. Kluwer, Boston, pp 211–241

Okonkwo SNC (1966) Studies on *Striga senegalensis. II*. Translocation of ^{14}C-labeled photosynthate, urea-^{14}C and sulphur^{35} between host and parasite. Am J Bot 53:142

Parker C, Riches CR (1993) Parasitic weeds of the world: biology and control. CAB International, Wallingford

Pate JS, Kuo J, Davidson NJ (1990) Morphology and anatomy of the haustorium of the root hemiparasite *Olax phyllanthi* (Olacaceae), with a special reference to the haustorial interface. Ann Bot 65:425–436

Pieterse CMJ, Dicke M (2007) Plant interactions with microbes and insects: from molecular mechanisms to ecology. Trends Plant Sci 12:564–569

Press MC (1995) Carbon and nitrogen relations. In: Press MC, Graves JD (eds) Parasitic plants. Chapman and Hall, London, pp. 103–124

Press MC, Tuohy JM, Stewart GR (1987) Gas exchange characteristics of the *Sorghum-Striga* host–parasite association. Plant Physiol 84:814–819

Press MC, Graves JD, Stewart GR (1990) Physiology of the interaction of angiosperm parasites and their higher plant hosts. Plant Cell Environ 13:91–104

Press MC, Whittaker JB (1993) Exploitation of the xylem stream by parasitic organisms. Philos Trans R Soc Lond B Biol Sci 341:101–111

Raka MM, Cynthia AG, Anne E, James H, J AD, Giles EDO, Sharon RL (2004) A Ca^{2+}/calmodulin-dependent protein kinase required for symbiotic nodule development: gene identification by transcript-based cloning. Proc Natl Acad Sci USA 101:4701–4705

Rank C, Rasmussen LS, Jensen SR, Pierce S, Press MC, Scholes JD (2004) Cytotoxic constituents of *Alectra* and *Striga* species. Weed Res 44:265–270

Richardson AO, Palmer JD (2006) Horizontal gene transfer in plants. J Exp Bot 58:1–9

Riopel JL, Baird WV (1987) Morphogenesis of the early development of primary haustoria in *Striga asiatica*. In: Musselman LJ (ed) Parasitic weeds in agriculture, vol. 1. CRC, Boca Raton, pp 107–125

Riopel JL, Timko MP (1995) Haustorial initiation and differentiation. In: Press MC, Graves JD (eds) Parasitic plants. Chapman and Hall, London, pp 39–79

Robert S, Simier P, Fer A (1999) Purification and characterization of mannose 6-phosphate reductase, a potential target for the control of *Striga hermonthica* and *Orobanche ramosa*. Aust J Plant Physiol 26:233–237

Roney JK, Khatibi PA, Westwood JH (2007) Cross-species translocation of mRNA from host plants into the parasitic plant dodder. Plant Physiol 143:1037–1043

Roth C, Rastogi S, Arvestad L, Dittmar K, Light S, Ekman D, Liberles DA (2007) Evolution after gene duplication: models, mechanisms, sequences, systems, and organisms. J Exp Zool 308B:58–73

Scholes JD, Press MC (2008) *Striga* infestation of cereal crops: an unsolved problem in resource limited agriculture. Curr Opin Plant Biol 11:180–186

Shah N, Smirnoff N, Stewart G (1987) Photosynthesis and stomatal characteristics of *Striga hermonthica* in relation to its parasitic habit. Physiol Plant 69:699–703

Shen H, Ye W, Hong L, Huang H, Wang Z, Deng X, Yang Q, Xu Z (2006) Progress in parasitic plant biology: host selection and nutrient transfer. Plant Biol 8:175–185

Smith CE, Ruttledge T, Zeng Z, O'Malley RC, Lynn DG (1996) A mechanism for inducing plant development: the genesis of a specific inhibitor. Proc Natl Acad Sci USA 93:6986–6991

Tank DC, Beardsley PM, Kelchner SA, Olmstead RG (2006) Review of the systematics of Scrophulariaceae s.l and their current disposition. Aust System Bot 19:289–307

Tennakoon KU, Pate JS, Stewart GR (1997) Haustorium-related uptake and metabolism of host xylem solutes by the root hemiparasitic shrub *Santalum acuminatum* (R. Br.) ADC (Santalaceae). Ann Bot 80:257–264

Tennakoon KU, Bolin JF, Musselman LJ, Maass E (2007) Structural attributes of the hypogeous holoparasite *Hydnora triceps* Drege & Meyer (Hydnoraceae). Am J Bot 94:1439–1449

Testa B (1995) The metabolism of drugs and other xenobiotics. Academic Press, New York

Thurman LD (1966) Genecological studies in *Orthocarpus* subgenus *Triphysaria*. University of California, Berkeley

Tomilov AA, Tomilova NB, Wroblewski T, Michelmore R, Yoder JI (2008) Trans-specific gene silencing between host and parasitic plants. Plant J 56(3):389–397

Torres MJ, Tomilov AA, Tomilova N, Reagan RL, Yoder JI (2005) Pscroph, a parasitic plant EST database enriched for parasite associated transcripts. BMC Plant Biol 24:24

Visser J, Dorr I (1987) The haustorium. In: Musselman LJ (ed) Parasitic weeds in agriculture, vol. 1. CRC, Boca Raton, pp 91–106

Wrobel RL, Matvienko M, Yoder JI (2002) Heterologous expression and biochemical characterization of an NAD(P)H: quinone oxidoreductase from the hemiparasitic plant *Triphysaria versicolor*. Plant Physiol Biochem 40:265–272

Young ND, Steiner KE, dePamphilis CW (1999) The evolution of parasitism in Scrophulariaceae/Orobanchaceae: plastid gene sequences refute an evolutionary transition series. Ann Mo Bot Gard V86:876–893

Zaneveld JR, Nemergut DR, Knight R (2008) Are all horizontal gene transfers created equal? Prospects for mechanism-based studies of HGT patterns. Microbiology 154:1–15

Zeng ZX, Cartwright CH, Lynn DG (1996) Cyclopropyl-*p*-benzoquinone: a specific organogenesis inhibitor in plants. J Am Chem Soc 118:1233–1234

Host Location and Selection by Holoparasitic Plants

Mark C. Mescher, Jordan Smith, and Consuelo M. De Moraes

Abstract Parasitic and carnivorous plants that adopt a heterotrophic lifestyle encounter novel environmental challenges that are shared with other heterotrophs, such as the need to locate hosts or lure prey and the need to overcome the defenses of their intended victims. These challenges are particularly acute for holoparasitic plants that depend entirely on their hosts for nutrients and other resources. In response to these challenges, holoparasitic plants employ a variety of strategies to locate and identify appropriate hosts. Root parasites such as *Striga* and *Orobanche* produce large numbers of tiny seeds that germinate only in response to host-derived chemical cues localized to the immediate vicinity of host roots. Other parasites, such as dodders (*Cuscuta*), produce relatively few large seeds that store sufficient resources for the parasitic seedling to "forage" for nearby hosts. Here we describe recent research on the mechanisms underlying these host-location strategies.

1 Introduction

1.1 Plant Behavior

If the concept of plant "behavior" is in some sense provocative, or even controversial, it is likely because behavior can easily seem, on first reflection, to be exactly the quality that animals possess and plants do not. A reasonable definition of the common-sense notion of behavior might be, "things that organisms do." And, to the casual observer, plants often don't seem to be doing much. Even Aristotle—who was manifestly not a casual observer—attributed to plants only the qualities of growth, reproduction, and decay, while reserving the powers of perception and locomotion for animals. More recent observers, aided by the tools of modern science, have shown that

M.C. Mescher (✉), J. Smith, and C.M. De Moraes
Center for Chemical Ecology, Department of Entomology, 539 ASI Building, The Pennsylvania State University, University Park, Pennsylvania, USA
e-mail: mcm19@psu.edu

F. Baluška (ed.), *Plant-Environment Interactions*, Signaling and Communication in Plants,
DOI: 10.1007/978-3-540-89230-4_6,

plants are not nearly so passive as they appear at first glance. Plants perceive the environments around them in myriad ways, as the examples described throughout this volume amply document. Plants also locomote, though over distances and timescales that are not always readily apparent to human observers.

Whether these activities of plants—or some subset of them—should be called behavior is a matter of intellectual perspective, the key question being whether such usage tends to illuminate the real and important commonalities between plants and animals or to obfuscate significant differences. The answer depends largely on which aspects of the phenomena we wish to emphasize. A mechanistic definition of behavior, drawing on work in animal systems, that makes explicit reference to muscles and nerves will necessarily exclude the actions of plants no matter how rapid or complex they might be. However, while such a definition might be criticized on grounds of utility or historical precedence, it cannot be argued that such a restrictive definition is incoherent, for there are obviously profound differences in the ways that plants and animals respond to and interact with their environment, and these distinctions are worth noting.

However, we prefer to emphasize the evolutionary function of behavior as an adaptive mechanism by which organisms achieve a better fit to dynamic and unpredictable environments by acquiring and responding to external information in ecological time. Thus, we are amenable to the recently proposed definition of plant behaviors as morphological or physiological responses to events or environmental changes that are rapid relative to the lifetime of an individual (Silvertown and Gordon 1989; Silvertown 1998; Karban 2008). As Karban (2008) points out, this definition is similar to commonly used descriptions of phenotypic plasticity in plants (Bradshaw 1965)—behavior under this definition being a form of phenotypic plasticity, occurring in response to a stimulus, that is relatively rapid and potentially reversible (Silvertown and Gordon 1989). It is likely, in fact, that plant responses occupy a continuum of rapidity and reversibility along which it may prove difficult to draw clear-cut distinctions. At one end of this continuum, the active foraging of the seedling of a parasitic dodder vine, for example, would likely satisfy even the most common-sense notion of behavior—if Aristotle had seen a time lapse video of a dodder seedling searching for a host he would likely have reconsidered the classification cited above. In contrast, the dependence of seed germination in other parasitic plants on exposure to chemical cues derived from the roots of host plants fits somewhat less easily with either an intuitive notion of behavior or with the technical definition described above. Nevertheless, it obviously makes sense to address these plant strategies together since, as we will discuss below, they serve fundamentally similar ecological functions as mechanisms of host location.

We should note, however, that such ambiguity is not unique to plant systems. The quiescence that eggs of some aquatic invertebrates exhibit—which we will discuss below as being analogous to the contingent germination strategies of plants—and from which they emerge only in response to environmental cues signaling the presence of favorable ecological conditions, also may not obviously fit with our commonsense notion of behavior. However, it is certainly the sort of thing that behavioral ecologists might study, which suggests another less technical but potentially useful definition of plant behavior: plant behaviors are those things that plants do which people who are interested in behavior might be keen to know about.

1.2 The Behavior of Parasitic Plants

Whatever definition we employ, we are likely to find that the behavior and ecology of plants most closely approaches those of animals in plant groups that adopt a parasitic or carnivorous habit. In their migration up the food chain, these plants encounter novel environmental challenges that are shared with other heterotrophs, such as the need to identify and locate organisms on which to feed and the need to overcome the defenses of their hosts or prey. This is especially the case for holoparasitic plants, which have forsaken the autotrophic habit entirely and derive their sustenance exclusively from their hosts. This similarity in the lifestyles of heterotrophic plants and animals was noted as early as the tenth century by an Arabian scholar, who described the actions of a parasitic plant, most likely a member of the genus *Cuscuta*, as corresponding "to those of the animal soul while its body remains that of a plant … for it attaches itself to trees, seeds, and thorns, and feeds itself as the worm from the juices of its host plant, thus with its soul carrying out the actions of animals" (Dieterici 1861 in Kuijt 1969).

In this chapter we will focus on the most distinctive "behavioral" characteristics of parasitic plants: their responses to environmental cues associated with the location and exploitation of host plants. Parasitic plants perceive and respond to cues from their host at many stages of development. In some cases, cues indicating the proximity of the host are required for the germination of seeds (Boumeester et al. 2003). Following germination, the radical of the parasitic seedling must grow toward and contact the body of the host plant, and this process also may be guided by the reception of chemical or other cues from the host (e.g., Runyon et al. 2006). Upon contacting the surface of the host, the attachment of the parasite and the penetration of host tissues (haustorium formation) are initiated and guided by the perception of host secondary metabolites (Yoder 2001; this volume). This chapter will focus primarily on the means by which parasites are able to find their hosts, as efficient host location is a particularly pressing problem for holoparasitic species, which depend entirely on the host for resources and thus must rapidly attach to a host following seed germination or else perish when the stored nutrients from the endosperm are exhausted (Butler 1995). The mechanisms underlying haustoria formation and the creation of a connection to the xylem of the host plant are addressed in more detail in the chapter on hemiparasitic plants.

2 The Lifestyle of Parasitic Plants

Approximately 4,500 flowering plant species, representing more than 1% of all angiosperms, have evolved the ability to parasitize other plants (Nickrent 2007; Parker and Riches 1993). Parasitism appears to have evolved independently as many as 11 times in angiosperms, and parasitic forms occur in diverse plant groups in approximately 270 genera across 22 families. The common feature that all parasitic plants share is their ability to acquire some or all of their nutrients and other resources from other plants through the production of a haustorium, a structure that is able to invade host plant tissues and act as the physiological bridge through which host resources are translocated to the parasite (Kuijt 1969; Press and Graves 1995).

A distinction can be drawn between holoparasitic plants, which lack chlorophyll and obtain all of their energy, water, and nutrients from the host, and hemiparasitic plants, which obtain some of their resources from the host but also carry out photosynthesis. However, this distinction is not always clear-cut (Musselman and Press 1995). Less than 10% of all parasitic species are strict holoparasites (Heide-Jørgensen 2008), but some other parasitic groups conduct only very limited photosynthesis. The genus *Cuscuta*, for example, contains some species that contain very small amounts of chlorophyll along with others that contain none at all. In still other groups, individuals may possess chlorophyll only at certain stages of their life cycle. For example, the root-parasitic species in the genus *Striga* are achlorophyllous when below ground and only become green and photosynthetic after their emergence above the soil surface (Musselman and Press 1995).

The ecology of holoparasitic or nearly holoparasitic species can be quite distinct from that of other plants (Heide-Jørgensen 2008), including more actively photosynthetic hemiparasites. Because the absence of chlorophyll frees holoparasitic species from a dependence on light, they can inhabit low-light environments and are able to evolve life histories in which most or all of the parasite's vegetative tissue remains underground or within the host plant. The vegetative bodies of parasites in the genus *Rafflesia*, for example, grow entirely within the tissues of the host, with only the flowers appearing externally. Holoparasitism also renders the absorptive root system superfluous, and it is absent in most strict holoparasites and greatly reduced in the Orobanchae. A further distinction is sometimes drawn between facultative and obligate parasites, but the biological relevance of this distinction is disputable, as it is not clear that any parasitic species routinely complete development without a host under natural conditions (Heide-Jørgensen 2008). A more meaningful distinction can be drawn between stem parasites, which attach to aboveground portions of their host plants, and root parasites, which make their attachments below ground. The latter account for approximately 60% of parasitic species.

There is great disparity in the extent to which the biology and ecology of parasitic plant species have been investigated, with a large majority of the research having been done on species that pose significant problems for agriculture. These include the witchweeds (*Striga* spp.) and broomrapes (*Orabanche* spp.), which parasitize host roots, and dodders (*Cuscuta* spp.), which make their aboveground attachments to plant shoots. We will focus our discussion primarily on these three taxa. The *Orabanche* are true holoparasites, while *Cuscuta*, as noted above, contains both strictly holoparasitic species and species with very limited photosynthetic ability. *Striga* are technically hemiparasitic, as they are chlorophyllic following emergence from the soil—though the photosynthetic capacity of isolated *Striga* chloroplasts is quite low, indicating strong dependence on the host (Tuquet et al. 1990). However, their subterranean seedlings, which must locate hosts and initiate parasitism, lack chlorophyll and are thus functionally holoparasitic (Parker and Riches 1993)—as we will discuss below, *Striga*, are also dependent on host-derived cues for germination and thus cannot mature under natural conditions in the absence of the host. Moreover, the germination and host location ecologies of *Striga* and *Orabanche* are quite similar, making it convenient to discuss these taxa together.

Despite accounting for a relatively small proportion of parasitic species, holoparasitic plants—or those that are functionally holoparasitic at the stage when parasitism is initiated—have a disproportionate impact on human agriculture. The root parasites *Striga*, *Orabanche*, and *Alectra* can be particularly pernicious pests, as they often inflict serious damage on host plants before the latter emerge from the soil, complicating control efforts (Runyon et al. 2008). *Striga* spp., for example, infest an estimated two-thirds of the cereals and legumes in sub-Saharan Africa, causing annual crop losses estimated at seven billion dollars and negatively impacting the lives of more than 300 million people (Berner et al. 1995; Musselman et al. 2001; Gressel et al. 2004; Press et al. 2001). The greatest economic costs are inflicted by *S. hermonthica* and *S. asiatica*, which between them cause major damage to many of the most important cereal crops, including maize, sorghum, millet, rice and sugar cane (Parker and Riches 1993).

3 Strategies for Seed Dispersal and Host Location

3.1 Seed Dispersal Strategies

Given the sedentary lifestyle of plants, angiosperm dispersal is accomplished primarily by the movement of seeds (although vegetative dispersal through growth or through the movement of vegetative tissue by wind or water is frequent in some species), and plants have evolved a wide array of strategies and mechanisms for effective seed dispersal (Butler 1995). For parasitic plants, a primary objective of seed dispersal strategies is to bring the seeds into the proximity of a host. For the reasons noted above, this is an especially pressing objective for holoparasites. Heide-Jørgensen (2008) described four primary seed dispersal strategies that are employed by parasitic plants:

(1) Some species produce large seeds with relatively high levels of stored resources (starch, fat, and protein) that will sustain the seedling for a limited period of time during which it will "forage" for a host. This is the strategy employed by *Cuscut*a spp., all the root-parasitic members of the Santalales, and several hemiparasitic Orobanchaceae (Kuijt 1969; Heide-Jørgensen 2008). This strategy entails the production of relatively large seeds, as the extent of seedling growth that can be supported by endospermic reserves present in the seed defines a critical distance beyond which a seedling has no possibility of reaching a host. For some species this distance can be quite large: seedlings of the shoot-parasitic dodder *C. gronovii* can search for hosts over distances of up to 35 cm (Costea and Tardif 2006). In contrast, *C. pentagona* seedlings rarely grow more than 10 cm before wilting (Runyon et al. 2006). For most root-parasitic holoparasites, the critical distance is probably on the order of millimeters (Salle et al.1998).

(2) A second strategy entails the production of sticky seeds that are dispersed by animals, primarily birds, and often deposited directly onto a branch of the host plant. As with the first strategy, this method of seed dispersal entails the production of relatively large seeds. This strategy is employed by the stem-parasitic loranths and mistletoes and is common among the Santales. The majority of the species that employ this strategy are hemiparasitic, and in some cases the endospermic tissues are capable of active photosynthesis, which is initiated immediately following germination. However, this strategy is also employed by the holoparasite *Tristerix aphyllus*, a member of the family Loranthaceae, which has a rather remarkable lifestyle (Heide-Jørgensen 2008): *T. aphyllus* exclusively parasitizes two columnar cacti from the southern Andes, *Echinopsis chilensis* and *Eulychnia acida*, and its seeds are dispersed by the Chilean mockingbird, *Mimus thenca* (Norton and Carpenter 1998; Gonzales et al. 2007). The seeds are typically deposited by the birds onto the spines of the cactus, where they adhere and then the newly germinated seedling grows up to 10 cm to bring the tip of the radicle into contact with the body wall of the cactus. After establishing itself on the host, *T. aphyllus* is entirely endophytic, with only its bright red inflorescences appearing on the exterior of the host, where they are pollinated by hummingbirds.
(3) A third strategy is similar to the second, but involves seeds that are brought into direct contact with the host by agents other than animals, including wind and water as well as self-dispersal. Seeds of *Arceuthobium*, for example, are covered with sticky viscin like those of other mistletoes, but rather than being carried by birds, their dispersal is achieved by the explosiveness of the fruits (Hinds and Hawksworth 1965; Garrison et al. 2000).
(4) The fourth strategy entails the production of seeds that are passively dispersed but that require exposure to stimulatory compounds from the host in order to initiate germination. This is the strategy employed by most of the holoparasitic root parasites, including *Orabanche*—in which the requirement for germination stimulants from the host was first observed in 1823 (Vaucher 1823)—as well as by *Striga*, on which a great deal of research has addressed the mechanisms underlying the stimulation of germination, as discussed in the next section. As a general rule, the host-derived exudates exploited for host recognition are active only within a few millimeters of the host roots. Consequently, this strategy entails the production of large numbers of small, long-lived seeds to enhance the probability that some seeds will come to rest in the immediate vicinity of a host.

4 Seed Germination

4.1 Seed Dormancy and Germination Requirements

Despite the dynamism outlined in this volume, plants cannot readily move large distances to find resources or escape harsh conditions and so must pursue other, more patient, strategies for coping with heterogeneous environments. Seed dormancy is an adaptative strategy, widely distributed among higher plants (Finch-Savage and Leubner-Metzger 2006), in which seeds enter a state of developmental quiescence, allowing time for the seeds to disperse and suspending growth until the seeds encounter a specific set of environmental conditions favorable to their development. Seed dormancy is a form of embryonic diapause, which exhibits widespread occurrence in both plants and animals. In many mammals, for example, fertilized eggs may enter a state of quiescence to await the presence of favorable conditions for development. This may occur as a matter of course, as in roe deer, where mating occurs in the fall but the development of fertilized eggs is delayed until the following spring (Sandell 1990). Or it may be contingent on specific ecological or social conditions. For example, in some mammals that produce multiple litters per year, the further development of fertilized eggs is suspended in response to the presence of physiological cues associated with lactation, indicating the presence of other dependent offspring (Lopes et al. 2004).

Diapause, embryonic or otherwise, is a common strategy employed by animals that inhabit highly variable or intermittently harsh environments. The planktonic crustacean *Daphnia* produces "resting" eggs that remain dormant to escape dry periods in temporary ponds or periods of intense predation in permanent ponds. The resumption of development is contingent upon exposure to environmental cues (e.g., photoperiod and temperature) associated with favorable ecological conditions (Hairston et al. 1995), and may possibly be inhibited by chemical cues indicating the presence of predatory fish (Lass et al. 2005), as has been reported for the reactivation of resting stages in dinoflagellates (Rengefors et al. 1998). Quiescent eggs of planktonic organisms may remain viable for many years, resulting in the accumulation in aquatic sediments of an "egg bank" analogous to the seed bank present in terrestrial soils (Hairston et al. 1995).

Among flowering plants, seed dormancy is the rule, and most seeds germinate only following exposure to one or more external stimuli signaling the presence of favorable growth conditions. For example, germination may depend on the presence of specific conditions relating to light, temperature, water, oxygen, and nutrients (Finch-Savage and Leubner-Metzger 2006). Parasitic plants also require permissive conditions with respect to these variables (Worsham 1987), but they face the additional challenge of needing to find a suitable host plant to parasitize—a particularly pressing issue for holoparasites and other obligately parasitic forms that must rapidly locate and attach to a host or perish. As a result, some parasitic forms are dependent on germination stimulants from the host. Even following germination, parasitic plants have been found to arrest development at a number of developmental stages, requiring signals from the host plant to continue growth. The stages at which development can be arrested include germination, haustorial initiation, host tissue penetration, physiological compatibility with the host, and apical meristem development (Nickrent et al. 1979; Boone et al. 1995). However, because seed germination is the critical first committed step in the developmental process, it can be the most discriminating in terms of host selection (Boone et al. 1995).

The details of the conditions required for germination appear to be highly variable across parasitic species, with most work having been done on the economically important species of *Stiga* and *Orabanche*, and especially on the important agricultural pests *S. asiatica* and *S. hermonthica*, which attack gramineous crops, and *S. gesneriodes*, which parasitizes legumes (Musselman 1980; Parker 1991). The seeds of Striga are very small, measuring around 0.15 × 0.3 mm, and therefore lack the reserves for sustained growth before host attachment—it is estimated that for successful host attachment germination must take place within 3–4 mm of the host root (Ramaiah et al. 1991). To compensate for these biological restrictions, *Striga* spp. may produce up to 450,000 seeds per plant, with a persistence in the soil of up to ten years (Eplee 1992). Prior to germination, *Striga* seedlings must undergo an after-ripening period during which seeds require a certain temperature and moisture regime for a period of about two weeks before they will respond to germination stimulants. This period may involve the breakdown of phenolic compounds that act as germination inhibitors (Musselman 1980). Following the after-ripening period, the seeds require a further conditioning period during which they are exposed to adequate levels of water and oxygen in the absence of light before exposure to germination stimulants can initiate germination. White light inhibits the germination of *S. asiatica* both before and immediately after exposure to germination stimulants (Egley 1972). However, beyond three hours after exposure to the maize germination stimulants, the developmental process is unresponsive to light. In the absence of host-derived stimulants, the seeds maintain dormancy and can remain viable through multiple preconditioning seasons. In *Orabanche,* seeds may remain viable for as long as 60 years (Heide-Jørgensen 2008). As discussed below, several classes of plant-derived compounds have been suggested to have germination-stimulating activity.

4.2 Germination Stimulants

4.2.1 Strigolactones

Strigol, the first germination-stimulating compound to be positively identified (Cook et al. 1966, 1972), was initially purified from hydroponically grown roots of cotton plants—a false host of *Striga* that stimulates seed germination but does not support development of the parasite—and was found to stimulate seed of *S. lutea*, eliciting 50% germination at concentrations as low as 10^{-5} ppm in water. Subsequently, a structural analog of strigol, sorgolactone, was isolated from sorghum, a true host of *Striga* (Hauck et al. 1992), while strigol itself was found to be present in the true hosts maize and millet (Siame et al. 1993). A chemically similar compound, alectrol, was identified from cowpea (Müller et al. 1992). Later, alectrol and another naturally occurring strigolactone, orabanchol, were found to serve as stimulants for Orabanche seed germination in response to root exudates of red clover.

Butler (1995) proposed the name "strigolactones" for this class of compounds. To date, nine naturally occurring strigolactones have been identified in plant root exudates (Akiyama and Hayashi 2008; Bouwmeester et al. 2007; Xie et al. 2007, 2008a, b; Matsuura et al. 2008). Strigol-like compounds have also been found in a number of medicinal plant species that are not known to be hosts or false hosts for parasitic weeds (Yasuda et al. 2003), suggesting that production of strigolactones may be widespread among plants. Strigolactones are typically present in root exudates in low quantities (cotton seedlings reportedly secreted ~15 pg of strigol per day; Sato et al. 2005), and several different strigalactones are present in most plants, with the ratios of compounds present varying from one species to another and even among varieties of individual species (Awad et al. 2006).

Structurally, a strigolactone comprises a tricyclic lactone that is connected, via an enol ether bond, to a methylbutenolide ring, and they were long regarded as sesquiterpenoids. However, Matusova et al. (2005) recently demonstrated the involvement of the carotenoid pathway in strigolactone biosynthesis, through a series of experiments employing carotenoid mutants of maize, and inhibitors of isoprenoid pathways on maize, sorghum and cowpea. Specifically, the tricyclic lactone was shown to be derived from the C40 carotenoids that originate from the plastidic, nonmevalonate methylerythritol phosphate (MEP) pathway.

Following the discovery of the role of strigol in stimulating the germination of parasitic plant seeds, a number of structural bioactivity studies aimed at elucidating the mode of action of strigolactones and developing synthetic analogs that might be used to induce "suicidal germination" of parasitic plant seeds in agricultural systems (e.g., Johnson et al 1981; Mangnus and Zwanenburg 1992; Mangus et al. 1992a, b; Bergmann et al. 1993; Kranz et al. 1996) led to the synthesis of a variety of synthetic strigolactone analogs, some of which stimulate germination in both *Striga* and *Orabanche* (Worsham 1987; Stewart and Press 1990, Bergmann et al. 1993). Among these were the so-called GR ("germination releaser") compounds that were first described by Johnson et al. (1976, 1981; see also Humphrey et al. 2006).

These structural analogs have variable rates of activity, with GR-7 and GR-24 having the strongest stimulatory effect on germination (Bergmann et al. 1993), and GR-24 came to be used as a standard positive control for studies of germination activity (Humphrey et al. 2006). Based on the results of numerous structure–activity studies, including those cited above, Mangnus and Zwanenburg (1992) proposed a tentative model for the molecular mechanism underlying the germination-stimulating activity of strigolactones. The model hypothesized a receptor-mediated process in which a nucleophilic group present at the receptor site attacks the enol bridge of the strigolactone molecule, with elimination of the D-ring serving as the mechanism for biological activation. This model is consistent with observed variation in the germination-stimulating activities of synthetic strigolactone analogs, but has not been confirmed by direct evidence as yet (Humphrey et al. 2006).

4.2.2 Strigolactones as Host-Location Cues for AM Fungi

An interesting question regarding host-derived germination stimulants is why plants produce them. These compounds presumably must have other functions that outweigh the cost to host plants of providing cues to their parasites. Recently, a number of studies have described a role for strigolactones as signals facilitating the colonization of plant roots by symbiotic abuscular arbuscular mycorrhizal (AM) fungi of the phylum Glomeromycota (Akiyama et al. 2005; Besserer et al. 2006). The symbiosis between AM fungi and plants evolved at least 460 million years ago, and more than 80% of land plants form symbioses with AM fungi (Akiyama and Hyashi 2008). Plants obtain water and mineral nutrients from their fungal partners, which are obligate symbionts dependent on carbon provided by the host plant to complete their life cycle.

Initiation of the symbiosis relies on the establishment of a network of connections between the roots of the host plant and the fungal hyphae, and entails extensive hyphal branching, presumably in response to chemical cues released by the host roots. Akiyama et al. (2005) demonstrated that the chemical factor responsible for inducing this branching is the strigolactone 5-deoxystrigol. Moreover, several other naturally occurring strigolactones, as well as GR24, were found to induce hyphal branching at similar concentrations.

It has been proposed that the emergence of strigolactone production during the evolution of strigolactone production as a host-location signal allowing AM fungi to find host roots may have provided an opportunity for later evolving parasitic weeds to co-opt it for their own ends (Bouwmeester et al. 2007, Akiyama and Hayashi 2008). This notion is supported by the observation that plant families where germination stimulant activity is relatively unreported tend to include plants which do not associate with AM fungi (Humphrey et al. 2006). The discovery of orobanchol in the root exudates of *Arabidopsis thaliana*, a nonhost of AM fungi but a host of *O. aegyptiaca* (Goldwasser et al. 2008), suggests, however, that strigolactones may be distributed beyond the host range of AM fungi.

4.2.3 Sesquiterpene Lactones

These compounds, which share some structural similarities with strigolactones, are widely distributed in plants and have been shown to have a variety of biological activities, including potential allelopathy (Macías et al. 2006). Several naturally occurring sesquiterpenes were shown to stimulate germination of *Striga* seeds (Fischer et al 1989). More recently, Macías and colleagues (2006) found that several sesquiterpene lactones induced germination of the seeds of *O. cumana* but not those of *O. crenata* or *O. ramosa* (de Luque et al. 2000; Galindo et al. 2002). *O. cumana* is a specialist parasite of sunflowers, which are known to contain large amounts of sesquiterpene lactones (Bouwmeester et al. 2003), and the response of *O. cumana* to these compounds (parthenolides) may represent a specific evolutionary response by this specialist parasite in addition to any naturally occurring recognition of strigolactones (Humphrey et al. 2006).

4.2.4 SXSg and the Debate Over Germination Stimulation by Sorghum

Following the discovery of strigol in cotton, but prior to the isolation of strigolactones from true hosts of *Striga* and *Orobanche*, a chemically different compound, the hydroquinine derivative dihydrosorgoleone, was isolated from sorghum root exudates and reported to have germination-stimulating activity (Chang et al. 1986). This compound is also commonly referred to as SXSg (*Sorghum xenognosin* of *Striga* germination). Lynn et al. (1981) introduced the term "xenognosis" to refer to the process of host recognition though the perception of host-derived chemical signals and "xenognosin" to refer to the signals by which recognition is achieved; however, the potential of this terminology for general utility appears to have been somewhat compromised by its subsequent close association with dihydrosorgoleone and with the position that this compound, to the specific exclusion of strigolactones, is "the" sorghum xenognosin (e.g., Boone et al. 1995; Palmer et al. 2004).

Early debate about the significance of dihydrosorgoleone relative to sorgolactone in sorghum and more generally about the nature of germination stimulants in natural soil systems (e.g., Boone et al. 1995; Wigchert and Zwanenburg 1999) focused on a number of issues, including the stability and diffusability of each compound and their distributions across host lines and species. Chang and Lynn (1986) followed by Lynne and colleagues (Boone et al. 1995) initially argued that the observed high activity of strigol and its relative stability were incompatible with its presumed function in limiting germination to the immediate vicinity of the host roots, in contrast to the electron-rich hydroquinone SXSg, which is readily autoxidized in soil and rapidly degrades. However, it was later reported that strigol and its analogs are much less stable in the soil, presumably because of hydrolytic degradation (Babiker et al. 1987, 1988). Moreover, Butler (1995) proposed a limited role for SXGs precisely because of its limited water solubility and rapid oxidation. Further arguments raised against the significance of SXSg (reviewed by Wigchert and Zwanenburg 1999) included the observation that variation in SXSg production among sorghum cultivars showed little correlation with the resistance or susceptibility of those cultivars to attack by *Striga* (Hess et al. 1992; Olivier and Leroux 1992), whereas the pattern of resistance is better correlated with strigolactone production (Wigchert and Zwanenburg 1999). Additionally, SXSg does not appear to be present in the root exudates of maize, which is highly susceptible to *Striga* (Housley et al. 1987).

Countervailing these arguments is the discovery of the compound resorcinol, a methylated analog of SXSg that reportedly acts as an autoxidation stabilizer (Fate and Lynn 1996), decreasing the effective concentrations of root exudates required for germination. Lynn and colleagues (e.g., Fate and Lynn 1996; Palmer et al. 2004) argued that the relative amounts of SXSg and resorcinol, taken together, accurately predict the germination zone of *S. asiatica* in several sorghum varieties. Germination in maize they attributed to the activity of a labile but as yet unidentified stimulant. A secondary debate centered on the viability of a model that attempts to explain the germination-stimulating activity of strigol based on the structural similarity of its D-ring to SXSg (e.g., Lynn and Boone 1993; Boone et al. 1995; Wigchert and Zwanenburg 1999; Palmer et al. 2004).

More recently, Matusova et al. (2005)—in the same study that demonstrated a carotenoid origin for strigolactones—reported that treating plants with the carotenoid biosynthesis inhibitor fluridone resulted in the complete inhibition of *Striga* germination in several plants including sorghum, suggesting that SXSg and other sorgoleone quinones are not directly involved in stimulating germination. It is unclear whether or how this result can be reconciled with previous reports that claim to demonstrate germination in response to SXSg (e.g., Chang et al. 1986; Fate and Lynn 1996).

Lynn and colleagues previously questioned whether strigolactones were plant-derived compounds at all, suggesting that they might rather be products of bacteria inhabiting the roots of plants grown hydroponically (Boone et al. 1995), but the subsequent identification of naturally occurring strigolactones from diverse plants (described above), and particularly the demonstration of their role in the colonization of plant roots by AM fungi, would seem to rule this out. Meanwhile, no corresponding body of evidence has emerged to support a similarly widespread role for sorgoleone quinines. Thus, more recent assertions that SXSg is "necessary and sufficient to induce seed germination in *Striga*" (Palmer et al. 2004) do not seem tenable, particularly in light of the recent findings regarding the effects of carotenoid inhibition on seed germination described above. Thus, the current weight of evidence seems to point toward strigolactones as the primary compounds stimulating the germination of parasitic weeds, while the significance of SXSg and related compounds is uncertain (Humphrey et al. 2006).

Nevertheless, the current literature on the relative significance of SXSg and strigolactones is somewhat muddled. For example, a recent text on the biology of parasitic plants devotes significant attention to the role of SXSg as a germination stimulant (Heide-Jørgensen 2008), and a recent review addressing the role of plant root exudates in interspecific interactions refers to SXSg as "the only plant-produced *Striga* germination inducer that has been identified and characterized" (Bais et al. 2006). It is likely that the apparent confusion on this point derives from an unfortunate tendency in some of the recent literature to describe either SXSg or strigolactones as "the" germination stimulants for parasitic plants, while providing little context regarding the controversy and conflicting data relating to the roles of the two compounds (e.g., Keyes et al. 2001; Palmer et al. 2004; Matusova and Bouwmeester 2006).

5 Host Location and Selection by Foraging Seedlings

In contrast to the fairly extensive work on the chemicals cues responsible for the germination of parasitic plant seeds described above, relatively little research has examined the cues responsible for guiding the growth of the seedling toward its host following germination. Though host location in the root parasites *Striga* and *Orabanche* is largely accomplished by restricting germination to the immediate vicinity of plant roots, Dube and Olivier (2001) postulated that the concentration gradients of germination stimulants may also guide radical growth toward the host's roots. However, this possibility has not yet been confirmed (Matusova et al. 2005).

Foraging and host selection by parasitic plants seedlings has been best studied in the cosmopolitan genus *Cuscuta* (dodders), which, like *Striga* and *Orabanche*, includes a number of important agricultural pests. The dodders are among the best known of the parasitic plants because their parasitism of host stems is readily observed and because of their "extraordinary appearance and behavior" (Kuijt 1969). Mature dodder vines, which contain little or no chlorophyll, are typically yellow or bright orange and can form an extensive interlaced mass of leafless stems; the total length of the reticulated branches of a single dodder plant may approach half a mile (Dean 1942). Unlike the seeds of *Striga* and *Orabanche*, those of *Cuscuta* have no specialized germination requirement and rather depend on foraging by the seedling to find a host (Parker and Riches 1993). The seeds do, however, possess a thick, impervious seed coat that must be eroded by mechanical abrasion in the soil prior to germination (Lyshede 1992) and may serve to distribute the germination of seeds over time. *Cuscuta* seeds can remain viable for up to 50 years under ideal conditions and for at least ten years in the soil (Menke 1954). Once the seedling has emerged, foraging occurs by circumnutation, a rotational movement pattern in which the growing seedling makes a counterclockwise rotation around its axis of growth on the order of once an hour. Upon contact with the stem of a potential host plant, the *Cuscuta* vine winds round tightly, making up to three complete coils prior to the initiation of haustoria formation (Parker and Riches 1993). While the swollen basal part of the seedling functions like a root in absorbing water and anchoring the plant, true roots are never produced (Kuijt 1969).

Evidence suggests that dodder vines are able to "choose" among potential hosts and are more likely to accept hosts of high nutritional quality (Kelly 1990, 1992; Kelly and Horning 1999; Koch et al. 2004). For example, Kelly (1992) found that individual stems of *C. europaea* transplanted onto various host plants were more likely to "accept" hosts of host of high nutritional status and to "reject" (grow away from) lower-quality hosts, although the cues that guide these preferences have not been established. The host preferences of *Cuscuta* spp. can induce changes in plant community structure and diversity where they become established (e.g., Pennings and Callaway 1996, 2002).

Runyon et al. (2006) recently demonstrated that foraging seedlings of *C. pentagona* use host plant-derived chemicals to locate their hosts. Chemotropism had previously been suggested to play a role in host location by *Cuscuta* (Bünning and Kaut 1956) but had never been firmly established. In the more recent study, seedlings were shown to exhibit directed growth toward blends of volatile chemicals emitted by the host plants tomato and impatiens as well as the nonhost wheat (*Cuscuta* spp. cannot successfully parasitize grasses). However, seedlings exhibited a preference for volatiles from tomato over those from wheat, suggesting a role for chemical cues in host discrimination. Seedlings were also found to exhibit a directed growth response to a number of individual compounds present in the tomato blend, including α-pinene, β-phellandrene, and β-myrcene (which was also present in the wheat blend). One compound from the wheat blend, (*Z*)-3-hexenyl acetate, was found to be repellent, inducing an aversive growth response.

Light cues have also been implicated in *Cuscuta* foraging. Because chlorophyll primarily absorbs red light, but reflects and transmits far-red light, foraging *Cuscuta* seedlings may perceive chlorophyllous neighbors as far-red objects or as regions with a high ratio of far-red to red light (Smith 1994). Orr et al. (1996) reported phototropism toward far-red light in (*C. planiflora*) seedlings. Benvenuti et al. (2004) documented a phototropic response of *C. pentagona* seedlings to light transmitted by leaves of sugar beet, and reported a stronger response to leaves with higher chlorophyll contents. Light cues have also been shown to influence the coiling of the dodder vine around the host and prehaustoria formation (e.g., Haidar et al. 1997).

References

Akiyama K, Matsuzaki K, Hayashi H (2005) Plant sesquiterpenes induce hyphal branching in arbuscular mycorrhizal fungi. Nature 435:824–827

Akiyama K, Hayashi H (2008) Plastid-derived strigolactones show the way to roots for symbionts and parasites. New Phytol 178:695–698

Awad AA, Sato D, Kusumoto D, Kamioka H, Takeuchi Y, Yoneyama K (2006) Characterization of strigolactones, germination stimulants for the root parasitic plants *Striga* and *Orobanche*, produced by maize, millet and sorghum. J Plant Growth Regul 48:221–227

Babiker AGT, Hamdoun AM, Rudwan A, Mansi NG, Faki HH (1987) Influence of soil-moisture on activity and persistence of the strigol analog Gr-24. Weed Res 27:173–178

Babiker AGT, Ibrahim NE, Edwards WG (1988) Persistence of Gr7 and *Striga* germination stimulant(s) from *Euphorbia aegyptiaca* Boiss in soils and in solutions. Weed Res 28:1–6

Bais HP, Weir TL, Perry LG, Gilroy S, Vivanco JM (2006) The role of root exudates in rhizosphere interations with plants and other organisms. Annu Rev Plant Biol 57:233–266

Benvenuti S, Dinelli G, Bonetti A (2004) Germination ecology of *Leptochloa chinensis*: a new weed in the Italian rice agro-environment. Weed Res 44:87–96

Bergmann C, Wegmann K, Frischmuth K, Samson E, Kranz A, Weigelt D, Koll P, Welzel P (1993) Stimulation of *Orobanche crenata* seed germination by (+)-strigol and structural analogs dependence on constitution and configuration of the germination stimulants. J Plant Physiol 142:338–342

Berner DK, Kling JG, Singh BB (1995) Striga research and control: a perspective from Africa. Plant Dis 79:652–660

Besserer A, Puech-Pages V, Kiefer P, Gomez-Roldan V, Jauneau A, Roy S, Portais JC, Roux C, Becard G, Sejalon-Delmas N (2006) Strigolactones stimulate arbuscular mycorrhizal fungi by activating mitochondria. PloS Biol 4:1239–1247

Boone LS, Fate G, Chang M, Lynn DG (1995) Seed germination. In: Graves JD, Press MC (eds) Parasitic plants. Chapman and Hall, London, pp 14–38

Bouwmeester HJ, Matusova R, Sun ZK, Beale MH (2003) Secondary metabolite signalling in host–parasitic plant interactions. Curr Opin Plant Biol 6:358–364

Bouwmeester HJ, Roux C, Lopez-Raez JA, Becard G (2007) Rhizosphere communication of plants, parasitic plants and AM fungi. Trends Plant Sci 12:224–230

Bradshaw AD (1965) Evolutionary significance of phenotypic plasticity in plants. Adv Genet 13:115–155

Bunning E, Kautt R (1956) On chemotropism of the seedlings of *C. europaea*. Biol Zentralbl 75:356–359

Butler LG (1995) Chemical communication between the parasitic weed *Striga* and its crop host: a new dimension in allelochemistry. In: Inderjit K, Dakshini MM, Einhellig FA (eds) Allelopathy: organisms, processes and application (ACS Symposium Series). ACS, Washington, DC, pp 158–168

Chang M, Netzly DH, Butler LG, and Lynn DG (1986) Chemical regulation of distance – Characterization of the 1st natural host germination stimulant for Striga asiatica. J. Am. Chem. Soc.108, 7858–7860

Cook CE, Whichard LP, Turner B, Wall ME (1966) Germination of witchweed (*Striga lutea* Lour): isolation and properties of a potent stimulant. Science 154:1189–1190

Cook CE, Coggon P, McPhail AT, Wall ME, Whichard LP, Egley GH, Luhan PA (1972) Germination stimulants. 2. Structure of strigol: potent seed germination stimulant for witchweed (*Striga lutea* Lour). J Am Chem Soc 94:6198–6199

Costea M, Tardif FJ (2006) The biology of Canadian weeds. 133. *Cuscuta campestris* Yuncker, C. *gronovii* Willd. ex Schult., *C. umbrosa* Beyr. ex Hook., C. epithymum (L.) L. and *C. epilinum* Weihe. Can J Plant Sci 86:293-316

de Luque AP, Galindo JCG, Macias FA, Jorrin J (2000) Sunflower sesquiterpene lactone models induce *Orobanche cumana* seed germination. Phytochemistry 53:45–50

Dean HL (1942) Total length of stem developed from a single seedling of *Cuscuta*. Proc Iowa Acad Sci 49:127–128

Dube MP, Olivier A (2001) *Striga gesnerioides* an its hosts, the cowpea: interaction and methods of control. Can J Bot 79:1225–1240

Egley GH (1972) Influence of seed envelope and growth regulators upon seed dormancy in witchweed (*Striga lutea* Lour). Ann Bot 36:755

Eplee RE (1992) Witchweed (*Striga asiatica*): an overview of management strategies in the USA. Crop Prot 11:3–7

Fate GD, Lynn DG (1996) Xenognosin methylation is critical in defining the chemical potential gradient that regulates the spatial distribution in *Striga* pathogenesis. J Am Chem Soc 118:11369–11376

Finch-Savage WE, Leubner-Metzger G (2006) Seed dormancy and the control of germination. New Phytol 171:501–523

Fischer NH, Weidenhamer JD, Bradow JM (1989) Dihydroparthenolide and other sesquiterpene lactones stimulate witchweed germination. Phytochemistry 28:2315–2317

Galindo JCG, de Luque AP, Jorrin J, Macias FA (2002) SAR studies of sesquiterpene lactones as *Orobanche cumana* seed germination stimulants. J Agric Food Chem 50:1911–1917

Garrison WJ, Miller GL, Raspet R (2000) Ballistic seed projection in two herbaceous species. Am J Bot 87:1257–1264

Goldwasser Y, Yoneyama K, Xie XA (2008) Production of strigolactones by *Arabidopsis thaliana* responsible for *Orobanche aegyptiaca* seed germination. Plant Growth Regul 55:21–28

Gonzales WL, Suarez LH, Guinez R, Medel R (2007) Phenotypic plasticity in the holoparasitic mistletoe *Tristerix aphyllus* (Loranthaceae): consequences of trait variation for successful establishment. Evol Ecol 21:431–444

Gressel J, Hanafi A, Head G, Marasas W, Obilana B, Ochanda J, Souissi T, Tzotzos G (2004) Major heretofore intractable biotic constraints to African food security that may be amenable to novel biotechnological solutions. Crop Protect 23:661–689

Haidar MA, Orr GL, Westra P (1997) Effects of light and mechanical stimulation on coiling and prehaustoria formation in *Cuscuta* spp. Weed Res 37:219–228

Hairston NG, Van Brundt RA, Kearns CM (1995) Age and survivorship of diapausing eggs in a sediment egg bank. Ecology 76:1706–1711

Hauck C, Muller S, Schildknecht H (1992) A germination stimulant for parasitic flowering plants from *Sorghum bicolor*, a genuine host plant. J Plant Physiol 139:474–478

Heide-Jørgensen HS (2008) Parasitic flowering plants. Brill, Leiden

Hess DE, Ejeta G, Butler LG (1992) Selecting sorghum genotypes expressing a quantitative biosynthetic trait that confers resistance to *Striga*. Phytochemistry 31:493–497

Hinds TE, Hawksworth FG (1965) Seed dispersal velocity in 4 dwarf mistletoes. Science 148:517–519

Housley TL, Ejeta G, Cherif-Ari O, Netzly DH, Butler LG (1987) Progress towards an understanding of sorghum resistance to *Striga*. In: Webe H, Forstreuter W (eds) Proceedings of the 4th International Symposium on Parasitic Flowering Plants. Philipps University, Marburg, Germany, pp 411–419

Humphrey AJ, Galster AM, Beale MH (2006) Strigolactones in chemical ecology: waste products or vital allelochemicals? Nat Prod Rep 23:592–614

Johnson AW, Rosebery G, Parker C (1976) Novel approach to *Striga* and *Orobanche* control using synthetic germination stimulants. Weed Res 16:223–227

Johnson AW, Gowda G, Hassanali A, Knox J, Monaco S, Razavi Z, Rosebery G (1981) The preparation of synthetic analogs of strigol. J Am Chem Soc Perkin Trans 1:1734–1743
Karban R (2008) Plant behaviour and communication. Ecol Lett 11:727–739
Kelly CK (1990) Plant foraging: a marginal value model and coiling response in *Cuscuta subinclusa*. Ecology 71:1916–1925
Kelly CK (1992) Resource choice in *Cuscuta europaea*. Proc Natl Acad Sci USA 89:12194–12197
Kelly CK, Horning K (1999) Acquisition order and resource value in *Cuscuta attenuata*. Proc Natl Acad Sci USA 96:13219–13222
Keyes WJ, Taylor JV, Apkarian RP, Lynn DG (2001) Dancing together. Social controls in parasitic plant development. Plant Physiol 127:1508–1512
Koch AM, Binder C, Sanders IR (2004) Does the generalist parasitic plant, *Cuscuta capestris,* selectively forage in heterogeneous plant communities? New Phytol 162:147–155
Kranz A, SamsonSchulz E, Hennig L, Welzel P, Muller D, MayerFigge H, Sheldrick WS (1996) Synthesis of new strigol analogues. Tetrahedron 52:14827–14840
Kuijt J (1969) The biology of parasitic flowering plants. University of California Press, Berkeley, CA
Lass S, Vos M, Wolinska J, Spaak P (2005) Hatching with the enemy: *Daphnia* diapausing eggs hatch in the presence of fish kairomones. Chemoecology 15:7–12
Lopes FL, Desmarais JA, Murphy BD (2004) Embryonic diapause and its regulation. Reproduction 128:669–678
Lynn DG, Boone LS (1993) Signaling germination in *Striga asiatica*. In: Schultz J, Raskin I (eds) Proceedings of the 8th Annual Penn State Symposium in Plant Physiology. Waverly Inc., Baltimore, MD, 47–53
Lynn DG, Steffens JC, Kamut VS, Graden DW, Shabanowitz J, Riopel JL (1981) Isolation and characterization of the first host recognition substance for parasitic angiosperms. J Am Chem Soc 103:1868–1870
Lyshede OB (1992) Studies on mature seeds of *Cuscuta pedicellata* and *C. campestris* by electron microscopy. Ann Bot 69:365–371
Macías FA, Fernandez A, Varela RM, Molinillo JMG, Torres A, Alves P (2006) Sesquiterpene lactones as allelochemicals. J Nat Prod 69:795–800
Mangnus EM, Zwanenburg B (1992) Tentative molecular mechanism for germination stimulation of *Striga* and *Orobanche* seeds by strigol and its synthetic analogues. J Agric Food Chem 40:1066–1070
Mangnus EM, Dommerholt FJ, Dejong RLP, Zwanenburg B (1992a) Improved synthesis of strigol analog Gr24 and evaluation of the biological activity of its diastereomers. J Agric Food Chem 40:1230–1235
Mangnus EM, Vanvliet LA, Vandenput DAL, Zwanenburg B (1992b) Structural modifications of strigol analogs: influence of the B and C rings on the bioactivity of the germination stimulant Gr24. J Agric Food Chem 40:1222–1229
Matsuura H, Ohashi K, Sasako H, Tagawa N, Takano Y, Ioka Y, Nabeta K, Yoshihara T (2008) Germination stimulant from root exudates of *Vigna unguiculata*. Plant Growth Regul 54:31–36
Matusova R, Bouwmeester HJ (2006) The effect of host-root-derived chemical signals on the germination of parasitic plants. In: Dicke M, Takken W (eds) Chemical ecology: from gene to ecosystem. Springer, Berlin, pp 39–54
Matusova R, Rani K, Verstappen FWA, Franssen MCR, Beale MH, Bouwmeester HJ (2005) The strigolactone germination stimulants of the plant-parasitic *Striga* and *Orobanche* spp. are derived from the carotenoid pathway. Plant Physiol 139:920–934
Menke HF (1954) Dodder infestation can halt certified seed production. West Feed Seed 9:24–37
Műller S, Hauck C, Schildknecht H (1992) Germination stimulants produced by *Vigna unguiculata* Walp Cv Saunders upright. J Plant Growth Regul 11:77–84
Musselman LJ (1980) The biology of *Striga*, *Orobanche*, and other root parasitic weeds. Annu Rev Phytopathol 18:463–489
Musselman LJ, Press MC (1995) Introduction to parasitic plants. In: Press MC, Graves JD (eds) Parasitic plants. Chapman and Hall, London, pp 1–13

Musselman LJ, Yoder JI, Westwood JH (2001) Parasitic plants major problem to food crops. Science 293:1434–1434

Norton DA, Carpenter MA (1998) Mistletoes as parasites: host specificity and speciation. Trends Ecol Evol 13:101–105

Nickrent DL (2007) Parasitic plant genera and species. Parasitic plant connection: http://www.parasiticplants.siu.edu/

Nickrent DL, Musselman LJ, Riopel JL, Eplee RE (1979) Haustorial initiation and non-host penetration in witchweed (*Striga asiatica*). Ann Bot 43:233–236

Olivier A, Leroux GD (1992) Root development and production of a witchweed (*Striga* Spp) germination stimulant in sorghum (*Sorghum bicolor*) cultivars. Weed Sci 40:542–545

Orr GL, Haidar MA, Orr DA (1996) Smallseed dodder (*Cuscuta planiflora*) gravitropism in red light and in red plus far-red. Weed Sci 44:795–796

Palmer AG, Gao R, Maresh J, Erbil WK, Lynn DG (2004) Chemical biology of multi-host/pathogen interactions: chemical perception and metabolic complementation. Annu Rev Phytopathol 42:439–464

Parker C (1991) Protection of crops against parasitic weeds. Crop Protect 10:6–22

Parker C, Riches CR (1993) Parasitic weeds of the world: biology and control. CAB, Wallingford

Pennings SC, Callaway RM (1996) Impact of a parasitic plant on the structure and dynamics of salt marsh vegetation. Ecology 77:1410–1419

Pennings SC, Callaway RM (2002) Parasitic plants: parallels and contrasts with herbivores. Oecologia 131:479–489

Press MC, Graves JD (1995) Parasitic plants. Chapman and Hall, London, pp 292

Press MC, Scholes JD, Riches CR (2001) Current status and future prospects for management of parasitic weeds (*Striga* and *Orobanche*). In: Riches CR, Farnham UK (eds) The world's worst weeds. British Crop Protection Council, London, pp 71–90

Ramaiah KV, Chidley VL, House LR (1991) A time-course study of early establishment stages of parasitic angiosperm *Striga asiatica* on susceptible sorghum roots. Ann Appl Biol 118:403–410

Rengefors K, Karlsson I, Hansson LA (1998) Algal cyst dormancy: a temporal escape from herbivory. Proc R Soc Lond B 265:1353–1358

Runyon JB, Mescher MC, De Moraes CM (2006) Volatile chemical cues guide host location and host selection by parasitic plants. Science 313:1964–1967

Runyon JB, Tooker JF, Mescher MC, De Moraes CM (2008) Parasitic plants in agriculture: chemical ecology of germination and host-plant location as targets for sustainable control. Sustainable Agriculture Reviews - Volume 1 – Springer

Salle G, Tuquet C, Raynal-Roques A (1998) Biology of flowering parasitic plants. Compt Rend Seanc Soc Biol Filial 192:9–36

Sandell M (1990) The evolution of seasonal delayed implantation. Q Rev Biol 65:23–42

Sato D, Awad AA, Takeuchi Y, Yoneyama K (2005) Confirmation and quantification of strigolactones, germination stimulants for root parasitic plants *Striga* and *Orobanche*, produced by cotton. Biosci Biotechnol Biochem 69:98–102

Siame BA, Weerasuriya Y, Wood K, Ejeta G, Butler LG (1993) Isolation of strigol, a germination stimulant for *Striga asiatica*, from host plants. J Agric Food Chem 41:1486–1491

Silvertown J (1998) Plant phenotypic plasticity and non-cognitive behaviour. Trends Ecol Evol 13:255–256

Silvertown J, Gordon DM (1989) A framework for plant behavior. Annu Rev Ecol Syst 20:349–366

Smith H (1994) Sensing the light environment the functions of the phytochrome family. In: Kendrick R, Kronenberg G (eds) Photomorphogenesis in plants, 2nd edn. Kluwer, Dordrecht,377–416

Stewart GR, Press MC (1990) The physiology and biochemistry of parasitic angiosperms. Annu Rev Plant Physiol Plant Mol Biol 41:127–151

Tuquet C, Farineau N, Salle G (1990) Biochemical composition and photosynthetic activity of chloroplasts from *Striga hermonthica* and *Striga aspera*, root parasites of field-grown cereals. Physiol Plant 78:574–582

Vaucher JP (1823) Mémoire sur la germination des orobanches. Mém MUS Hist Nat Paris 10:261–273
Worsham AD (1987) Germination of witchweed seeds. In: Musselman LJ (ed) Parasitic weeds in agriculture. CRC, Boca Raton, FL, pp 45–61
Wigchert SCM, Zwanenburg B (1999) A critical account on the inception of *Striga* seed germination. J Agric Food Chem 47:1320–1325
Xie X, Kusumoto D, Takeuchi Y, Yoneyama K, Yamada Y (2007) 2'-Epi-orobanchol and solanacol, two unique strigolactones, germination stimulants for root parasitic weeds, produced by tobacco. J Agric Food Chem 55:8067–8072
Xie X, Yoneyama K, Kusumoto D, Yamada Y, Yokota T, Takeuchi Y (2008a) Isolation and identification of alectrol as (+)-orobanchyl acetate, a germination stimulant for root parasitic plants. Phytochemistry 69:427–431
Xie XN, Yoneyama K, Kusumoto D, Yamada Y, Takeuchi Y, Sugimoto Y (2008b) Sorgomol, germination stimulant for root parasitic plants, produced by *Sorghum bicolor*. Tetrahedron Lett 49:2066–2068
Yasuda N, Sugimoto Y, Kato M, Inanaga S, Yoneyama K (2003) (+)-Strigol, a witchweed seed germination stimulant, from *Menispermum dauricum* root culture. Phytochemistry 62:1115–1119
Yoder JI (2001) Host-plant recognition by parasitic Scrophulariaceae. Curr Opin Plant Biol 4:359–365

Plant Innate Immunity

Jacqueline Monaghan, Tabea Weihmann, and Xin Li

Abstract Plants possess an elaborate multi-layered defense system that relies on the intrinsic ability of plant cells to perceive the presence of pathogens and trigger local and systemic responses. Transmembrane receptors detect highly conserved microbial features and activate signaling cascades that induce defense gene expression. Pathogens deliver effector proteins into plant cells that suppress these responses by interfering with signaling components. Plants, in turn, evolved intracellular resistance (R) protein receptors to recognize these effector proteins or their activities in the plant cell. Activated R proteins trigger a series of physiological changes in the infected cell that restrict pathogen growth locally and resonate systemically to enhance immunity throughout the plant. In this chapter we summarize our current understanding of defense responses employed by plants during pathogen infection.

1 Introduction

There are numerous examples of human suffering caused by the failure of crops due to plant disease. One of the most commonly cited examples is the great potato famine that hit Ireland in the middle of the nineteenth century as the result of potato late blight caused by *Phytophthora infestans*. This disease not only caused the deaths of an estimated one million people, but it also led to a mass emigration out of Ireland into North America, and has been credited as the linchpin that sparked a

J. Monaghan, T. Weihmann, and X. Li (✉)
Michael Smith Laboratories, University of British Columbia, Vancouver, BC, Canada
Department of Botany, University of British Columbia,
e-mail: monagh9@interchange.ubc.ca;
e-mail: weihmann@interchange.ubc.ca
e-mail: xinli@interchange.ubc.ca

F. Baluška (ed.), *Plant-Environment Interactions*, Signaling and Communication in Plants, 119
DOI: 10.1007/978-3-540-89230-4_7,

real interest in plant pathology as a scientific discipline (Holub 2001; Judelson and Blanco 2005). Plant diseases cost farmers billions of dollars each year due to crop loss or disease prevention strategies. Just one of many recent examples is the rice blast fungus, *Magnaporthe grisea*, which affects most rice-producing areas in the world and is estimated to ruin enough crops to feed 60 million mouths each year (Dean et al. 2005). The ability of plant pathogens to spread rapidly through crop fields and cause huge damage is exacerbated by the modern practice of monoculture farming, where single cultivars are planted over large areas of land year after year. However, most plants are resistant to most potential pathogens, and there has been a worldwide effort to understand the innate mechanisms that underlie this ability. A clear understanding of the interplay between plants and their pathogens is fundamental to the development of environmentally friendly management approaches of plant diseases.

Even though plants are host to every type of microbial pathogen (including fungi, oomycetes, bacteria, and viruses), they are not infected easily. Plants present microbes with a number of obstacles to overcome before they can successfully infect plant cells. Examples include cuticular waxes, antimicrobial enzymes and other secondary metabolites, as well as plant cell walls (Thordal-Christensen 2003). Microbes that have adapted to certain plants have found ways to circumvent these barriers and cause disease, whereas nonadapted microbes are unable to overcome these defenses. Plant species that can be colonized by a pathogen become "hosts" for that pathogen, whereas resistant species are "nonhosts." However, individual plant cultivars within a host species can become resistant to pathogen infection once they have evolved specific defense genes. The genetic relationship between host plants and their pathogens was first described in detail by Harold Flor in the 1940s and 1950s. Flor meticulously studied the genetic relationship between races of flax rust fungus and a number of flax varieties with respect to host susceptibility and resistance (Flor 1971). Based on his work, Flor hypothesized that resistance is the consequence of the correct combination of single genetic loci in the host and the pathogen. He proposed that the products of plant *Resistance* (*R*) genes interact with pathogenic *Avirulence* (*Avr*) gene products in a corresponding gene-for-gene manner. These pathogenic proteins are called "avirulent" because, instead of contributing to virulence, their recognition by R proteins leads to plant resistance. Rather than existing solely to reveal their identity, many Avr proteins have been shown to contribute to virulence in susceptible plants (Jones and Dangl 2006).

The advent of molecular biology and the use of genetic models have led to a number of recent discoveries outlining signaling components in plant defense pathways. *Arabidopsis thaliana*, which emerged as a model for plant molecular biology in the mid-1980s, has been widely adopted by plant pathologists studying defense signaling networks. The labs of Frederick Ausubel, Brian Staskawicz, and Jeffery Dangl characterized the interaction between arabidopsis and the bacterial pathogen *Pseudomonas syringae* which causes leaf spots, and revealed a highly amendable plant–pathogen system that is now widely used to dissect genetic components of plant defense (Whalen et al. 1991; Dong et al. 1991; Katagiri et al. 2002). This system has been instrumental to the field, and both genomes are now fully sequenced (Buell et al. 2003; The Arabidopsis Genome Initiative 2000). Arabidopsis is also host to the water mold *Hyaloperonospora parasitica* which causes downy mildew on leaves (Slusarenko and Schlaich 2003), and this system, established largely by Eric Holub, Jonathan Jones, and Jane Parker, has been extremely useful in the study of plant defense. In addition to these Arabidopsis systems, agriculturally important plant–pathogen systems are also widely studied as models, such as powdery mildew of barley led by Paul Shulze–Lefert's group, bacterial blight of rice pioneered by Pamela Ronald's team, bacterial spot of tomato and pepper largely studied by Greg Martin and Ulla Bonas' groups, leaf rust of flax by Jeff Ellis' team, and leaf mold of tomato led by Jonathan Jones' and Pierre de Wit's groups. Together, the establishment of these model systems has enabled researchers to identify key players in host immune responses and pathogen virulence at the molecular level.

We now know that signaling in plant disease resistance shares many conceptual features with mammalian innate immunity (Nürnberger et al. 2004), although there are several lines of evidence to suggest that these pathways evolved convergently (Ausubel 2005). Though plants lack an adaptive immune system like that found in vertebrates, plant cells are equipped with a number of extra- and intracellular immune receptors that detect the presence of pathogenic microbes and activate defense responses. Plants have a set of receptors that detect highly conserved and slowly evolving features of whole groups of microbes such as flagellin, the major protein found in bacterial flagella (Gómez-Gómez and Boller 2002). The activation of these receptors induces defense gene expression, ion fluxes, and the production of reactive oxygen species in the plant cell that limit microbial growth. Successful pathogens have either adapted to evade recognition by plants, or have evolved ways of interfering with or suppressing defense signaling, mostly through the expression of effectors delivered into host cells during an infection (Jones and Dangl 2006). In an elegant example of coevolution, plants have, in turn, evolved intracellular R proteins to recognize specific pathogenic effectors and activate signaling cascades leading to massive cellular reprogramming that eventually restricts pathogen growth (Dangl and Jones 2001; Jones and Dangl 2006). Pathogens can evolve additional effectors to overcome plant defense, and thus, the "arms race" between host and pathogen goes on.

This chapter will outline some of the signaling events triggered in plant cells following the recognition of biotrophic pathogens which feed on living plant tissue. This defense response is distinct from that used to combat herbivory and necrotizing pathogens, which will not be covered here (interested readers are referred to the following excellent reviews: Schilmiller and Howe 2005; Farmer et al. 2003). We start with a discussion on the recognition of highly conserved microbial patterns at the plant cell surface and some of the defense responses that follow. Certain ways that pathogen effectors have adapted to suppress these events is briefly elaborated. Intracellular R proteins that respond to the presence of these effectors and some of the signaling components involved in R protein-mediated defenses form the remainder and bulk of the chapter.

2 Recognition and Response at the Plant Cell Surface

2.1 *Microbe-Associated Molecular Patterns and Pattern Recognition Receptors*

Like animals, plants are able to recognize highly conserved features of microbes known as microbe-associated molecular patterns (MAMPs). MAMPs are typically necessary for and integral to microbial lifestyles and are therefore not easily lost or mutated, making them ideal targets for detection by plant immune receptors. For example, both plants and animals can detect the presence of Gram-negative bacteria through the perception of lipopolysaccharides (LPSs) found in their outer membrane (Dow et al. 2000). Plants respond to other MAMPs including peptides or motifs characteristic to bacterial proteins such as flagellin, elongation factor Tu (EF-Tu), and cold shock proteins, as well as to sugars found in bacterial and fungal cell walls (peptidoglycan and chitin, respectively; reviewed in Nürnberger et al. 2004). Thus, plants have evolved the ability to differentiate between self and non-self as part of an early warning system against potential pathogen infection.

MAMPs are recognized in mammals by transmembrane Toll-like receptors (TLRs) and cytosolic Nod proteins (Akira et al. 2006), collectively referred to as pattern or pathogen recognition receptors (PRRs). In plants, transmembrane receptor-like kinases (RLKs) play an integral role in MAMP perception and signal relay. Two PRRs that have been well characterized in plants include FLS2 (FLAGELLIN-SENSITIVE2; Gómez-Gómez and Boller 2000), and EFR (EF-Tu RECEPTOR; Zipfel et al. 2006), which recognize bacterial flagellin and EF-Tu, respectively. FLS2 and EFR have an extracellular leucine-rich repeat (LRR) domain and a cytosolic serine/threonine kinase domain, and likely represent members of a larger group of RLKs involved in MAMP perception (Zipfel 2008). Plants respond to MAMPs rapidly with pronounced changes in gene expression, cell wall alterations, accumulation of antimicrobial proteins and compounds, and changes in apoplastic pH levels that hinder the growth of microbial populations to some extent but are only slightly effective at preventing the growth of virulent pathogens (Gómez-Gómez and Boller 2000).

Bacterial pathogenesis is largely reliant on the ability to move into and within the plant apoplast, and this motility is provided by flagella. FLS2 proteins in Arabidopsis, tomato, and tobacco recognize and respond to bacterial flagellin, indicating that this recognition module is conserved across plant species (Zipfel 2008). FLS2 in Arabidopsis binds a small but highly conserved 22-amino acid epitope, flg22, from the N-terminus of the flagellin protein (Chinchilla et al.2006). Arabidopsis plants treated with flg22 one day prior to infection with virulent bacteria exhibit a reduction in bacterial growth compared to plants that are not pre-treated; conversely, *fls2* mutants are unable to perceive and respond to the flg22 elicitor, which is reflected in the higher susceptibility of these plants to bacterial infection (Zipfel et al. 2004). The same phenomenon has been observed using the EF-Tu elf18 epitope as an elicitor (Zipfel et al. 2006), demonstrating that the detection of single MAMPs can prime cells against further attack.

PRR activation and downstream signaling are tightly controlled. FLS2 is negatively regulated by the kinase-associated protein phosphatase KAPP (Gómez-Gómez et al. 2001) at the plasma membrane, and is internalized following flg22 binding by vesicle-mediated endocytosis as part of a negative feedback regulation scheme (Robatzek et al. 2006). Both FLS2 and EFR are positively regulated by another RLK, BAK1 (brassinosteroid-associated kinase 1; Chinchilla et al. 2007; Hesse et al. 2007). Interestingly, both tobacco and Arabidopsis mutants with compromised FLS2 activity become susceptible to nonadapted pathogens (Zipfel 2008), suggesting that PRRs are integral to both host and nonhost resistance. Flagellin from the legume-associated nitrogen-fixing symbiont *Rhizobium* is not recognized in Arabidopsis by FLS2; nor is flagellin from the plant pathogen *Agrobacterium* (Felix et al.1999), indicating that microbes are under evolutionary pressure to alter MAMPs to avoid recognition by the host PRR surveillance system.

2.2 Signaling Downstream of PRR Activation

The perception of MAMPs is relayed through finely tuned mitogen-activated protein kinase (MAPK) signaling cascades. MAPKs are used as signal transducers in all eukaryotes, and are an integral part of both mammalian and plant immunity (Nakagami et al. 2005; Nürnberger et al. 2004). These cascades are composed of at least a MAPK kinase kinase (MAPKKK), a MAPK kinase (MAPKK), and a MAPK, activated by phosphorylation in that order. MAPK cascades that act both positively and negatively on resistance are activated following PRR activation. The Arabidopsis MAPKs MPK3, MPK4, and MPK6 are activated early in the FLS2-mediated pathway (Nakagami et al. 2005). Interestingly, whereas the phosphorylation cascade leading to MPK3 and MPK6 activation promotes resistance, the cascade involved in MPK4 activation plays an inhibitory role (Suarez-Rodriguez et al. 2006). This has been supported genetically, as *mpk4* knock-out mutants constitutively activate defense markers and have enhanced resistance to pathogen infection (Petersen et al. 2000), whereas silencing *MPK6* causes heightened susceptibility to pathogens (Menke et al. 2004). The details of these pathways have not yet been fully elucidated, but it is presumed that the simultaneous activation of both positive and negative regulators allows resistance outputs to be carefully balanced according to the nature of the signal (Suarez-Rodriguez et al. 2006).

The activation of MPK3 and MPK6 induces the transcription of defense-related genes via WRKY transcription factors (Asai et al. 2002). The WRKY family comprises a large group of plant-specific transcription factors (TFs) with a WRKY DNA-binding domain that binds W-box (C/TTGACC/T) promoter elements (Ülker and Somssich 2004). W-boxes are found in the promoters of many defense-related genes, and WKRYs have been implicated in transcriptional reprogramming in response to biotic stresses such as MAMP perception and pathogen infection (Euglem and Somssich 2007). There are over 70 WRKY TFs encoded in the Arabidopsis genome, and 90 in rice (Ülker and Somssich 2004), and there is a high level of functional redundancy among the members of this large gene family (Euglem and Somssich 2007). The functional homologs WRKY22 and WRKY29 have been shown to be downstream targets of MPK3 and MPK6 activated in response to bacterial and fungal pathogens (Asai et al. 2002).

Plants are able to sense and respond to the presence of potential pathogenic microbes in their immediate environment. For pathogens to successfully colonize and exploit plant cells, they must avoid detection by the host. Phytopathogens (and animal pathogens; Finlay and McFadden 2006) employ a number of strategies to evade host surveillance that, for the most part, interfere with or suppress host defense signaling in one way or other (Göhre and Robatzek 2008; Zhou and Chai 2008). Evasion and/or virulence are accomplished through the expression and delivery of pathogenic effector proteins into host cells during an infection. Thus, in addition to transmembrane PRRs, plants are also equipped with intracellular surveillance elements known collectively as R proteins to sense and respond to the activities of these effectors.

3 Immune Responses Mediated by Plant Resistance Proteins

3.1 Pathogen Virulence Through the Delivery of Effectors

Phytopathogens require access to plant cells to acquire photosynthate and other metabolites, and to accomplish this they have evolved various mechanisms to deliver effectors into the apoplast and/or directly into plant cells. The most widely studied bacterial delivery system used during plant infection is the type three secretion system (T3SS) employed by many Gram-negative bacteria to gain access to plant tissue. This secretion system is characterized by an assembled protein pilus that extends from the bacterium and punctures the cell membrane in a syringe-like manner, releasing a battery of effectors directly into the host cell (Jin and He 2001). The pilus is essential to pathogenicity, as bacterial mutants lacking pilus components lose virulence and cannot cause disease on normally susceptible host plants (Alfano and Collmer 1996). In addition to bacterial effectors, some fungal and oomycete effectors have been detected intracellularly (Birch et al. 2008). There is accumulating evidence to suggest that oomycetes secrete and translocate effectors into plant cells by hijacking the host endocytic pathway, a mechanism similar to that used by the human malaria parasite (Birch et al. 2008).

Several effectors in plant pathogens have been cloned (most of which are bacterial and delivered by the T3SS), and their virulence functions are now being characterized. The roles of most effectors have proven elusive, as their primary amino acid sequences provide few clues regarding protein function. Recent structural and functional analysis, however, has revealed that some effectors mimic eukaryotic host signaling proteins, including transcription factors, proteases and phosphatases, to alter immune responses (Kay et al. 2007; Göhre and Robatzek 2008). The structure of the *P. syringae* effector AvrPtoB was recently shown to bear a striking resemblance to E3 ubiquitin ligases (Janjusevic et al. 2006), and was also found to have intrinsic E3 enzymatic activity (Abramovitch et al. 2006). As AvrPtoB requires this enzymatic function for virulence on susceptible plants, it is thought to suppress positive regulators of immunity via protein degradation (Janjusevic et al. 2006; Abramovitch et al. 2006). Another *P. syringae* effector, AvrPto, was recently shown to bind the PRRs FLS2 and EFR, preventing their phosphorylation and thus suppressing downstream MAPK signaling and defense outputs in susceptible plants (Xiang et al. 2008; He et al. 2006). AvrPto also inhibits another kinase, the R protein Pto, contributing to virulence in susceptible plants (Xing et al. 2007). In addition, defense-related MAPK cascades can be directly targeted by pathogenic effectors (Shan et al. 2007). Together, these examples demonstrate that successful pathogens evolved specific effectors to evade host perception and suppress host defense responses.

3.2 Resistance Proteins

Although used by pathogens to promote virulence in susceptible plants, some effector proteins can render infections avirulent if they are recognized in resistant plants by R proteins. The activation of R proteins triggers immune responses that are far more effective than those triggered by PRRs. The activation of R proteins leads to substantial ion fluxes, the induction of pathogenesis-related (*PR*) genes, the accumulation of the signaling molecule salicylic acid (SA), and an oxidative burst that leads to the accumulation of reactive oxygen species. Not only do these physiological changes create an unfavorable environment for pathogen growth, they are also often associated with a form of localized programmed cell death known as the hypersensitive response (HR), in which threatened cells commit suicide to restrict pathogen growth. The HR is particularly effective against pathogens requiring living tissue, as it confines them to dead cells where they are deprived of essential nutrients.

There are several classes of R proteins in plants. By far the most prominent class comprises intracellular NB–LRR proteins, which possess a central nucleotide-binding site (NB), and a C-terminal highly variable leucine-rich repeats (LRR) domain. This group can be further subdivided based on two structural variations at the N-terminus: CC–NB–LRRs possess a putative coiled-coil (CC) domain and TIR–NB–LRRs possess a region similar to the Toll and interleukin 1 receptor domain (TIR) found in mammalian immune receptors (Takeda et al. 2003). There are an estimated 125 NB–LRR proteins encoded in the Arabidopsis genome (Jones and Dangl 2006), although most of these genes have not yet been shown to function in pathogen defense. Proteins with LRRs function in a diverse array of cellular processes in addition to plant immunity. LRRs have been shown to mediate protein–protein interactions in eukaryotes (Kobe and Deisenhofer 1995), and, in the case of R proteins, are thought to determine Avr recognition specificity (Martin et al. 2003). NB domains are found in a number of proteins, including ATPases, G-proteins, and, notably, in animal apoptosis regulators and proteins involved in innate immunity (Takken et al. 2006). This domain is thought to regulate the activity of R protein activation through the binding and hydrolysis of ATP (Tameling et al. 2006). The CC and TIR domains likely function in signaling, as CC- and TIR–NB–LRRs signal through distinct downstream pathways (Aarts et al. 1998), although it is also possible that these domains function in recognition specificity, as is the case with the R protein N in tobacco (Burch-Smith et al. 2007).

In addition to NB–LRRs, there are other classes of R proteins in plants. A large class of R proteins in tomato includes the Cf proteins, effective against infection by the leaf mold *Cladisporium fulvum* (Rivas and Thomas 2005). These proteins span the plasma membrane and have an extracellular LRR domain and a small cytosolic domain of unknown function. The R protein Xa-21 in rice encodes an RLK similar to FLS2 and EFR that confers resistance against bacterial *Xanthomonas* species (Song et al. 1995), and tomato Pto is a cytosolic serine/threonine kinase required for resistance to *P. syringae* pv. *tomato* (Martin et al. 1993). Interestingly, no cloned Arabidopsis *R* genes encode proteins that clearly resemble Pto, Xa-21 or the Cf proteins, highlighting the importance of studying resistance mechanisms in a number of species (Martin et al. 2003). There are also some rather unusual R proteins found in Arabidopsis. RRS1 (resistance to *R. solanacearum* 1), required for resistance to *Ralstonia solanacearum*, is a TIR–NB–LRR with a C-terminal nuclear localization sequence (NLS) and a WRKY domain, merging a defense receptor with a transcriptional regulator (Deslandes et al. 2002). *RPW8* (*RESISTANCE TO POWDERY MILDEW8*) confers resistance to a broad-range of powdery mildew strains and encodes a protein with a predicted N-terminal transmembrane domain and a CC domain (Xiao et al. 2001).

3.3 Recognition of Pathogen Effectors

Although several cognate R–Avr pairs have been identified, the relationship between these pairs is not always well understood at the molecular level. The simplest model predicts that R proteins are receptors for Avr ligands. For example, it has been shown that the R protein Pto interacts directly with its cognate effector AvrPto, and that this interaction is necessary for resistance (Tang et al. 1996). Although there are a few other cases, most attempts to show direct interactions between R and Avr proteins have not been fruitful, suggesting that additional host proteins are involved in effector recognition. In 1998, Eric Van der Biezen and Jonathan Jones introduced the idea that, as opposed to directly interacting with effector proteins, R proteins might guard or monitor the integrity of effector targets (Van der Biezen and Jones 1998); an idea that was later articulated as the "guard hypothesis" (Dangl and Jones 2001). In this model, R proteins screen for pathogen-induced modifications in host proteins to trigger immune signaling.

A well-established example of such a pathogen-modified protein in plants is RIN4 (RPM1-INTERACTING PROTEIN4). RIN4 is localized to the plasma membrane, and is monitored by the likewise localized CC–NB–LRR R proteins RPM1 (RESISTANCE

TO *P. syringae* pv. *maculicola*1) and RPS2 (RESISTANT TO *P. syringae*2). During infection, *P. syringae* releases several effectors into plant cells, including AvrRpm1, AvrB, and AvrRpt2, which are thought to target a number of host proteins as part of a virulence strategy. AvrRpt2, for example, is a cysteine protease (Coaker et al. 2005) that modifies plant auxin levels to promote virulence and pathogen growth (Chen et al. 2007). Although most virulence targets of these effectors have not been identified, it has been shown that AvrRpm1, AvrB, and AvrRpt2 interact with and modify RIN4 either by phosphorylation or cleavage (Mackey et al. 2002; Axtell et al. 2003). Intriguingly, these interactions with RIN4 do not promote virulence and are not required for successful infection (Belkhadir et al. 2004). Instead, RIN4 phosphorylation is monitored by RPM1 and its cleavage is monitored by RPS2, and either event leads to plant resistance (Mackey et al. 2002; Kim et al. 2005). RIN4 physically interacts with and represses both RPM1 and RPS2 (Mackey et al. 2002, 2003). The inhibitory function of RIN4 has been shown genetically, as partial loss-of-function *rin4* mutant plants have heightened resistance to virulent pathogens, suggesting a negative role in immunity (Mackey et al. 2002). Also, *rin4* phenotypes are fully suppressed in *rin4 rpm1 rps2* triple mutants, indicating that RIN4 is indeed a negative regulator of these R proteins (Belkhadir et al. 2004). Another example is AvrPto, which, as mentioned before, targets the PRRs FLS2 and EFR to suppress plant immunity. AvrPto also binds and inhibits the kinase Pto (Xing et al. 2007), but unlike binding FLS2 and EFR, this interaction activates the NB–LRR protein Prf (Pseudomonas resistance and fenthion sensitivity) and leads to resistance (Mucyn et al. 2006). Thus, Pto might have evolved to compete with FLS2 and ERF binding to initiate defense (Zipfel and Rathjen 2008). The guard hypothesis predicts that R proteins evolved to keep a watchful eye on a subset of proteins that are modified by pathogen effectors (including some plant proteins that may mimic virulence targets; Xing et al. 2007). It is likely that most effector modifications augment virulence in some way; however, the detection of even one of these events in a plant expressing the appropriate R protein can lead to an immune response and render the pathogen avirulent (Belkhadir et al. 2004).

3.4 R Protein Activation

There are significant physiological and metabolic consequences associated with deregulated R proteins in plants. For example, point mutation in a predicted TIR–NB–LRR type *R* gene *SNC1* (*suppressor of npr1, constitutive 1*) renders the protein constitutively active, causing immune signaling components to be turned on in the absence of pathogen infection (Zhang et al. 2003a). Although *snc1* plants display enhanced resistance to pathogens, there are considerable costs to fitness. Partly as the result of heightened levels of SA, *snc1* plants exhibit severe dwarfism (Li et al. 2001), as do many mutants in which SA-mediated resistance signaling is overactive (Durrant and Dong 2004). Other mutations associated with deregulated disease resistance pathways result in spontaneous HR-like lesion formation, as is the case with a gain-of-function mutation in the TIR–NB–LRR R protein *SSI4* (*SUPPRESSOR OF*

SA-INSENSITIVITY OF npr1–5, 4; Shirano et al. 2002). To avoid these extreme costs to plant health, defense pathways are tightly regulated.

R proteins are thought to exist in a repressed form in the absence of pathogens, either through inhibitory folding or interaction with negative regulators (Marathe and Dinesh-Kumar 2003). Analysis of Rx, a potato CC–NB–LRR type R protein, indicated that the CC and NB–LRR protein domains physically interact with each other in a nonthreatening environment, but that these interactions dissipate in the presence of the cognate pathogen effector (Moffett et al. 2002). It is reasonable to expect that other NB–LRR R proteins undergo conformational changes in response to pathogen infection, and that they normally exist in an inhibitory conformation to avoid unwarranted activation. In addition, a number of NB–LRR R proteins associate with cytosolic HSP90 (HEAT-SHOCK PROTEIN90) and its co-chaperones RAR1 (REQUIRED FOR MLA12 RESISTANCE1), SGT1 (SUPPRESSOR OF THE G2 ALLELE OF *skp1*), and HSC70 (CYTOSOLIC HEAT SHOCK COGNATE70; Shirasu and Schulze-Lefert 2003; Noël et al. 2007). It is thought that this association facilitates the formation of R protein complexes and/or helps maintain R protein stability during the transition from a signal-incompetent to a signal-competent state (Shirasu and Schulze-Lefert 2003).

This chaperone complex might also mediate the localization and movement of R proteins within the cell (Seo et al. 2008). Recent convincing evidence indicates that some NB–LRR R proteins likely shuttle from the cytoplasm to the nucleus. This finding was somewhat unexpected, as many NB–LRR R proteins are predicted to be cytosolic (Dangl and Jones 2001). However, some pathogen effectors are thought to be targeted to the nucleus, so it is conceivable that R proteins might also be present in the nucleus to monitor their activities. The R proteins MLA10 (MILDEW A 10) in barley, N in tobacco, and RPS4 (RESISTANT TO *P. syringae* 4) in Arabidopsis were shown to localize to both the cytoplasm and the nucleus, and their nuclear localization and accumulation is necessary for downstream signaling and immunity to avirulent pathogens (Burch-Smith et al. 2007; Shen et al. 2007; Wirthmueller et al. 2007). In this regard, it is not surprising that certain mutations in components of the nucleocytoplasmic trafficking machinery have detrimental effects on defense responses (Wiermer et al. 2007). In the nucleus, MLA10 interacts directly with a subset of WRKY TFs that repress MAMP-mediated gene expression, suggesting that this R protein induces the expression of defense genes by sequestering negative regulators (Shen et al. 2007). Importantly, this finding also provides direct evidence that plants respond to MAMPs and effectors using some of the same resistance programs. The ability of R proteins to shuttle into the nucleus might afford plants an alternative and more direct route to modulate defense outputs when threatened by avirulent pathogens (Shen et al. 2007).

3.5 R Protein-Mediated Signaling

R protein signaling channels through several pathways that later converge and activate a common suite of defense outputs (summarized in Fig. 1b). NB–LRR proteins

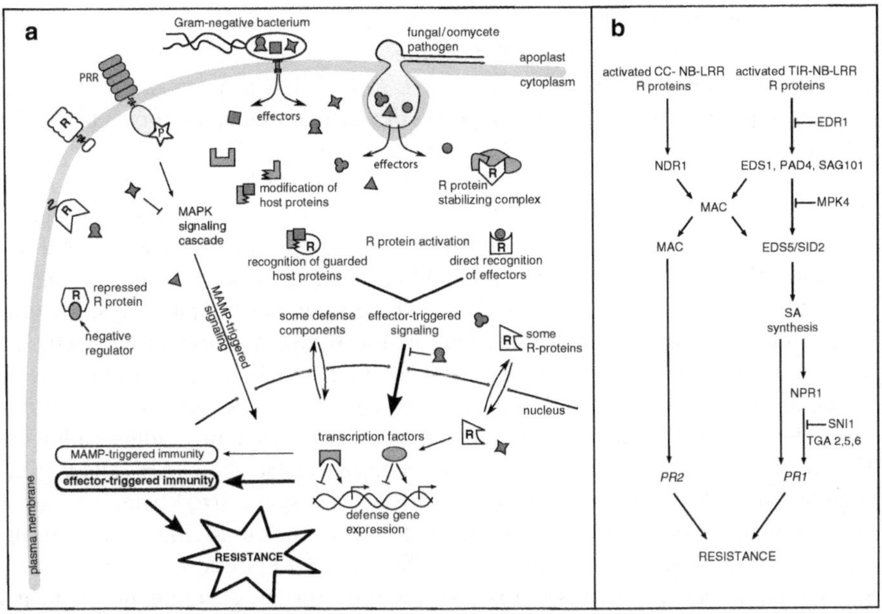

Fig. 1 Signaling events involved in plant innate immunity. **a** Plants have evolved the ability to perceive highly conserved microbe-associated molecular patterns (MAMPs) via transmembrane pattern recognition receptors (PRRs). PRR activation triggers mitogen-activated protein kinase (MAPK) signaling cascades that induce defense gene expression and hinder the growth of some microbial populations. During infection, pathogenic microbes deliver effector proteins into host cells, where they function to suppress or interfere with MAMP-triggered immunity and other defense responses. In resistant plants, cytoplasmic and membrane-associated resistance (R) proteins recognize effectors either directly or indirectly through the surveillance of guarded plant proteins and trigger effector-triggered immunity. Activated R proteins result in genetic reprogramming and pronounced physiological changes in the infected plant cell that ultimately result in resistance. **b** Genetic representation of some key signaling components activated during CC- and TIR–NB–LRR R protein-mediated resistance. Please see text for more details

of the CC type signal through the plasma membrane-associated protein NDR1 (NON-SPECIFIC DISEASE RESISTANCE1), whereas those of the TIR-type signal through the lipase-like protein EDS1 (ENHANCED DISEASE SUSCEPTIBILITY1) and its interacting partners PAD4 (PHYTOALEXIN-DEFICIENT4) and SAG101 (SENESCENCE-ASSOCIATED GENE101; Aarts et al. 1998; Feys et al. 2005). Importantly, there are two known *R* genes, *RPP7* and *RPP8* (*RESISTANCE TO P. parasitica 7* and *8*), that do not require NDR1 or EDS1 for downstream signaling, suggesting that additional transduction modules exist in defense signaling (McDowell et al. 2000). Aside from the fact that NDR1 works cooperatively with RIN4 to activate CC–NB–LRR R proteins such as RPM1 and RPS2 (Day et al. 2006), the molecular function of NDR1 and its specific downstream signaling components remain elusive. EDS1 interacts with PAD4 and SAG101 in distinct protein complexes in the cytosol and the nucleus (Feys et al. 2005), and is essential for the accumulation of SA and the transduction of signals derived from reactive oxygen species during infection (Wiermer et al. 2005).

There are a number of additional negative regulators that suppress EDS1 induction, suggesting that EDS1-activated pathways are strictly controlled (Glazebrook 2001). For example, *EDR1* (*ENHANCED DISEASE RESISTANCE1*) encodes a MAPKKK that functions upstream of EDS1 to suppress downstream signaling (Frye et al. 2000). Similarly, MPK4, one of the MAPKs induced following FLS2 activation, negatively regulates EDS1-activated SA signaling (Petersen et al. 2000).

Whereas jasmonic acid (JA) and ethylene are integral to resistance against herbivores and necrotrophic pathogens in plants, SA has long been associated with resistance to biotrophic pathogens. JA and SA signaling networks generally antagonize one another, but there is some cross-talk between the two pathways. Infection by avirulent biotrophic pathogens leads to local accumulation of SA, which is thought to mobilize a long-distance signal. In response to this mobile signal, systemic cells accumulate SA and express defense genes, effectively guarding themselves against potential further attack by a broad range of virulent pathogens. This phenomenon is known as systemic acquired resistance (SAR; Durrant and Dong 2004). SA production induced by infection is synthesized from chorismate by the enzyme isochorismate synthase (ICS1, also known as SID2; Nawrath and Métraux 1999; Wildermuth et al. 2001). The protein EDS5 (also known as SID1; SA INDUCTION-DEFICIENT1) is also required for SA accumulation, and it is most likely involved in transporting an SA precursor (Nawrath and Métraux 1999; Nawrath et al. 2002). Mutants unable to synthesize or accumulate SA become more susceptible to pathogen infection (Nawrath and Métraux 1999), highlighting the importance of this molecule in plant defense.

SA accumulation is associated with a buildup of reactive oxygen species that causes significant changes in cellular redox levels. These redox changes are sensed in the cytosol by the key defense protein NPR1 (NON-EXPRESSOR OF PR GENES1; Dong 2004). NPR1 is thought to exist in an inactive state as an oligomer that responds to redox alterations by monomerizing and relocating to the nucleus, where it interacts with multiple basic leucine zipper TGA transcription factors to induce the expression of the defense gene *PR1* (Mou et al.2003). The transcription factors TGA2, TGA5, and TGA6 have redundant functions and were shown to play both positive and negative roles in the regulation of SAR (Zhang et al. 2003b). The activation of *PR1* also requires derepression of its negative regulator, SNI1 (SUPPRESSOR OF *npr1*, INDUCIBLE1; Li et al. 1999). In addition to *PR1,* there are several other *PR* genes activated during defense. These include chitinases, glucanases, proteinases, and RNases that were shown to have antimicrobial activities in vitro.

EDS1 activates additional pathways in an infected cell that are independent of SA accumulation and NPR1. While *PR1* is downregulated in *eds5* mutants, expression of another *PR* gene, the β-1,3-glucanase *PR2*, appears to be unaffected (Nawrath and Métraux 1999). Also, *PR2* is constitutively expressed in *snc1* gain-of-function plants and is not suppressed by *npr1* and is only partially suppressed by *eds5* (Zhang et al. 2003a). Together, these data reveal that SA- and NPR1-independent defense signaling pathways are activated downstream of R proteins. A handful of components have been shown to function in these pathways, including a recently characterized multiprotein nuclear complex called the MAC (MOS4-associated complex; Palma et al. 2007). SA accumulation following pathogen infection is unaffected in MAC mutants, and epistasis analysis with *npr1* suggests that the MAC functions separately from NPR1 (Palma et al. 2007). Interestingly, the MAC seems to be required for resistance conditioned by both CC- and TIR–NB–LRR R proteins, representing a possible convergence point between the NDR1 and EDS1-activated pathways (Palma et al. 2007). Loss-of-function mutations in any of the known MAC components lead to higher susceptibility to virulent pathogen infection (Palma et al. 2007), suggesting that the SA-independent pathway is necessary for both basal and R protein-mediated defenses.

4 Concluding Remarks

The past decade has seen great advances in our understanding of the plant immune system. Plants, which are under constant threat of pathogen infection, rely on an intricate network of signaling components to effectively fend off microbial colonization. The first level of defense is carried out at the plant cell surface, where PRRs detect highly conserved MAMPs and activate low-level resistance responses. The detection of menacing effector proteins then activates R proteins that trigger more effective defense responses, often ending in an HR to restrict the growth of biotrophic pathogens. The overall theme of an evolutionary arms race between plants and pathogens is presented in Fig. 1a.

However, there are still many probing questions that are currently unanswered. (1) What interplay occurs between MAMP- and R-mediated resistance? (2) How, mechanistically, and in which subcellular compartments do R proteins recognize their cognate effectors? (3) How are multiple signaling pathways coordinated, and what cross-talk is present between the distinct signaling pathways? Future work in these areas will truly enlighten our knowledge of plant innate immunity to microbial infection.

Acknowledgments We thank Dr. Marcel Wiermer for helpful discussions and critical reading of the manuscript, and we are grateful to Yuti Cheng for research assistance.

References

Aarts N, Metz M, Holub E, Staskawicz BJ, Daniels MJ, Parker JE (1998) Different requirements for *EDS1* and *NDR1* by disease resistance genes define at least two *R* gene-mediated signaling pathways in *Arabidopsis*. Proc Nat Acad Sci USA 95:10306–10311

Abramovitch RB, Janjusevic R, Stebbins CE, Martin G (2006) Type III effector AvrPtoB requires intrinsic E3 ubiquitin ligase activity to suppress plant cell death and immunity. Proc Nat Acad Sci USA 103:2851–2856

Akira S, Uematsu S, Takeuchi O (2006) Pathogen recognition and innate immunity. Cell 124:783–801

Alfano JR, Collmer A (1996) Bacterial pathogenesis in plants: life up against the wall. Plant Cell 9:1683–1698

Asai T, Tena G, Plotnikova J, Willmann MR, Chiu W-L, Gómez-Gómez L, Boller T, Ausubel FM, Sheen J (2002) MAP kinase signaling cascade in *Arabidopsis* innate immunity. Nature 415:977–983

Ausubel FM (2005) Are innate immune signaling pathways in plants and animals conserved? Nature Immunol 6:973–979

Axtell MJ, Chisholm ST, Dahlbeck D, Staskawicz BJ (2003) Genetic and molecular evidence that the *Pseudomonas syringae* type III effector protein AvrRpt2 is a cysteine protease. Mol Microbiol 49:1537–1546

Belkhadir Y, Nimchuk Z, Hubert DA, Mackey D, Dangl JL (2004) Arabidopsis RIN4 negatively regulates disease resistance mediated by RPS2 and PRM1 downstream or independent of the NDR1 signal modulator and is not required for the virulence functions of bacterial type III effectors AvrRpt2 or AvrRpm1. Plant Cell 16:2822–2835

Birch PRJ, Boevink PC, Gilroy EM, Hein I, Pritchard L, Whisson SC (2008) Oomycete RXLR effectors: delivery, functional redundancy and durable disease resistance. Curr Opin Plant Biol 11:1–7

Buell CR, Joardar V, Lindeberg M, Selengut J, Paulsen IT, Gwinn ML, Dodson RJ, Deboy RT, Durkin AS, Kolonay JF, Madupu R, Daugherty S, Brinkac L, Beanan MJ, Haft DH, Nelson WC, Davidsen T, Zafar N, Zhou L, Liu J, Yuan Q, Khouri H, Fedorova N, Tran B, Russell D, Berry K, Utterback T, Van Aken SE, Feldblyum TV, D'Ascenzo M, Deng WL, Ramos AR, Alfano JR, Cartinhour S, Chatterjee AK, Delaney TP, Lazarowitz SG, Martin GB, Schneider DJ, Tang X, Bender CL, White O, Fraser CM, Collmer A (2003) The complete genome sequence of the Arabidopsis and tomato pathogen *Pseudomonas syringae* pv. *tomato* DC3000. Proc Nat Acad Sci USA 100:10181–10186

Burch-Smith TM, Schiff M, Caplan JL, Tsao J, Czymmek K, Dinesh-Kumar SP (2007) A novel role for the TIR domain in association with pathogen-derived elicitors. PLoS Biol 5:501–514

Chen Z, Agnew JL, Cohen JD, He P, Shan L, Sheen J, Kunkel BN (2007) *Pseudomonas syringae* type III effector AvrRpt2 alters Arabidopsis thaliana auxin physiology. Proc Nat Acad Sci U S A 204:20131–20136

Chinchilla D, Bauer Z, Regenass M, Boller T, Felix G (2006) The Arabidopsis receptor kinase FLS2 binds flg22 and determines the specificity of flagellin perception. Plant Cell 2:465–476

Chinchilla D, Zipfel C, Robatzek S, Kemmerling B, Nürnberger T, Jones JDG, Felix G, Boller T (2007) A flagellin-induced complex of the receptor FLS2 and BAK1 initiates plant defence. Nature 448:497–500

Coaker G, Falick A, Staskawicz B (2005) Activation of a phytopathogenic bacterial effector protein by a eukaryotic cyclophilin. Science 308:548–550

Dangl JL, Jones JDG (2001) Plant pathogens and integrated defense responses to infection. Nature 411:826–833

Day B, Dahlbeck D, Staskawicz BJ (2006) NDR1 interaction with RIN4 mediates the differential activation of multiple disease resistance pathways in *Arabidopsis*. Plant Cell 18:2782–2791

Dean RA, Talbot NJ, Ebbole DJ, Farman ML, Mitchell TK, Orbach MJ, Thon M, Kulkarni R, Xu J-R, Pan H, Read ND, Lee Y-W, Carbone I, Brown D, Oh YY, Donofrio N, Jeong JS, Soanes DM, Djonovic S, Kolomiets E, Rehmeyer C, Li W, Harding M, Kim S, Lebrun M-H, Bohnert H, Coughlan S, Butler J, Calvo S, Ma L-J, Nicol R, Purcell S, Nusbaum C, Galagan JE, Birren BW (2005) The genome sequence of the rice blast fungus *Magnaporthe grisea*. Nature 434:980–986

Deslandes L, Olivier J, Theulieres F, Hirsch J, Feng DX, Bittner-Eddy P, Beyon J, Marco Y (2002) Resistance to Ralstonia solanacerum in *Arabidopsis thaliana* is conferred by the recessive RRS1-R gene, a member of a novel family of resistance genes. Proc Nat Acad USA 99:2404–2409

Dong X (2004) NPR1, all things considered. Curr Opin Plant Biol 7:547–552

Dong X, Mindrinos M, Davis KR, Ausubel FM (1991) Induction of *Arabidopsis* defense genes by virulent and avirulent *Pseudomonas syringae* strains and by a cloned avirulence gene. Plant Cell 3:61–72

Dow M, Newman MA, von Roepenack E (2000) The induction and modulation of plant defense responses by bacterial lipopolysaccharides. Annu Rev Phytopathol 38:241–261

Durrant WE, Dong X (2004) Systemic acquired resistance. Annu Rev Phytopathol 42:185–209

Euglem T, Somssich IE (2007) Networks of WRKY transcription factors in defense signaling. Curr Opin Plant Biol 10:366–371

Farmer EE, Almeeras E, Krishnamurthy V (2003) Jasmonates and related oxylipins in plant responses to pathogenesis and herbivory. Curr Opin Plant Biol 6:372–378

Felix G, Duran JD, Volko S, Boller T (1999) Plants have a sensitive perception system for the most conserved domain of bacterial flagellin. Plant J 18:265–276

Feys BJ, Wiermer M, Bhat RA, Moisan LJ, Medina-Escobar N, Neu C, Cabral A, Parker JE (2005) Arabidopsis SENESCENCE-ASSOCIATED GENE101 stabilizes and signals within an ENHANCED DISEASE SUSCEPTIBILITY1 complex in plant innate immunity. Plant Cell 17:2601–2613

Finlay BB, McFadden G (2006) Anti-immunology: evasion of the host immune system by bacterial and viral pathogens. Cell 124:767–782

Flor HH (1971) Current status of the gene-for-gene concept. Annu Rev Phytopathol 9:275–296

Frye CA, Tang D, Innes RW (2000) Negative regulation of defense responses in plants by a conserved MAPKK kinase. Proc Nat Acad Sci USA 98:373–378

Glazebrook J (2001) Genes controlling expression of defense responses in *Arabidopsis*: 2001 status. Curr Opin Plant Biol 4:301–308

Göhre V, Robatzek S (2008) Breaking the barriers: microbial effector molecules subvert plant immunity. Annu Rev Phytopathol 46:189–215

Gómez-Gómez L, Boller T (2000) FLS2: an LRR receptor-like kinase involved in the perception of the bacterial elicitor flagellin in Arabidopsis. Mol Cell 5:1003–1011

Gómez-Gómez L, Boller T (2002) Flagellin perception: a paradigm for innate immunity. Trends Plant Sci 7:251–256

Gómez-Gómez L, Bauer Z, Boller T (2001) Both the extracellular leucine-rich repeat domain and the kinase activity of FSL2 are required for flagellin binding and signaling in Arabidopsis. Plant Cell 13:1155–1163

He P, Shan L, Lin N-C, Martin GB, Kemmerling B, Nürnberger T, Sheen J (2006) Specific bacterial suppressors of MAMP signaling upstream of MAPKKK in *Arabidopsis* innate immunity. Cell 125:563–575

Hesse A, Hann DR, Simenes-Ibanez S, Jones AM, He K, Li J, Schroeder JI, Peck SC, Rathjen JP (2007) The receptor-like kinase SERK3/BAK1 is a central regulator of innate immunity in plants. Proc Nat Acad Sci USA 104:12217–12222

Holub EB (2001) The arms race is ancient history in *Arabidopsis*, the wildflower. Nat Rev Gen 2:516–527

Janjusevic R, Abramovitch RB, Martin G, Stebbins CE (2006) A bacterial inhibitor of host programmed cell death defenses is an E3 ubiquitin ligase. Science 311:222–226

Jin Q, He S-Y (2001) Role of the Hrp pilus in type III protein secretion in *Pseudomonas syringae*. Science 294: 2556–2558

Jones JDG, Dangl JL (2006) The plant immune system. Nature 444:323–329

Judelson HS, Blanco FA (2005) The spores of *Phytophthora*: weapons of the plant destroyer. Nat Rev Microbiol 3:47–58

Katagiri F, Thilmony R, He S-Y (2002) The *Arabidopsis thaliana-Pseudomonas syringae* Interaction. In: The Arabidopsis Book. American Society of Plant Biologists, Rockville. Available via DIALOG. www.aspb.org/publications/arabidopsis/ Cited 30 September 2002

Kay S, Hahn S, Marois E, Hause G, Bonas U (2007) A bacterial effector acts as a plant transcription factor and induces a cell size regulator. Science 318:648–651

Kim HS, Desveaux D, Singer AU, Patel P, Sondek J, Dangl JL (2005) The *Pseudomonas syringae* effector AvrRpt2 cleaves its C-terminally acylated target, RIN4, from Arabidopsis membranes to block RPM1 activation. Proc Nat Acad Sci USA 102:6496–6501

Kobe B, Deisenhofer J (1995) Proteins with leucine-rich repeats. Curr Opin Struct Biol 5:409–416

Li X, Zhang Y, Clarke JD, Li Y, Dong X (1999) Identification and cloning of a negative regulator of systemic acquire resistance, SNI1, through a screen for suppressors of NPR1-1. Cell 98:329–339

Li X, Clark JD, Zhang Y, Dong X (2001) Activation of an EDS1-mediated R-gene pathway in the snc1 mutant leads to constitutive, NPR1-independent pathogen resistance. Mol Plant Microbe Interact 14:1131–1139

Mackey D, Holt BF III, Wiig A, Dangl JL (2002) RIN4 interacts with *Pseudomonas syringae* type III effector molecules and is required for RPM1-mediated resistance in Arabidopsis. Cell 108:743–754

Mackey D, Belkhadir Y, Alonso JM, Ecker JR, Staskawicz BJ (2003) Arabidopsis RIN4 is a target of the type III virulence effector AvrRpt2 and modulates RPS2-mediated resistance. Cell 7:379–389

Marathe R, Dinesh-Kumar SP (2003) Plant defense: one post, multiple guards?! Mol Cell 11:284–286

Martin GB, Brommonschenkel SH, Chunwongse J, Frary A, Ganal MW, Spivey R, Wu T, Earle ED, Tanksley SD (1993) Map-based cloning of a protein kinase gene conferring disease resistance in tomato. Science 262:1432–1436

Martin GB, Bogdanove AJ, Sessa G (2003) Understanding the functions of plant disease resistance proteins. Annu Rev Plant Biol 54:23–61

McDowell JM, Cuzick A, Can C, Beyon J, Dangl JL, Holub EB (2000) Downy mildew (*Peronospora parasitica*) resistance genes in *Arabidopsis* vary in functional requirements for NDR1, EDS1, NPR1, and salicylic acid accumulation. Plant J 22:523–529

Menke FL, Van Pelt JA, Pieterse CM, Klessig DF (2004) Silencing of the mitogen-activated protein kinase MPK6 compromises disease resistance in Arabidopsis. Plant Cell 16:897–907

Moffett P, Farnham G, Peart J, Baulcombe DC (2002) Interaction between domains of a plant NBS-LRR protein in disease resistance-related cell death. EMBO J 21:4511–4519

Mou Z, Fan W, Dong X (2003) Inducers of plant systemic acquired resistance regulate NPR1 function through redox changes. Cell 113:935–944

Mucyn TS, Clemente A, Andriotis VM, Balmuth AL, Oldroyd GE, Staskawicz BJ, Rathjen JP (2006) The tomato NBARC-LRR protein Prf interacts with Pto kinase in vivo to regulate specific plant immunity. Plant Cell 18:2792–2806

Nakagami H, Pitzschke A, Hirt H (2005) Emerging MAP kinase pathways in plant stress signaling. Trends Plant Sci 10:339–346

Nawrath C, Métraux J-P (1999) Salicylic acid induction-deficient mutants of Arabidopsis express *PR-2* and *PR-5* and accumulate high levels of camalexin after pathogen inoculation. Plant Cell 11:1393–1404

Nawrath C, Heck S, Parinthawong N, Métraux J-P (2002) EDS5, an essential component of salicylic acid-dependent signaling for disease resistance in Arabidopsis, is a member of the MATE transporter family. Plant Cell 14:275–286

Noël LD, Cagna G, Stuttmann J, Wirthmüler L, Betsuyaku S, Witte CP, Bhat R, Pochon N, Colby T, Parker JE (2007) Interaction between SGT1 and cytosolic/nuclear HSC70 chaperones regulates Arabidopsis immune responses. Plant Cell 19:4061–4076

Nürnberger T, Brunner F, Kemmerling B, Piater L (2004) Innate immunity in plants and animals: striking similarities and obvious differences. Immunol Rev 198:249–266

Palma K, Zhao Q, Cheng Y, Bi D, Monaghan J, Cheng W, Zhang Y, Li X (2007) Regulation of plant innate immunity by three proteins in a complex conserved across the plant and animal kingdoms. Genes Dev 21:1484–1493

Petersen M, Brodersen A, Naested H, Andreasson E, Lindhart U, Johansen B, Nielsen HB, Lacy M, Austin MJ, Parker JE, Sharma SB, Klessig DF, Martienssen R, Mattsson O, Jensen AB, Mundy J (2000) Arabidopsis MAP Kinase 4 negatively regulates systemic acquired resistance. Cell 103:111–1120

Rivas S, Thomas CM (2005) Molecular interactions between tomato and the leaf mold pathogen *Cladosporium fulvum*. Annu Rev Phytopathol 43:395–436

Robatzek S, Chinchilla D, Boller T (2006) Ligand-induced endocytosis of the pattern recognition receptor FLS2 in Arabidopsis. Genes Dev 20:537–542

Schilmiller AL, Howe GA (2005) Systemic signaling in the wound response. Curr Opin Plant Biol 8:369–377

Seo Y-S, Lee S-K, Song M-Y, Suh J-P, Han T-R, Ronald P, Jeon J-S (2008) The HSP90-SCT1-RAR1 molecular chaperone complex: a core modulator in plant immunity. J Plant Biol 51:1–10

Shan L, He P, Sheen J (2007) Intercepting host MAPK signaling cascades by bacterial type III effectors. Cell Host Microbe 1:167–174

Shen Q-H, Saijo Y, Mauch S, Christoph B, Bieri S, Beat K, Hikary S, Üker B, Somssich IE, Schulze-Lefert P (2007) Nuclear activity of MLA immune receptors links isolate-specific and basal disease-resistance responses. Science 315:1098–1103

Shirano Y, Kachroo P, Shah J, Klessig DF (2002) A gain-of-function mutation in an Arabidopsis Toll-Interleukin1 receptor-nucleotide binding site-leucine-rich repeat type R gene triggers defense responses and results in enhanced disease resistance. Plant Cell 14:3149–3162

Shirasu K, Schulze-Lefert P (2003) Complex formation, promiscuity and multi-functionality: protein interactions in disease-resistance pathways. Trends Plant Sci 8:252–258

Slusarenko AJ, Schlaich NL (2003) Pathogen profile: downy mildew of *Arabidopsis thaliana* caused by *Hyaloperonospora parasitica* (formerly *Peronospora parasitica*). Mol Plant Pathol 4:159–170

Song WY, Wang GL, Chen LL, Kim HS, Pi LY, Holsten T, Gardner J, Wang B, Zhai WX, Zhu LH, Fauquet C, Ronald P (1995) A receptor kinase-like protein encoded by the rice disease resistance gene, Xa21. Science 270:1804–1806

Suarez-Rodriguez MC, Adams-Phillips L, Liu Y, Wang H, Su S-H, Jester PJ, Zhang S, Bent AF, Krysan PJ (2006) MEKK1 is required for flg22-induced MPK4 activation in Arabidopsis plants. Plant Physiol 143:661–669

Takeda K, Kaisho T, Akira S (2003) Toll-like receptors. Annu Rev Immunol 21:335–376

Takken FLW, Albrecht M, Tameling WIL (2006) Resistance proteins: molecular switches of plant defence. Curr Opin Plant Biol 9:1–8

Tameling WIL, Vossen JH, Albrecht M, Lengauer T, Berden JA, Haring MA, Cornelissen BJC, Takken FLW (2006) Mutations in the NB-ARC domain of I-2 that impair ATP hydrolysis cause autoactivation. Plant Physiol 140:1233–1245

Tang X, Frederick RD, Zhou J, Halterman DA, Jia Y, Martin GB (1996) Initiation of plant disease resistance by physical interaction of AvrPto and Pto kinase. Science 274:2060–2063

The Arabidopsis Genome Initiative (2000) Analysis of the genome sequence of the flowering plant *Arabidopsis thaliana*. Nature 408:796–815

Thordal-Christensen H (2003) Fresh insights into processes of nonhost resistance. Curr Opin Plant Biol 6:351–357

Ülker B, Somssich IE (2004) WRKY transcription factors: from DNA binding towards biological function. Curr Opin Plant Biol 7:491–498

Van Der Biezen EA, Jones JDG (1998) Plant disease resistance proteins and the gene-for-gene concept. Trends Biochem Sci 23:454–456

Whalen MC, Innes RW, Bent AF, Staskawicz BJ (1991) Identification of *Pseudomonas syringae* pathogens of *Arabidopsis* and a bacterial locus determining avirulence on both *Arabidopsis* and soybean. Plant Cell 2:49–59

Wiermer M, Feys BJ, Parker JE (2005) Plant immunity: the EDS1 regulatory node. Curr Opin Plant Biol 8:383–389

Wiermer M, Palma K, Zhang Y, Li X (2007) Should I stay or should I go? Nucleocytoplasmic trafficking in plant innate immunity. Cell Microbiol 9:1880–1890

Wildermuth MC, Dewdney J, Wu G, Ausubel FM (2001) Isochorismate synthase is required to synthesize salicylic acid for plant defence. Nature 414:562–565

Wirthmueller L, Zhang Y, Jones JDG, Parker JE (2007) Nuclear accumulation of the Arabidopsis immune receptor RPS4 is necessary for triggering EDS1-dependent defense. Curr Biol 17:2023–2029

Xiang T, Zong N, Zou Y, Wu Y, Zhang J, Xing W, Li Y, Tang X, Zhu L, Chai J, Zhou JM (2008) *Pseudomonas syringae* effector AvrPto blocks innate immunity by targeting receptor kinases. Curr Biol 18:74–80

Xiao S, Ellwood S, Calis O, Patrick E, Li T, Coleman M, Turner JG (2001) Broad-spectrum mildew resistance in *Arabidopsis thaliana* mediated by RPW8. Science 291:118–120

Xing W, Zou Y, Liu Q, Liu J, Luo X, Huang Q, Chen S, Zhu L, Bi R, Hao Q, Wu J-W, Zhou J-M, Chai J (2007) The structural basis for activation of plant immunity by bacterial effector protein AvrPto. Nature 449:243–247

Zhang Y, Goritschnig S, Dong X, Li X (2003a) A gain-of-function mutation in a plant disease resistance gene leads to constitutive activation of downstream signal transduction pathways in suppressor of npr1–1, constitutive 1. Plant Cell 15:2636–2646

Zhang Y, Tessaro MJ, Lassner M, Li X (2003b) Knockout analysis of Arabidopsis transcription factors TGA2, TGA5, and TGA6 reveals their redundant and essential roles in systemic acquired resistance. Plant Cell 15:2647–2653

Zhou J-M, Chai J (2008) Plant pathogenic bacterial type III effectors subdue host responses. Curr Opin Microbiol 11:179–185

Zipfel C (2008) Pattern-recognition receptors in plant innate immunity. Curr Opin Immunol 20:10–16

Zipfel C, Rathjen JP (2008) Plant immunity: AvrPto targets the frontline. Curr Opin Plant Biol 18:R218–R220

Zipfel C, Robatzek S, Navarro L, Oakeley EJ, Jones JDG, Felix G, Boller T (2004) Bacterial disease resistance in Arabidopsis through flagellin perception. Nature 428:764–767

Zipfel C, Kunze G, Chinchilla D, Caniard A, Jones JDG, Boller T, Felix G (2006) Perception of the bacterial PAMP EF-Tu by the receptor EFR restricts *Agrobacterium*-mediated transformation. Cell 125:749–760

Airborne Induction and Priming of Defenses

Martin Heil

Abstract At first glance, the idea of "talking trees" goes against common sense, but we now know that plants can indeed perceive volatile organic compounds (VOCs) or specific light reflected from or transmitted by their neighbors, and that this perception triggers specific responses. Airborne plant–plant communication usually affects the resistance phenotype of a plant growing close to an attacked neighbor. An explanation for this "information parasitism" appears to be that VOCs serve many purposes, including airborne within-plant signaling. Airborne systemic resistance induction is faster than signaling via the vascular system, independent of orthostichy, and it allows distant plant parts to be primed in order to achieve an optimized systemic defense expression. Plants do need to be able to emit and perceive VOCs and so it is difficult for them to stop their neighbors from "eavesdropping." Plant–plant communication via VOCs has become an accepted phenomenon, but further studies are required to estimate its true importance under ecologically realistic conditions.

1 Introduction

Many novels deal with plants—or plant-like organisms—that talk to each other or to humans. Usually, the occurrence of such plants in a novel implies that it can be classified as fantasy, science fiction, exaggeratedly esoteric, or—at best—that it deals with dreams. Most plants cannot even move in a visible way, and they do not produce any noise, so how can plants communicate? Although plant behavior has been intensively and controversially discussed since the time of Charles Darwin, many found the concept of plant communication a difficult one to swallow (Karban 2008). Similarly, the related phenomenon of "talking trees" did not enter the scientific literature until 1983. That year, Rhoades (1983) reported increased levels of anti-herbivore resistance in undamaged Sitka willow trees growing close to herbivore-infested

M. Heil
Dpto. de Ingeniería Genética, CINVESTAV—Irapuato, Km. 9.6 Libramiento Norte, CP 36821, Irapuato, Guanajuato, México
e-mail: mheil@ira.cinvestav.mx

F. Baluška (ed.), *Plant-Environment Interactions*, Signaling and Communication in Plants,
DOI: 10.1007/978-3-540-89230-4_8,

conspecific plants, and Baldwin and Schultz (1983) found that when undamaged poplar and sugar maple saplings shared the same air as damaged plants their chemical defenses were enhanced. Apparently, the attacked plants managed to warn their neighbors somehow.

Assuming they do exist, do these phenomena represent true communication? The answer to this question depends on how we define "communication." If we mean "an intentional exchange of information among individuals," the answer is "no," since plants lack any conscious behavior. However, this definition appears to be too narrow, as it also excludes most well-accepted forms of communication among animals, and intentionality cannot be proven for most (if not all) species besides humans. Richard Karban therefore suggested a two-step definition that applies the term "plant communication" to situations where cues that are emitted from plants cause rapid responses in a receiver organism, and where the emission of these cues is plastic and conditional (Karban 2008).

Does airborne communication exist among plants? The present chapter tries to answer this question and uses Karban's definition. I will first describe the phenomenon of plants responding to airborne signals that are released from damaged plants, present the signals involved (in so far as they are known), and then mention other situations where the exchange of airborne information among plants elicits rapid responses in the receiver. I finally discuss the ecological and evolutionary consequences of the exchange of information among plants.

2 Airborne Plant–Plant Signaling

2.1 Induced Defenses Against Pathogens and Herbivores

Plants respond to attacks by pathogens or herbivores with extensive changes in gene expression that lead to induced resistance phenomena: various traits are then expressed de novo or at much higher intensities to reduce or prevent further damage (Karban and Baldwin 1997; Sticher et al. 1997; Walling 2000; Durrant and Dong 2004). As both pathogens and herbivores are mobile, such responses are usually not restricted to the damaged tissue but are expressed systemically; in as-yet undamaged organs too. Hormones such as jasmonic acid (JA), salicylic acid (SA) and their derivatives are produced at the site of attack and spread throughout the plant. Since plant vascular bundles represent a highly sophisticated system for long-distance transport (Le Hir et al. 2008), early research on the translocated signals focused on—and found—signaling compounds that are transported via the phloem and the xylem (Métraux et al. 1990; Dicke and Dijkman 1992; Constabel et al. 1995; Zhang and Baldwin 1997; Thorpe et al. 2007; for reviews, see Starck 2006; Wasternack 2007; Heil and Ton 2008).

Recent studies have revealed, however, that long-distance signaling can also be mediated by volatile compounds that move in the headspace outside the plants (Heil and Ton 2008; *see* Fig.1). In particular, green-leaf volatiles (GLVs) and other

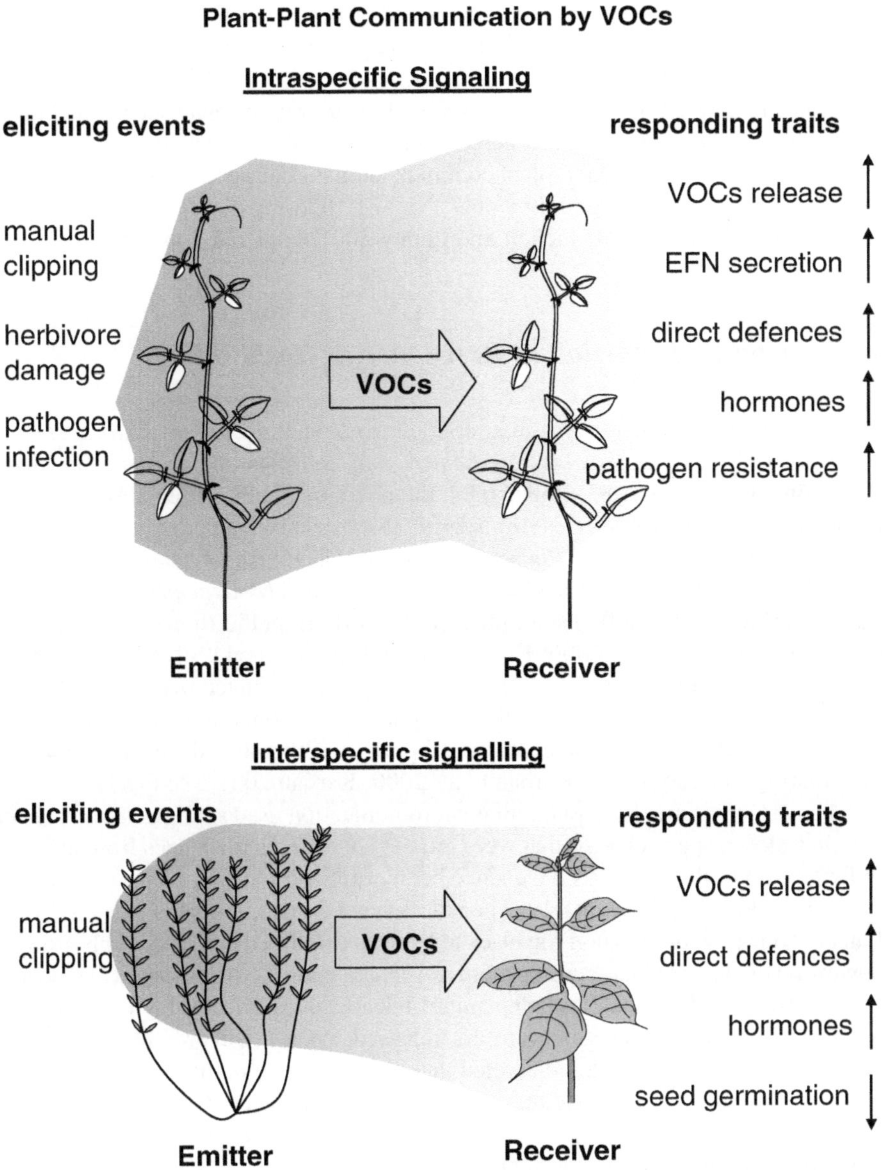

Fig. 1 Airborne plant–plant signaling. VOC-mediated plant–plant communication occurs both among plants of the same species and among individuals belonging to different species. Intraspecific communication has been reported for black alder, corn, lima bean, sagebrush, sugar maple and tobacco. It can be elicited by manual clipping, natural herbivore damage or pathogen infection, and may affect direct defenses against herbivores such as proteinase inhibitors and leaf phenolics, indirect defenses such the release of VOCs and the secretion of extrafloral nectar, the production of the signaling hormones SA and JA, and plant pathogen resistance. Interspecific communication has so far only been reported in the case of manually clipped sagebrush, which enhanced direct and indirect herbivore resistance in neighboring tobacco plants and reduced seed germination rates in its direct vicinity

herbivore-induced volatile organic compounds (VOCs) have been reported in the context of systemic plant responses to local damage (Xu et al. 1994; Birkett et al. 2000; Ellis and Turner 2001; Voelckel et al. 2001; Schmelz et al. 2003; Kishimoto et al. 2005; Karban et al. 2006). Since such volatiles are released from the plant surface and move through the air, they can also affect neighboring plants and thus mediate a phenomenon associated with airborne plant–plant communication—the expression of resistance in intact plants, which is triggered by cues from neighboring plants that are currently under attack (Farmer 2001; Pickett and Poppy 2001; Heil and Ton 2008).

2.2 Airborne Induction of Resistance to Herbivores

Plant–plant communication was first reported from the Sitka willow *Salix sitchensis* (Rhoades 1983), poplar *Populus* × *euroamericana*, and sugar maple *Acer saccharum* (Baldwin and Schultz 1983). Since then, the phenomenon has been detected in the taxonomically unrelated species *Arabidopsis thaliana*, black alder (*Alnus glutinosa*), corn (*Zea mays*), lima bean (*Phaseolus lunatus*), sagebrush (*Artemisia tridentata*) and wild and cultivated tobacco (*Nicotiana attenuata* and *N. tabacum*) (Shulaev et al. 1997; Karban et al. 2000; Tscharntke et al. 2001; Engelberth et al. 2004; Choh et al. 2006; Karban et al. 2006; Kost and Heil 2006; Heil and Silva Bueno 2007b; Ton et al. 2007; Godard et al. 2008). Most of these cases related to signaling among plants that belong to the same species, but plant–plant communication even occurs among different species; for example, clipping sagebrush induced resistance in neighboring tobacco plants (Karban et al. 2000; Karban 2001; *see* Fig.1).

The first experiments on plant–plant communication used saplings that had been kept in a closed space or trees that were growing at different distances from attacked individuals, but these reports have since been criticized for their lack of ecological realism (Baldwin and Schultz 1983) or their lack of true controls (Rhoades 1983). Later on, however, observations on black alder trees confirmed that individuals growing downwind of clipped plants became more resistant to future herbivore attack (Tscharntke et al. 2001). While manual clipping might release unrealistically high amounts of VOCs, or even compounds that are not released when herbivores feed on plants, field studies on lima bean demonstrated that plant–plant communication also works under ecologically realistic conditions: receivers that were otherwise untreated suffered less from herbivory when they were exposed to the air that came from beetle-damaged emitters (Heil and Silva Bueno 2007b).

2.3 Airborne Induction of Resistance to Pathogens

Plant–plant communication appears to commonly trigger herbivore resistance. Are similar phenomena also involved in pathogen resistance? Salicylic acid is a hormone that plays a central role in the systemic acquired resistance of plants to biotrophic pathogens, and its volatile derivative, methyl salicylate (MeSA), has been put forward as the most likely mobile signal (Park et al. 2007). In tobacco, MeSA is enzymatically converted back to SA by SA-binding protein 2 (SABP2), and SA then forms the active resistance-inducing compound (Kumar and Klessig 2003; Forouhar et al. 2005). In principle, this also opens up the possibility of airborne signaling in the context of pathogen resistance. Resistance expression has indeed been reported in tobacco plants that were exposed to the MeSA-rich air from infected plants (Shulaev et al. 1997) and in lima bean plants exposed to VOCs released from resistance-expressing conspecifics (Yi, Ryu, Heil, unpublished data).

Moreover, several aspects of pathogen resistance appear to depend on oxylipins rather than SA, at least in arabidopsis (Pieterse et al. 1998; Truman et al. 2007), and oxylipins were involved in the resistance of *Vicia faba* to bean rust fungus (*Uromyces fabae*) (Walters et al. 2006). Some oxylipin-derived green-leaf volatiles (GLVs) exhibit antimicrobial activity (Nakamura and Hatanaka 2002; Dilantha Fernando et al. 2005; Matsui 2006; Shiojiri et al. 2006) and may thus also mediate airborne pathogen resistance. Indeed, exposure to GLVs such as *trans*-2-hexenal, *cis*-3-hexenal or *cis*-3-hexenol enhanced the resistance of arabidopsis to the fungal pathogen *Botrytis cinerea* (Kishimoto et al. 2005). Plant–plant communication mediated by volatile compounds may therefore be a common phenomenon in the context of plant pathogen resistance too.

3 Mechanisms of Plant–Plant Communication

3.1 VOCs Prime and Induce Defense Responses in Intact Plants

Studies aimed at a mechanistic understanding of VOC-mediated resistance in intact plants reported changes in the expression of defense-related genes (Arimura et al. 2000; Farag et al. 2005; Paschold et al. 2006; Ton et al. 2007; Godard et al. 2008), increased production rates of MeJA (Godard et al. 2008), JA or defensive compounds (Baldwin and Schultz 1983; Farmer and Ryan 1990; Engelberth et al. 2004; Ruther and Fürstenau 2005), and increased production of indirect defenses such as VOCs (Ton et al. 2007) and extrafloral nectar (Choh et al. 2006; Kost and Heil 2006).

In many cases, exposure to lower concentrations of volatiles that failed to activate plant defenses to their full extent could still affect plant resistance by sensitizing the plant's defense arsenal (Engelberth et al. 2004; Choh and Takabayashi 2006; Heil and Kost 2006; Kessler et al. 2006; Heil and Silva Bueno 2007b; Ton et al. 2007). This so-called *priming* affects various induced-resistance phenomena (Conrath et al. 2006; Goellner and Conrath 2008). Priming prepares the plant to respond more rapidly and/or effectively to subsequent attack and is normally activated at much lower concentrations of the resistance-inducing signal than required for full induction of active defenses (van Hulten et al. 2006; Bruce et al. 2007). Hence, primed plants do not show enhanced defense activity, but they do respond much more rapidly or strongly to wounding or infection than unprimed plants.

Compounds that trigger defensive responses in as-yet undamaged plants are still being discovered, but most of the volatiles that have been identified so far in this context are either the gaseous derivatives of jasmonic acid and salicylic acid (MeJA and MeSA) or GLVs (Arimura et al. 2000; Engelberth et al. 2004; Ruther and Fürstenau 2005; Ruther and Kleier 2005; Kost and Heil 2006). Green-leaf volatiles are C_6 compounds that are rapidly released upon tissue damage, since they are synthesized by pre-existing enzymes from precursors that already exist in the undamaged cell (Turlings and Wäckers 2004). GLVs that have been observed to prime or induce herbivore resistance at the genetic, biochemical or phenotypic levels include, for example, *cis*-3-hexenyl acetate (corn and lima bean: see Engelberth et al. 2004; Kost and Heil 2006; Heil et al. 2008) and *cis*-3-hexen-1-ol, *trans*-2-hexenal, *cis*-3-hexenal, *trans*-2-pentenal and *trans*-2-heptenal (corn: see Engelberth et al. 2004; Ruther and Fürstenau 2005).

Another candidate is the gaseous hormone ethylene, which plays a modulating role in plant defensive reactions to pathogens (van Loon et al. 2006) and herbivores (Xu et al. 1994; von Dahl and Baldwin 2007). For example, ethylene perception in the pathogen-infected leaf is required for the expression of systemic pathogen resistance (Verberne et al. 2003). Ethylene also augmented induced volatile production of maize upon exposure to *cis*-3-hexenol, but exposure to ethylene alone had no effect (Ruther and Kleier 2005). Apparently, ethylene increases the plant's response to GLVs but does not serve as a primary signal. Apart from MeSA, MeJA, ethylene and GLVs, *cis*-jasmone can trigger defensive responses via airborne transport. However, this herbivore-induced volatile activates different sets of genes than MeJA (Birkett et al. 2000; Bruce et al. 2007), which suggests a different mode of action. Correspondingly, *cis*-jasmone failed to induce extrafloral nectar secretion in lima bean, a JA-responsive trait that can be elicited by *cis*-3-hexenyl acetate (Kost and Heil 2006).

3.2 The Unknown Receptor: Where Do Plants Keep Their Noses?

Plants perceive various VOCs that are released from their neighbors and respond with changes in gene expression to augment their resistance to pathogens or herbivores. Plants have odors, and the components of these odors have been investigated in detail, but how do plants smell? Where should we search for the "noses" of plants? Elucidating the mechanisms by which plants perceive volatile signals is a major challenge for future research.

Both MeSA and MeJA can be converted back into the respective active plant hormone, a physiological mechanism that provides an obvious explanation for plant responses to these two particular volatiles. In contrast, little is known about signaling responses to GLVs. The high structural diversity of resistance-inducing GLVs makes a specific, receptor protein-mediated mechanism of perception unlikely. Although it has been suggested that GLVs with an α,β-unsaturated carbonyl groups can trigger plant defense due to their activities as reactive electrophile species (Almeras et al. 2003), other biologically active GLVs lack this motif (Ruther and Kleier 2005; Kost and Heil 2006; Heil et al. 2008). Changes in transmembrane potentials are involved in early signaling events in the cellular response to stress (Maffei et al. 2007), and exposure to GLVs such as *cis*-hexenyl acetate changed membrane potentials in intact lima bean leaves (M. Maffei, pers. commun.). It is therefore tempting to speculate that the dissolution of volatiles in the membranes leads to changes in transmembrane potentials or somehow disintegrates the membrane and thereby induces gene activity. However, much more research will be required before we gain an understanding of the mechanisms by which plants perceive GLVs and other resistance-related plant odors.

3.3 *Far-Red-Mediated Perception of Neighboring Plants*

While we are still searching for the plant olfaction system, other receptors that plants use to sense their neighbors are well known. Light that has been transmitted through or reflected from plant surfaces has a higher ratio of far red to red light than full sunlight (Smith 2000), and plants have evolved specific photoreceptors (phytochrome B) that act as far-red sensing systems (Ballaré and Scopel 1997) which allow them to detect the presence of putative competitors. In response, they shape their morphology and future growth accordingly (Fig. 2). For example, far-red

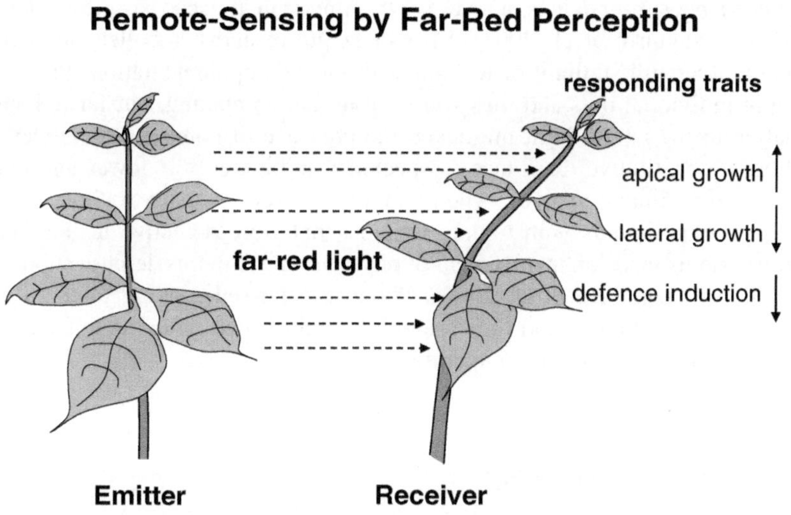

Fig. 2 Remote sensing by far-red reception. Light that is reflected by or transmitted through plant tissue has a higher far-red content than full sunlight, and phytochrome B and other receptors in plants can sense this change. Far-red light means the presence of putative competitors, particularly when it is perceived by vertical plant structures (and hence when it comes from the side). Common responses to this include increased apical growth at the expense of lateral growth and "shade avoidance:" growth away from the far-red source usually enables plants to grow into more open spaces. As resistance expression is costly, native South American tobacco has even been reported to reduce its levels of herbivore resistance when receiving lateral far-red light; i.e., when it apparently is, or soon will be, exposed to intensive competition by neighbors

sensing leads to stronger apical growth at the expense of lateral shoot production and thus helps plants to overgrow their competitors (Ballaré 1999). Although the emission of far-red light from plant tissues is not likely to be very plastic and thus violates the second part of Karban's (2008) definition of communication, far-red perception mediates highly sophisticated information transfer events among plants and triggers apparently adaptive responses.

Competition for light, water, space and nutrients is a central issue in the lives of plants. Successfully sensing a future competitor even before any shortage in these resources actually limits growth may therefore have significantly beneficial effects (Ballaré 1999). As well as future growth, resource shortages eventually limit reproduction. Resistance traits are costly (Agrawal et al. 1999; Agrawal 2000; Heil 2002; Heil and Baldwin 2002), and plants therefore suffer from a "growth–differentiation" dilemma, i.e., they can invest their limited resources in either growth or defense, but not both (Herms and Mattson 1992). As a consequence, the level of resistance that is expressed in response to a defined induction event (Cipollini and Bergelson 2001; Dietrich et al. 2004) and the net costs of resistance induction (Heil et al. 2000; Cipollini 2002; Cipollini et al. 2003; Dietrich et al. 2005) can depend on resource availability and competition.

Due to these constraints, plants may obtain fitness benefits by reducing their defense investments in situations when future competition is likely (Cipollini 2004). Surprisingly enough, a connection between far-red sensing and defense induction has indeed been found in a native South American tobacco species, *Nicotiana longiflora* (Izaguirre et al. 2006). Miriam Izaguirre and her colleagues exposed plants to either full sunlight or to light with far-red supplementation. Plants were grown in individual pots and thus were not in fact competing, but far-red supplementation to the lateral light mimicked the presence of competitors. Under these conditions, constitutive resistance to specialist herbivores was lower and defense induction was impaired even when the plants were actually being damaged. Additional experiments with tomato mutants that were defective in their far-red sensory systems made an involvement of phytochrome B in this defense suppression highly likely (Izaguirre et al. 2006). Plants use far-red sensing to monitor the presence of other plants, and they are able to adjust their actual defensive efforts according to the presence of competitors.

3.4 Airborne Allelopathy

A receiver that responds when its neighbor is attacked by immobile enemies invests in a resistance that it never needs. VOC release may thus principally inhibit the growth of neighboring plants. In fact, native annuals germinated at lower rates when seeded beneath clipped sagebrush plants, and additional experiments made an involvement of VOCs in this phenomenon likely (Karban 2007). The underlying physiological mechanisms remain unknown. However, mechanically clipped sagebrush emits high amounts of MeJA (Karban et al. 2000), and jasmonates can have tremendously negative effects on anabolism and may even induce senescence;

in fact, JA was originally described as a "growth inhibitor" (Creelman and Mullet 1997). Similarly, *cis*-jasmonate was said to be an alleopathic agent (Pickett et al. 2007). It is thus tempting to speculate that allelopathy by herbivore-induced VOCs might simply result from exposing neighbors to an overdose of jasmonates.

4 Ecological and Evolutionary Considerations

Communication requires an emitter and a receiver, but the net effects of this interaction can differ dramatically among the interacting partners. While communication among animals often benefits both sender and receiver, the few scattered reports in which effects have been considered at all lead to the interpretation that airborne plant–plant communication is much more unidirectional: it benefits the receiver at the cost of the sender or, less commonly, the sender at the cost of the receiver. Moreover, most experimental designs have to some extent violated the prerequisite of ecological realism. Does plant–plant communication have any ecological relevance in nature, and how does it affect the partners involved? The following paragraphs present the little information on this topic that exists to date, and discusses some of the questions that are still open to debate.

4.1 Does It Actually Exist?

As mentioned above, the first reports on VOC-mediated plant–plant communication have been criticized for a lack of true controls or for missing ecological relevance. Since then, not that much has changed, unfortunately: the majority of studies are still being conducted under laboratory or greenhouse conditions (without natural air movements that might dilute the cues and without growing the plants in mixed stands). Most field studies have used manual clipping treatments and so probably did not work with quantitatively and qualitatively realistic VOC bouquets. However, recent studies on lima bean conducted at the plant's native area in the coastal region of Southern México have used beetle-damaged emitter shoots and found reduced herbivore damage on receivers (Heil and Silva Bueno 2007b). In other words, plant–plant communication can indeed occur under ecologically realistic conditions!

Heil and Silva Bueno (2007b) did, however, intertwine senders and receivers to mimic the natural growth of lima bean, a liana. Similarly, most other studies on plant–plant communication have searched for effects in plants growing very close to the emitter. Resistance induction was found in wild tobacco plants growing 15 cm downwind from clipped sagebrushes (Karban et al. 2000) and in black alders growing at a distance of 1 m from clipped trees (Tscharntke et al. 2001). VOC-mediated allelopathic effects occurred when receivers were seeded directly underneath the clipped sagebrushes (Karban 2007).

Distance is a crucial parameter in plant–plant communication, and positive reports are restricted to plants that were grown very close to the emitter. Systematically investigating the maximum distance over which cues can be exchanged is important but is a difficult task: volatiles diffuse in the air and move by eddy current dispersal, so the distances over which they can affect other plants thus depend strongly on wind speed, air humidity and temperature.

Although generalizations have proven impossible so far, it appears safe to assume that VOC-mediated plant–plant communication only functions over short distances. This would also solve the old question that arose from the observation of VOC-induced production of VOCs: how do plants avoid endless autoinduction? When present at lower doses, VOCs usually prime rather than fully induce resistance responses (Engelberth et al. 2004; Heil and Kost 2006; Kessler et al. 2006; Frost et al. 2007, 2008; Heil and Silva Bueno 2007b; Ton et al. 2007). Due to the rapid diffusion that occurs under natural conditions, it is plausible that resistance-inducing volatiles are normally diluted to priming or—at larger distances—completely inactive concentrations.

4.2 Evolutionary Considerations

Although at least some mechanistic aspects of airborne resistance induction are now well understood at the physiological and even the genetic levels, our knowledge about the fitness effects on both partners involved is surprisingly restricted, and information on the evolutionary consequences of plant–plant communication as well as on its evolutionary origins is apparently lacking. Surprisingly enough, it appears that this interaction usually benefits the receiver, and normally even at the cost of the emitter. Most plants use the information on the presence of (damaged) neighbors to adapt their growth and defensive phenotype according to current environmental conditions (i.e., enemy pressure and competition). Airborne plant–plant communication thus benefits only the receiver in most cases. Since plants usually compete with each other for light, space, water and nutrients, it can even be expected that this communication has detrimental effects on the emitter, which is already damaged.

Why should plants warn their neighbors that enemies are around? And why are there no mutualistic forms of plant–plant communication? The most widely appreciated cases of mutualistic communication of plants with other organisms are the signals that flowers emit to attract their pollinators. Plants can communicate to enable or stabilize mutualisms, so why have most of the described cases of plant–plant communication indicated detrimental effects on one of the partners, and why is it usually the emitter of the cues that suffers? Are plants egoistic, and if they are, why don't they simply stop emitting the active cues?

4.2.1 Why Are There No Mutualistic Forms of Plant–Plant Communication?

Although there is no definitive answer to this question, the most likely explanation for a lack of mutually beneficial plant–plant communication appears to be that plants do not have too many positive messages for each other. In fact, plants usually establish mutualisms with organisms from other kingdoms, but hardly ever with other plants. Mutualisms aid the exchange of resources and services that one partner can provide easily and that is difficult for the other partner to produce/achieve (Bronstein 1994). Plants are sessile, autotrophic organisms; the normal mutualisms of plants are thus established either with highly mobile animals (which are then in charge of the transport of pollen or seeds, or which indirectly defend the plant) or with physiologically very different microorganisms (such as N-fixing *Rhizobia* or mycorrhizal fungi). The normal interaction of a plant with other plants, in contrast, is competition. The same remains true for interactions within a single species: while many animals cooperate with each other, social cooperation is almost absent from the plant kingdom. It appears that plants are too similar to each other to establish mutually beneficial interactions. Plants mainly interact negatively rather than positively, and this pattern shows up in plant–plant communication as well.

4.2.2 Since Emitters Usually Suffer Due to Communication, Why Don't They Stop Emitting Cues?

The most likely answer to this question is that plants cannot simply avoid emitting the cues that other plants use as the source of information. Even the emission of those VOCs that most commonly induce resistance in neighbors appears to come with so many beneficial effects that it cannot be ceased easily. VOCs have direct inhibitory effects on microbes (Nakamura and Hatanaka 2002; Dilantha Fernando et al. 2005; Matsui 2006; Shiojiri et al. 2006) and they serve to attract the third trophic level in the context of indirect defense (Dicke 1986; Dicke et al. 1990; Turlings et al. 1990; Tumlinson et al. 1999; Heil 2008). Moreover, as predicted by Edward Farmer and Colin Orians (Farmer 2001; Orians 2005), VOCs serve as hormones and mediate signaling among different parts of the same individual plant (Karban et al. 2006; Frost et al. 2007; Heil and Silva Bueno 2007b).

Compared to signaling via the vascular system, airborne within-plant signaling is faster and independent of orthostichy, and airborne signals can move independently, unlike the unidirectional flow in phloem and xylem (Heil and Ton 2008). Herbivores and pathogens are highly mobile and do not necessarily move according to plant anatomy. Particularly in anatomically complex plants such as lianae (lima bean) or shrubs (sagebrush), an internal signal would be less efficient than an airborne signal, since leaves that are very close spatially may be connected to different shoots and may therefore be separated by an anatomical distance of several meters. VOCs can serve as a cue to trigger defense responses in exactly those parts of a plant where resistance is actually required: in the parts that are spatial (but not necessarily anatomical) neighbors (Heil and Silva Bueno 2007a).

Finally, even self-priming by herbivore-induced volatiles has been described in the context of airborne within-plant signaling (Frost et al. 2007; Heil and Silva Bueno 2007b). This observation led to the prediction of a two-step regulatory system in which airborne signals prime distal tissues to respond more efficiently to vascular signals or direct attack (Heil and Ton 2008). VOC-exposed tissues, then, would only respond with full resistance expression when the confirming vascular signal reaches the distal parts of a locally damaged plant or when true herbivore damage occurs. This allows plants to achieve additional fine-tuning of the systemic resistance response. VOC-mediated long-distance signaling within plants can facilitate detailed tailoring of plant systemic responses to local damage and can be expected to be the rule rather than the exception. Plants need to be able to both emit and perceive VOCs, and just cannot completely avoid the dangers of "eavesdropping" neighbors. However, dosage-dependent effects should strongly reduce this putative ecological cost of external signaling. Volatile-mediated signaling works over short distances, and the probability that the leaf nearest to an attacked one belongs to the same plant is high. As a result, the chance that eavesdropping by competing plants will become a significant problem remains relatively low (Heil and Ton 2008).

5 Conclusions

Plants emit volatile organic compounds that transport detailed information on their status of attack, and they reflect high amounts of far-red light, which signals their mere presence. Both types of information are perceived and used by neighboring plants to adjust their growth rate, morphology or resistance phenotype accordingly. VOCs that are released in response to herbivore feeding or pathogen infection are controlled by the emitting plant, and airborne plant–plant signaling thus fulfills all the requirements of being a true form of communication. The evolutionary origins of this phenomenon appear to reside in the internal functions that VOCs fulfill as hormones in systemic resistance induction. To perform this function, plants need all of the traits that are required for the production and emission of VOCs as well as for their reception, and communication among different individuals is likely an inevitable by-product of within-plant signaling "worn on the outside."

Most plant–plant signaling events aid the receiver, even at the cost of the emitter, with the only exception being VOC-mediated allelopathy, which supposedly benefits the emitter at the cost of the receivers. Unfortunately, no true generalizations have been elucidated as yet, since most studies on plant–plant communication have been conducted under highly controlled rather than ecologically realistic conditions, and since fitness effects on the partners have apparently never been considered. Even the question of how far a plant can be from the emitter and still perceive its signal has never been investigated.

Plant–plant communication exists in nature. The plant species for which the phenomenon has been described comprise both monocotyledons, such as corn, and several distantly related dicotyledons, and thus the phenomenon can be regarded as being taxonomically common. However, much more must be done before we can determine how common plant–plant communication is under realistic conditions, how it usually affects the sender and the receiver of the information, and how it ultimately shapes the structure of plant communities and the evolution of plant species.

Acknowledgements I thank Dale Walters and Carlos Ballaré for helpful comments on earlier versions of this manuscript, and CONACyT (Consejo Nacional de Ciencia y Tecnología de México) for financial support.

References

Agrawal AA (2000) Benefits and costs of induced plant defense for *Lepidium virginicum* (Brassicaceae). Ecology 81:1804–181

Agrawal AA, Strauss SY, Stout MJ (1999) Costs of induced responses and tolerance to herbivory in male and female fitness components of wild radish. Evolution 53:1093–1104

Almeras E, Stolz S, Vollenweider S, Reymond P, Mene-Saffrane L, Farmer EE (2003) Reactive electrophile species activate defense gene expression in *Arabidopsis*. Plant J 34:202–216

Arimura G-I, Ozawa R, Shimoda T, Nishioka T, Boland W, Takabayashi J (2000) Herbivory-induced volatiles elicit defence genes in lima bean leaves. Nature 406:512–515

Baldwin IT, Schultz JC (1983) Rapid changes in tree leaf chemistry induced by damage: evidence for communication between plants. Science 221:277–279

Ballaré CL (1999) Keeping up with the neighbours: phytochrome sensing and other signalling mechanisms. Trends Plant Sci 4:97–102

Ballaré CL, Scopel AL (1997) Phytochrome signalling in plant canopies: testing its population-level implications with photoreceptor mutants of *Arabidopsis*. Funct Ecol 11:441–450

Birkett MA, Campbell CAM, Chamberlain K, Guerrieri E, Hick AJ, Martin JL, Matthes M, Napier JA, Pettersson J, Pickett JA, Poppy GM, Pow EM, Pye BJ, Smart LE, Wadhams GH, Wadhams LJ, Woodcock CM (2000) New roles for *cis*-jasmone as an insect semiochemical and in plant defense. Proc Natl Acad Sci USA 97:9329–9334

Bronstein JL (1994) Our current understanding of mutualism. Quart Rev Biol 69:31–51

Bruce TJA, Matthes MC, Napier JA, Pickett JA (2007) Stressful "memories" of plants: evidence and possible mechanisms. Plant Sci 173:603–608

Choh Y, Takabayashi J (2006) Herbivore-induced extrafloral nectar production in lima bean plants enhanced by previous exposure to volatiles from infested conspecifics. J Chem Ecol 32:2073–2077

Choh Y, Kugimiya S, Takabayashi J (2006) Induced production of extrafloral nectar in intact lima bean plants in response to volatiles from spider mite-infested conspecific plants as a possible indirect defense against spider mites. Oecologia 147:455–460

Cipollini DF (2002) Does competition magnify the fitness costs of induced resistance in *Arabidopsis thaliana*? A manipulative approach. Oecologia 131:514–520

Cipollini DF (2004) Stretching the limits of plasticity: can a plant defend against both competitors and herbivores? Ecology 85:28–37

Cipollini DF, Bergelson J (2001) Plant density and nutrient availability constrain constitutive and wound-induced expression of trypsin inhibitors in *Brassica napus*. J Chem Ecol 27:593–610

Cipollini DF, Purrington CB, Bergelson J (2003) Costs of induced responses in plants. Basic Appl Ecol 4:79–89

Conrath U, Beckers GJM, Flors V, Garcia-Agustin P, Jakab G, Mauch F, Newman MA, Pieterse CMJ, Poinssot B, Pozo MJ, Pugin A, Schaffrath U, Ton J, Wendehenne D, Zimmerli L, Mauch-Mani B (2006) Priming: getting ready for battle. Mol Plant Microbe Interact 19:1062–1071

Constabel CP, Bergey DR, Ryan CA (1995) Systemin activates synthesis of wound-inducible tomato leaf polyphenol oxidase via the octadecanoid defense signaling pathway. Proc Natl Acad Sci USA 92:407–411

Creelman RA, Mullet JE (1997) Biosynthesis and action of jasmonates in plants. Annu Rev Plant Physiol Plant Mol Pathol 48:355–381

Dicke M (1986) Volatile spider-mite pheromone and host-plant kairomone, involved in spaced-out gregariousness in the spider mite *Tetranychus urticae*. Physiol Entomol 11:251–262

Dicke M, Dijkman H (1992) Induced defence in detached uninfested plant-leaves: effects on behavior of herbivores and their predators. Oecologia 91:554–560

Dicke M, van Beek TA, Posthumus MA, Ben Dom N, van Bokhoven H, de Groot AE (1990) Isolation and identification of volatile kairomone that affects *Acarine* predator–prey interactions. Involvement of the host plant in its production. J Chem Ecol 16:381–395

Dietrich R, Ploss K, Heil M (2004) Constitutive and induced resistance to pathogens in *Arabidopsis thaliana* depend on nitrogen supply. Plant Cell Environ 27:896–906

Dietrich R, Ploss K, Heil M (2005) Growth responses and fitness costs after induction of pathogen resistance depend on environmental conditions. Plant Cell Environ 28:211–222

Dilantha Fernando WG, Ramarathnam R, Krishnamoorthy AS, Savchuk SC (2005) Identification and use of potential bacterial organic antifungal volatiles in biocontrol. Soil Biol Biochem 37:955–964

Durrant WE, Dong X (2004) Systemic acquired resistance. Annu Rev Phytopathol 42:185–209

Ellis C, Turner JG (2001) The *Arabidopsis* mutant cev1 has constitutively active jasmonate and ethylene signal pathways and enhanced resistance to pathogens. Plant Cell 13:1025–1033

Engelberth J, Alborn HT, Schmelz EA, Tumlinson JH (2004) Airborne signals prime plants against insect herbivore attack. Proc Natl Acad Sci USA 101:1781–1785

Farag MA, Fokar M, Zhang HA, Allen RD, Paré PW (2005) (Z)-3-Hexenol induces defense genes and downstream metabolites in maize. Planta 220:900–909

Farmer EE (2001) Surface-to-air signals. Nature 411:854–856

Farmer EE, Ryan CA (1990) Interplant communication: airborne methyl jasmonate induces synthesis of proteinase inhibitors in plant leaves. Proc Natl Acad Sci USA 87:7713–7716

Forouhar F, Yang Y, Kumar D, Chen Y, Fridman E, Park SW, Chiang Y, Acton TB, Montelione GT, Pichersky E, Klessig DF, Tong L (2005) Structural and biochemical studies identify tobacco SABP2 as a methyl salicylate esterase and implicate it in plant innate immunity. Proc Natl Acad Sci USA 102:1773–1778

Frost C, Appel H, Carlson J, De Moraes C, Mescher M, Schultz J (2007) Within-plant signalling by volatiles overcomes vascular constraints on systemic signalling and primes responses against herbivores. Ecol Lett 10:490–498

Frost C, Mescher MC, Carlson J, De Moraes CM (2008) Plant defense priming against herbivores: getting ready for a different battle. Plant Physiol 146:818–824

Godard KA, White R, Bohlmann J (2008) Monoterpene-induced molecular responses in *Arabidopsis thaliana*. Phytochemistry 69:1838–1849

Goellner K, Conrath U (2008) Priming: it's all the world to induced disease resistance. Eur J Plant Pathol 121:233–242

Heil M (2002) Ecological costs of induced resistance. Curr Opin Plant Biol 5:345–350

Heil M (2008) Indirect defence via tritrophic interactions. New Phytol 178:41–61

Heil M, Baldwin IT (2002) Fitness costs of induced resistance: emerging experimental support for a slippery concept. Trends Plant Sci 7:61–67

Heil M, Kost C (2006) Priming of indirect defences. Ecol Lett 9:813–817

Heil M, Silva Bueno JC (2007a) Herbivore-induced volatiles as rapid signals in systemic plant responses. Plant Signal Behav 2:191–193

Heil M, Silva Bueno JC (2007b) Within-plant signaling by volatiles leads to induction and priming of an indirect plant defense in nature. Proc Natl Acad Sci USA 104:5467–5472

Heil M, Ton J (2008) Long-distance signalling in plant defence. Trends Plant Sci 13:264–272

Heil M, Hilpert A, Kaiser W, Linsenmair KE (2000) Reduced growth and seed set following chemical induction of pathogen defence: does systemic acquired resistance (SAR) incur allocation costs? J Ecol 88:645–654

Heil M, Lion U, Boland W (2008) Defence-inducing volatiles: in search for the active motif. J Chem Ecol 34:601–604

Herms DA, Mattson WJ (1992) The dilemma of plants: to grow or to defend. Quart Rev Biol 67:283–335

Izaguirre MM, Mazza CA, Biondini M, Baldwin IT, Ballaré CL (2006) Remote sensing of future competitors: impacts on plant defenses. Proc Natl Acad Sci USA 103:7170–7174

Karban R (2001) Communication between sagebrush and wild tobacco in the field. Biochem Syst Ecol 29:995–1005

Karban R (2007) Experimental clipping of sagebrush inhibits seed germination of neighbours. Ecol Lett 10:791–797

Karban R (2008) Plant behaviour and communication. Ecol Lett 11:727–739

Karban R, Baldwin IT (1997) Induced responses to herbivory. University of Chicago Press, Chicago

Karban R, Baldwin I, Baxter K, Laue G, Felton G (2000) Communication between plants: induced resistance in wild tobacco plants following clipping of neighboring sagebrush. Oecologia 125:66–71

Karban R, Shiojiri K, Huntzinger M, McCall AC (2006) Damage-induced resistance in sagebrush: volatiles are key to intra- and interplant communication. Ecology 87:922–930

Kessler A, Halitschke R, Diezel C, Baldwin IT (2006) Priming of plant defense responses in nature by airborne signaling between *Artemisia tridentata* and *Nicotiana attenuata*. Oecologia 148:280–292

Kishimoto K, Matsui K, Ozawa R, Takabayashi J (2005) Volatile C6-aldehydes and allo-ocimene activate defense genes and induce resistance against *Botrytis cinerea* in *Arabidopsis thaliana*. Plant Cell Physiol 46:1093–1102

Kost C, Heil M (2006) Herbivore-induced plant volatiles induce an indirect defence in neighbouring plants. J Ecol 94:619–628

Kumar D, Klessig DF (2003) High-affinity salicylic acid-binding protein 2 is required for plant innate immunity and has salicylic acid-stimulated lipase activity. Proc Natl Acad Sci USA 100:16101–16106

Le Hir R, Beneteau J, Bellini C, Vilaine F, Dinant S (2008) Gene expression profiling: keys for investigating phloem functions. Trends Plant Sci 13:273–280

Maffei ME, Mithöfer A, Boland W (2007) Before gene expression: early events in plant-insect interaction. Trends Plant Sci 12:310–316

Matsui K (2006) Green leaf volatiles: hydroperoxide lyase pathway of oxylipin metabolism. Curr Opin Plant Biol 9:274–280

Métraux J-P, Signer H, Ryals J, Ward E, Wyss-Benz M, Gaudin J, Raschdorf K, Schmid E, Blum W, Inverardi B (1990) Increase in salicylic acid at the onset of systemic acquired resistance. Science 250:1004–1006

Nakamura S, Hatanaka A (2002) Green-leaf-derived C6-aroma compounds with potent antibacterial action that act on both gram-negative and gram-positive bacteria. J Agric Food Chem 50:7639–7644

Orians C (2005) Herbivores, vascular pathways, and systemic induction: facts and artifacts. J Chem Ecol 31:2231–2242

Park SW, Kaimoyo E, Kumar D, Mosher S, Klessig DF (2007) Methyl salicylate is a critical mobile signal for plant systemic acquired resistance. Science 318:113–116

Paschold A, Halitschke R, Baldwin IT (2006) Using 'mute' plants to translate volatile signals. Plant J 45:275–291

Pickett JA, Poppy GM (2001) Switching on plant genes by external chemical signals. Trends Plant Sci 6:137–139

Pickett JA, Birkett MA, Moraes MCB, Bruce TJA, Chamberlain K, Gordon-Weeks R, Matthes MC, Napier JA, Smart LE, Wadhams LJ, Woodcock CM (2007) *cis*-Jasmone as allelopathic agent in inducing plant defence. Allelopathy J 19:109–117

Pieterse CMJ, van Wees SCM, van Pelt JA, Knoester M, Laan R, Gerrits N, Weisbeek PJ, van Loon LC (1998) A novel signaling pathway controlling induced systemic resistance in Arabidopsis. Plant Cell 10:1571–1580

Rhoades DF (1983) Responses of alder and willdow to attack by tent caterpillars and webworms: evidence for pheromonal sensitivity of willows. In: Hedin PA (ed) Plant resistance to insects. American Chemical Society, Washington, 55–68

Ruther J, Fürstenau B (2005) Emission of herbivore-induced volatiles in absence of a herbivore. Response of *Zea mays* to green leaf volatiles and terpenoids. Zeitschr Naturforsch C 60:743–756

Ruther J, Kleier S (2005) Plant-plant signaling: ethylene synergizes volatile emission in *Zea mays* induced by exposure to (Z)-3-Hexen-1-ol. J Chem Ecol 31:2217–2222

Schmelz EA, Alborn HT, Tumlinson JH (2003) Synergistic interactions between volicitin, jasmonic acid and ethylene mediate insect-induced volatile emission in *Zea mays*. Physiol Plant 117:403–412

Shiojiri K, Kishimoto K, Ozawa R, Kugimiya S, Urashimo S, Arimura G, Horiuchi J, Nishioka T, Matsui K, Takabayashi J (2006) Changing green leaf volatile biosynthesis in plants: an approach for improving plant resistance against both herbivores and pathogens. Proc Natl Acad Sci USA 103:16672–16676

Shulaev V, Silverman P, Raskin I (1997) Airborne signalling by methyl salicylate in plant pathogen resistance. Nature 385:718–721

Smith H (2000) Phytochromes and light signal perception by plants—an emerging synthesis. Nature 407:585–591

Starck Z (2006) Role of conducting systems in the transduction of long-distance stress signals. Acta Physiol Plant 28:289–301

Sticher L, Mauch-Mani B, Métraux J-P (1997) Systemic acquired resistance. Annu Rev Phytopathol 35:235–270

Thorpe MR, Ferrieri AP, Herth MM, Ferrieri RA (2007) ^{11}C-imaging: methyl jasmonate moves in both phloem and xylem, promotes transport of jasmonate, and of photoassimilate even after proton transport is decoupled. Planta 226:541–551

Ton J, D'Allesandro M, Jourdie V, Jakab G, Karlen D, Held M, Mauch-Mani B, Turlings TCJ (2007) Priming by airborne signals boosts direct and indirect resistance in maize. Plant J 49:16–26

Truman W, Bennettt MH, Kubigsteltig I, Turnbull C, Grant M (2007) Arabidopsis systemic immunity uses conserved defense signaling pathways and is mediated by jasmonates. Proc Natl Acad Sci USA 104:1075–1080

Tscharntke T, Thiessen S, Dolch R, Boland W (2001) Herbivory, induced resistance, and interplant signal transfer in *Alnus glutinosa*. Biochem Syst Ecol 29:1025–1047

Tumlinson JH, Paré PW, Turlings TCJ (1999) Plant production of volatile semiochemicals in response to insect-derived elicitors. In: Chadwick DJ, Goode JA (eds) Insect–plant interactions and induced plant defence. Wiley,Chichester 95–109

Turlings TCJ, Wäckers FL (2004) Recruitment of predators and parasitoids by herbivore-injured plants. In: Cardés RT, Millar JG (eds) Advances in insect chemical ecology. Cambridge University Press Cambridge 21–75

Turlings TCJ, Tumlinson JH, Lewis WJ (1990) Exploitation of herbivore-induced plant odors by host -seeking parasitic wasps. Science 250:1251–1253

van Hulten M, Pelser M, van Loon LC, Pieterse CMJ, Ton J (2006) Costs and benefits of priming for defense in *Arabidopsis*. Proc Natl Acad Sci USA 103:5602–5607

van Loon LC, Geraats BPJ, Linthorst HJM (2006) Ethylene as a modulator of disease resistance in plants. Trends Plant Sci 11:184–191

Verberne MC, Hoekstra J, Bol JF, Linthorst HJM (2003) Signaling of systemic acquired resistance in tobacco depends on ethylene perception. Plant J 35:27–32

Voelckel C, Schittko U, Baldwin IT (2001) Herbivore-induced ethylene burst reduces fitness costs of jasmonate- and oral secretion-induced defenses in *Nicotiana attenuata*. Oecologia 127:274–280

von Dahl CC, Baldwin IT (2007) Deciphering the role of ethylene in plant–herbivore interactions. J Plant Growth Regul 26:201–209

Walling LL (2000) The myriad plant responses to herbivores. J Plant Growth Regul 19:195–216

Walters DR, Cowley T, Weber H (2006) Rapid accumulation of trihydroxy oxylipins and resistance to the bean rust pathogen *Uromyces fabae* following wounding in *Vicia faba*. Ann Bot 97:779–784

Wasternack C (2007) Jasmonates: an update on biosynthesis, signal transduction and action in plant stress response, growth and development. Ann Bot 100:681–697

Xu Y, Chang PFL, Liu D, Narasimhan ML, Raghothama KG, Hasegawa PM, Bressan RA (1994) Plant defense genes are synergistically induced by ethylene and methyl jasmonate. Plant Cell 6:1077–1085

Zhang Z-P, Baldwin IT (1997) Transport of [2-^{14}C]jasmonic acid from leaves to roots mimics wound-induced changes in endogenous jasmonic acid pools in *Nicotiana sylvestris*. Planta 203:436–441

Chemical Signaling During Induced Leaf Movements

Minoru Ueda and Yoko Nakamura

Abstract Chemical aspects of the circadian leaf movement known as nyctinasty are discussed in this chapter. Each nyctinastic plant from the five different genera examined so far contained a pair of factors, one of which induces leaf closure while the other induces leaf opening. Changes in the relative contents of the closing and opening factors correlated with nyctinastic leaf movement. The use of fluorescence-labeled and photoaffinity-labeled factors revealed that the leaf-closing factor binds to a 38-kDa membrane protein of motor cells.

1 Introduction

In general, plants are rooted and are unable to move from place to place by themselves. However, some plants are able to move in certain ways. Leguminous plants are known to open their leaves in daytime and to "sleep" at night with their leaves folded (Fig. 1). This leaf movement follows a circadian rhythm and is regulated by a biological clock with a cycle of about 24h. This phenomenon, known as nyctinasty, has been of great interest to scientists for centuries, with the oldest records dating from the time of Alexander the Great (Kirchner 1874).

It was Charles Darwin, well known for his theory of evolution, who established the science of plant movement in his later years. In 1880, Darwin published an invaluable book entitled *The Power of Movement in Plants*, which was based on experiments using more than 300 different kinds of plants (Darwin 1875). However, despite the advances in science that have been made since Darwin's time, it is still difficult to establish the molecular basis for these processes. Our study focused on the chemical mechanisms of Darwin's observations.

M. Ueda (✉) and Y. Nakamura
Laboratory of Organic Chemistry, Department of Chemistry,
Tohoku University, Japan
e-mail: ueda@mail.tains.tohoku.ac.jp

F. Baluška (ed.), *Plant-Environment Interactions*, Signaling and Communication in Plants, 153
DOI: 10.1007/978-3-540-89230-4_9, © Springer-Verlag Berlin Heidelberg 2009

Fig. 1 Nyctinastic leaf movement of *Albizzia saman* (*Samanea saman*)

The physiological mechanism of nyctinasty has been investigated extensively by Satter et al. (1981, 1990), Moran (2007a, b), Moshelion et al. (2000, 2002), etc., on a leguminous plant, *Albizzia saman* (*Samanea saman*) (Lee 1990). Nyctinastic leaf movement is induced by the swelling and shrinking of motor cells in the pulvinus, a joint-like thickening located at the base of the petiole. Motor cells play a key role in plant leaf movement. A flux of potassium ions across the plasma membranes of the motor cells is followed by a massive flux of water, which results in the swelling or shrinking of these cells (Satter et al. 1981). An issue of great interest is the circadian rhythmic regulation of the opening and closing of the potassium channels involved in nyctinastic leaf movement. Chemical studies on nyctinasty have also been carried out, and many attempts have been made to isolate the endogenous factors that are involved in the control of nyctinasty (Schildknecht 1983).

2 Leaf-Closing and -Opening Substances in Nyctinastic Plants

Nyctinastic plants contain two types of endogenous bioactive substances: leaf-opening and leaf-closing factors, which possibly mediate nyctinastic leaf movement (Ueda and Yamamura 2000c; Ueda and Nakamura 2006). When the leaves of a leguminous plant were separated from its stem, their leaflets continued to move according to the circadian rhythm: they were open in the daytime and closed at night (Fig. 2). To date, we have identified five sets of leaf-closing and leaf-opening factors (**1–10**) in five nyctinastic plant species (Fig. 3) (Miyoshi et al. 1987; Shigemori et al. 1989; Ueda et al. 1995, 1997a, b, 1998a, b, 1999a, b, c, 2000a). All of these factors were effective at concentrations of 10^{-5} to 10^{-6}M when applied exogenously. This bioactivity is similar to those of known phytohormones such as IAA, gibberellin, etc. It was also shown that each nyctinastic plant uses a specific set of leaf-movement factors,

a b

Fig. 2a–b Bioassays of leaf movement factors: **a** leaf-closing factor causes the plant leaf to fold; **b** leaf-opening factor causes the plant leaf to open

and that the set of factors are conserved within the same genus. None of the factors were effective in the plant belonging to other genuses, even at concentrations that were 10,000- 100,000 -fold higher than normal (Yamamura and Ueda 2000 ; Ueda and Nakamura 2006). For example, 1 is effective as a leaf-opening factor for Albizzia julibrissin Durazz. at 10^{-5} M, but it was not effective for other genera, such as Mimosa, Cassia and Phyllanthus, even at 10^{-1} M (Ueda et al 2000a). These findings suggest that nyctinasty is controlled by genus-specific chemical factors (Ueda et al. 2000b).

3 Bioorganic Studies of Nyctinasty Using Functionalized Leaf-Movement Factors as Molecular Probes

3.1 Leaf Movement Factors for the Genus Albizzia

Most of the physiological studies on nyctinasty have been carried out in plants belonging to the genus *Albizzia* (Satter et al.1981, 1990; Moran 2007b; Moshelion et al. 2002). Considering that each nyctinastic plant has a pair of leaf-movement factors whose bioactivities are specific to the plant genus (Ueda et al. 2000b), bioorganic studies of nyctinasty performed using *Albizzia* plants are important. We revealed that **1** (a closing factor) (Ueda et al. 2000) and **2** (an opening factor) (Ueda et al. 1997) are common leaf-movement factors among three *Albizzia* plants; furthermore, **1** and **2** were shown to be ineffective for plants belonging to other plant genera, such as *Mimosa*, *Phyllanthus*, *Cassia*, etc (Ueda et al. 1997a, 2000b). We focused on the mode of action of **1** in *A. saman* in order to study the bioorganic chemistry of nyctinasty. Synthetic molecular probes that are designed to mimic **1**, such as fluorescence-labeled **1** and photoaffinity-labeled **1**, provide powerful tools for such studies (Kotzyba-Hilbert et al. 1995).

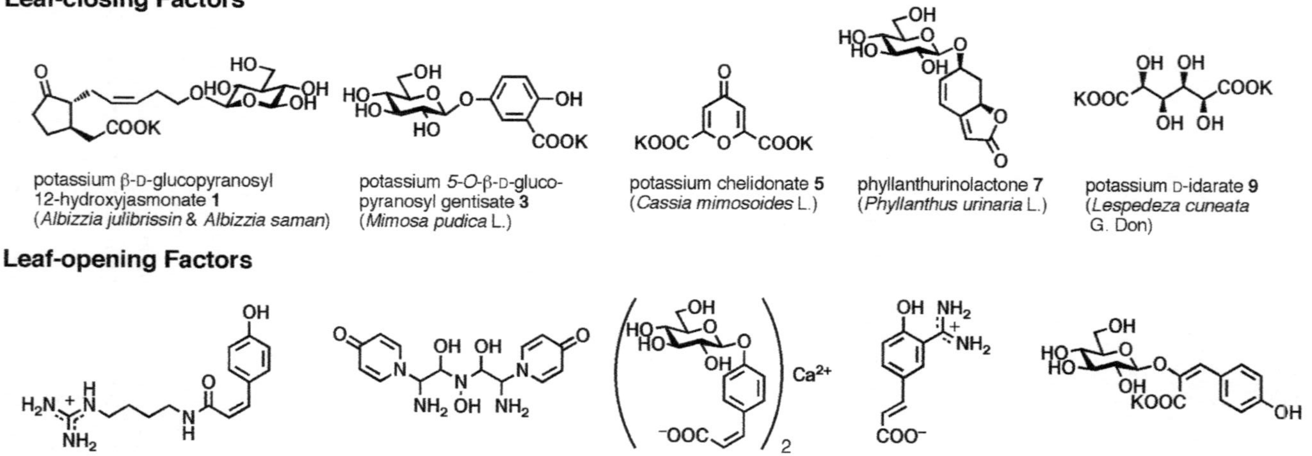

Leaf-closing Factors

potassium β-D-glucopyranosyl 12-hydroxyjasmonate **1** (*Albizzia julibrissin* & *Albizzia saman*)

potassium 5-*O*-β-D-gluco-pyranosyl gentisate **3** (*Mimosa pudica* L.)

potassium chelidonate **5** (*Cassia mimosoides* L.)

phyllanthurinolactone **7** (*Phyllanthus urinaria* L.)

potassium D-idarate **9** (*Lespedeza cuneata* G. Don)

Leaf-opening Factors

cis-p-coumaroylagmatine **2** (*Albizzia julibrissin* & *Albizzia saman*)

mimopudine **4** (*Mimosa pudica* L.)

calcium 4-*O*-β-D-gluco-pyranosyl*cis-p*-coumarate **6** (*Cassia mimosoides* L.)

phyllurine **8** (*Phyllanthus urinaria* L.)

potassium lespedezate **10** (*Lespedeza cuneata* G. Don)

Fig. 3 Leaf-closing and -opening factors from five nyctinastic plants

3.2 *The Enantiodifferential Approach to Identifying the Target Cell and Target Protein of the Leaf-Closing Factor*

Unfortunately, many difficulties usually accompany the molecular identification of a target protein using a functional probe. The most serious of these is nonspecific binding between the probe and multiple proteins, which are usually observed along with the specific recognition between the probe and its true target protein. Nonspecific binding arises due to noncovalent association between the probe and protein, which mainly occurs for two reasons: probe hydrophilicity (Tamura et al. 2003), and electrostatic interactions between the probe and protein (Wilchek et al. 1984) due to the acid dissociation properties of their carboxylate and ammonium groups (Fig. 4). Competitive inhibition is usually used to confirm the specific binding in experiments using probes: the binding of probe to the target protein is competitively inhibited in the presence of excess unlabeled ligand. However, when a ligand has carboxylate or ammonium groups that are easily dissociated, competitive binding experiments yield misleading results, because any nonspecific binding between the probe and proteins due to electrostatic interactions is also inhibited competitively by the unlabeled ligand. This phenomenon is well known in affinity chromatography using charged ligands (Wilchek et al. 1984). Thus, a more reliable method is necessary to confirm the specificity of binding between probe and target protein.

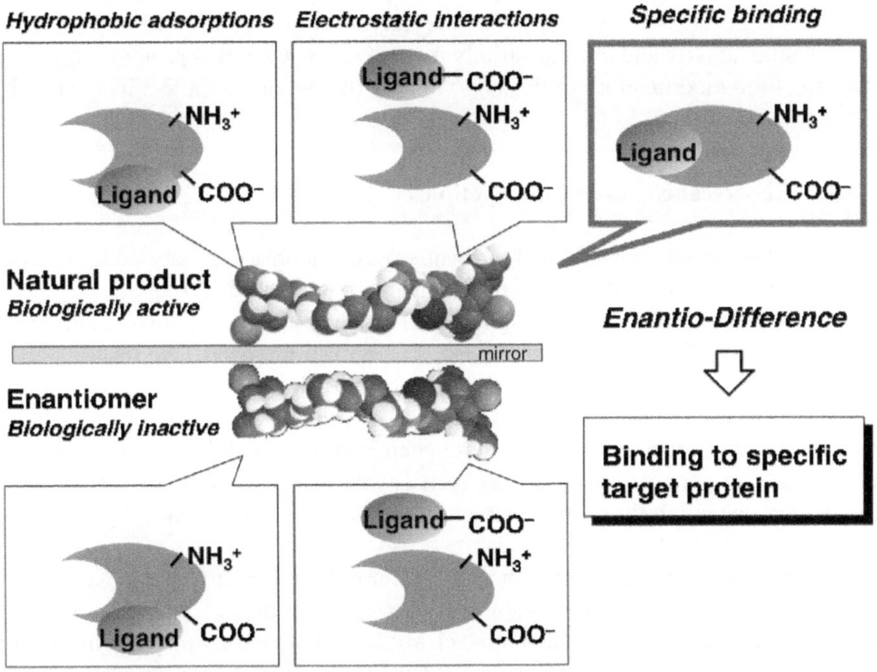

Fig. 4 Concept of the enantiodifferential approach

Enantio-pairs of chiral natural products have almost the same physical properties, with the exception of their optical rotations and affinities for chiral molecules, such as proteins. Both enantiomers exhibit the same nonspecific binding to proteins based on noncovalent associations or electrostatic interactions (Fig. 4). A clear difference is observed, however, in any specific binding based on the stereospecific molecular recognition of a ligand by its target protein. We used an enantiomer of a chiral natural product as a control in bioorganic studies using probe compounds. We applied this enantiodifferential approach to the identification of the target protein of **1**, a chemical factor for leaf-closing activity in the leaf of *A. saman*. We used an enantio-pair type of molecular probe designed for **1** to confirm the specific recognition between **1** and its target protein.

3.3 *Structure–Activity Relationship Studies on the Leaf-Closing Factor for the Genus Albizzia*

Important information on the molecular design of molecular probes was obtained from a structure–activity relationship study on **1** using an enantiomer of **1** (**11**), a D-galactoside derivative (**12**), a *cis*-analog (**13**), and a potassium *epi*-tuberonate (**14**) (Fig. 5). The leaf-closing activity of **12** in *A. saman* leaves was as strong as that of **1** (5×10^{-4}M), but **11**, **13**, and **14** did not exhibit any leaf-closing activity, even at 1×10^{-3}M. These results showed that the aglycone moiety of **1** is important for leaf-closing activity and must be strictly recognized by the target protein, suggesting that structural modifications to the sugar moiety of **1** would not affect its bioactivity.

3.3.1 Fluorescence Studies on Nyctinasty

Based on these results, we designed and synthesized a fluorescence-labeled leaf-closing factor from a pair of optically pure enantiomers of methyl jasmonate (Asamitsu et al. 2006). The probes were designed as D-galactosides to circumvent enzymatic hydrolysis by endogenous β-glucosidase. To confirm the result, we used a pair of diastereomer-type probes (**15** and **16**) in which each enantiomeric aglycone was connected to the D-galactose moiety. A pair of probes (**15** and **16**) were selected because proteins recognizing the stereochemistry of a galactose moiety, such as membrane transporters or glycosidases such as galactosidase, would also be detected by a difference in binding between the two enantiomers when a pair of enantiomer-type probes are used.

The enantiodifferential approach clearly demonstrated the involvement of a target protein of **1** in the motor cell located in the adaxial side of pulvini, which is dominated by extensor cells (Fig. 6) (Nakamura et al. 2006a). Upon comparing the results obtained for the fluorescence-labeled leaf-closing factor probe **15** and its enantiomer **16**, it was clearly apparent that fluorescence occurred in the extensor cell due to the

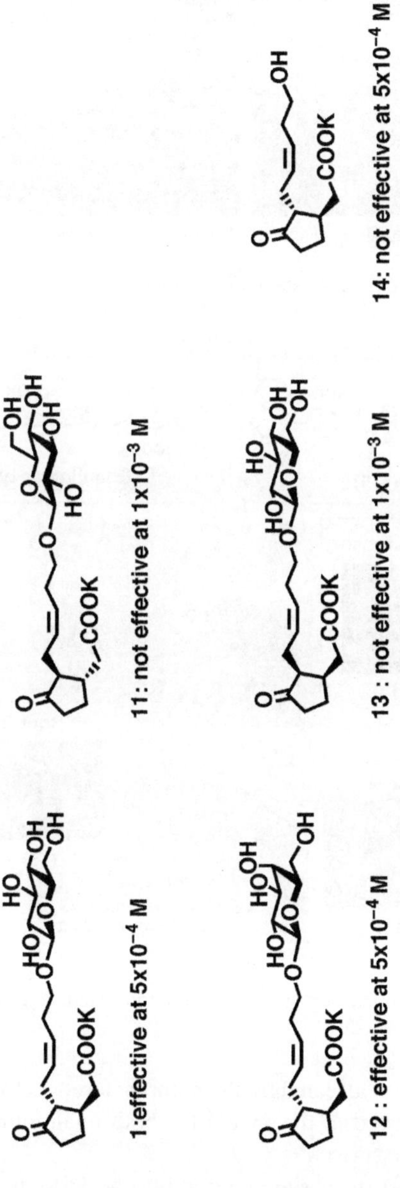

Fig. 5 SAR study of the leaf-closing factor of *Albizzia* plants

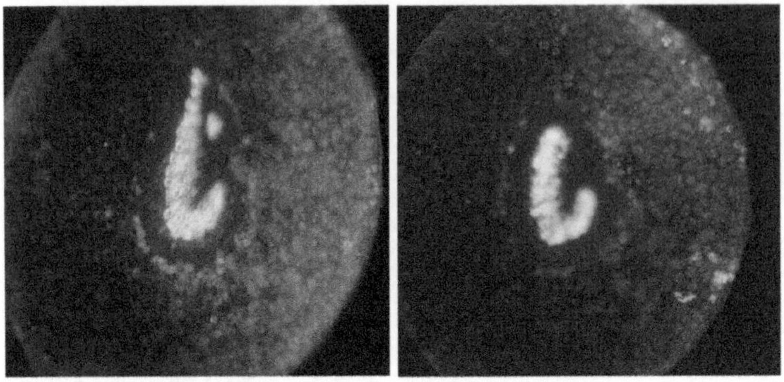

Fig. 6 Enantiodifferential fluorescence staining of pulvinus

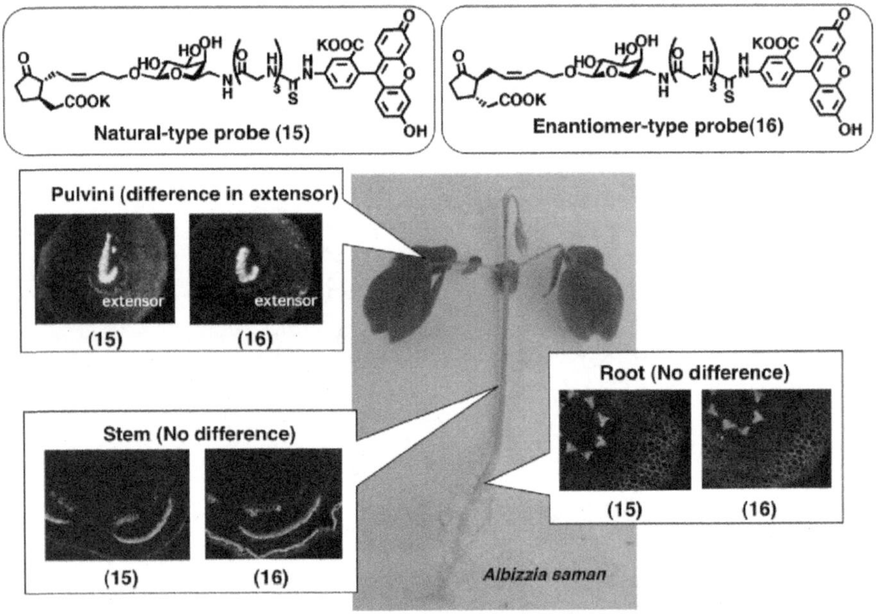

Fig. 7 Enantiodifferential fluorescence staining of pulvinus, stem and root

specific binding, which was affected by the natural stereochemistry. In addition, the strong fluorescence observed in the xylem for both enantiomers was attributed to nonspecific binding of the probes.

Moreover, no other part of *A. saman* bound probe **15** stereospecifically (Fig. 7) (Nakamura et al. 2008). Thus, the actual target cell for the leaf-closing factor was confirmed to be the motor cell. These results strongly suggested the involvement of some specific target protein in the motor cell.

3.3.2 Photoaffinity Labeling of the Target Protein for the Leaf-Closing Factor

We designed and synthesized photoaffinity probe **17** with a benzophenone and a biotinyl group in the sugar moiety (Nakamura et al. 2008). An enantiodifferential photoaffinity labeling experiment was carried out using **17** and **18** against protoplasts of motor cells (Nakamura et al. 2008) that were prepared from *A. saman* leaves according to Satter's method (Fig. 8) (Gorton and Satter 1984). Protoplasts were prepared from the ca. 200 leaflet pulvini collected, and photocrosslinking was carried out on the cell surface using probe **17** or **18**. After treatment with streptavidin–FITC conjugate, labeled protoplasts with the biologically active probe **17** gave green fluorescence due to fluorescein on the plasma membrane of protoplasts (Fig. 9). This result strongly suggested that the target protein that recognizes the stereochemistry of the aglycone in probe **17** is associated with the plasma membrane of motor cells.

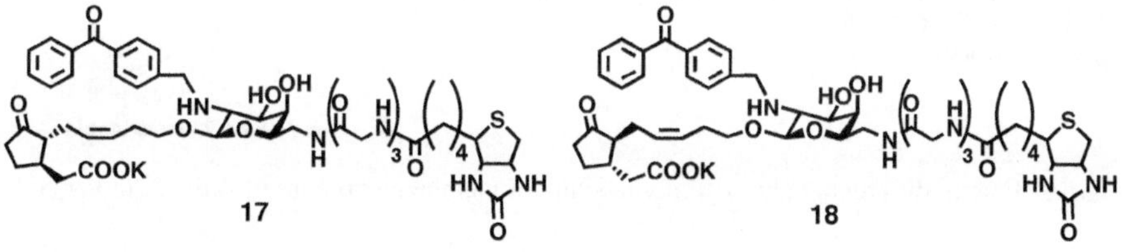

Fig. 8 Photoaffinity probes based on the leaf-closing factor of *Albizzia* plants

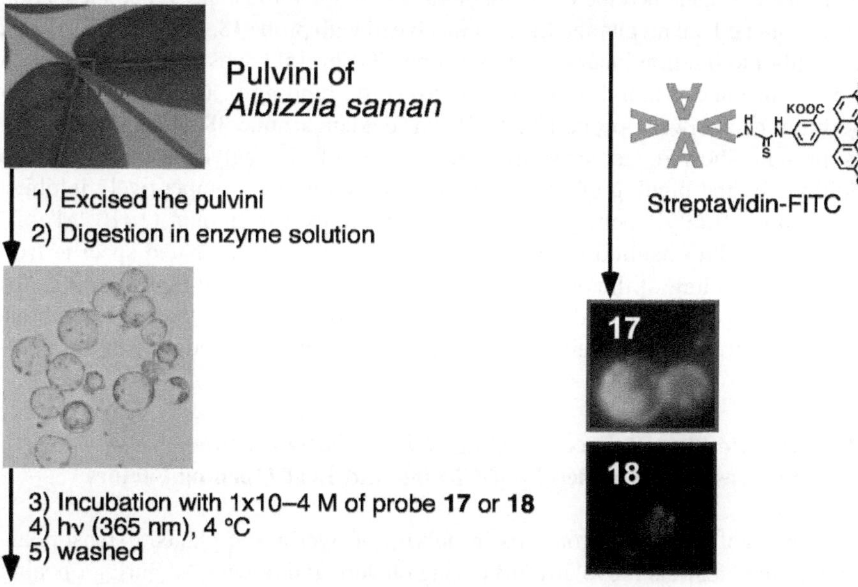

Fig. 9 Enantiodifferential photoaffinity labeling of *Albizzia* motor cells

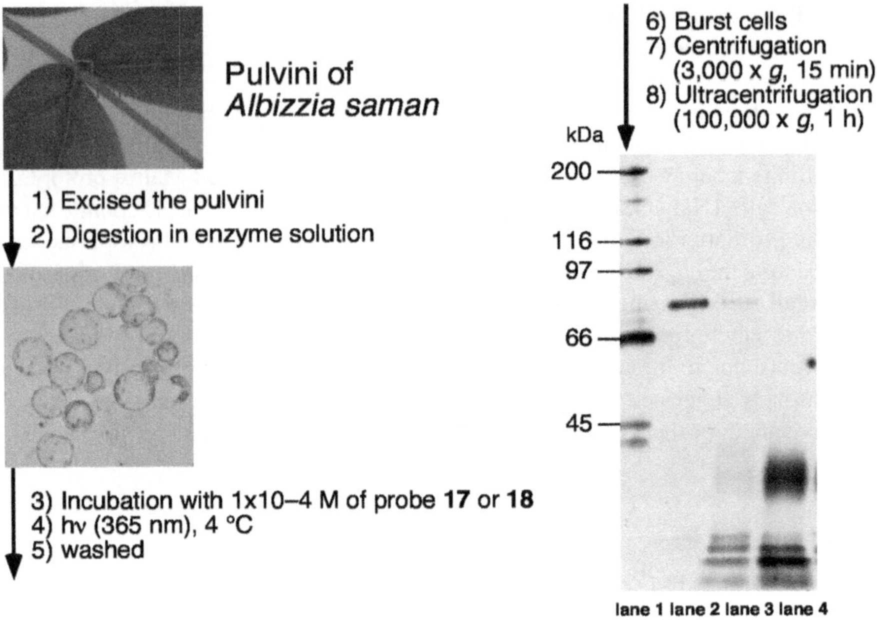

Fig. 10 Enantiodifferential photoaffinity labeling of membrane proteins of *Albizzia* motor cells

SDS–PAGE analysis and chemiluminescence analysis of photocrosslinked membrane proteins of protoplasts were carried out (Fig. 10) (Nakamura et al. 2008). In Fig. 10, lane 2 contained the crude membrane fraction without any probe incubation, lane 3 contained the membrane fraction incubated with probe **18**, and lane 4 contained the membrane fraction incubated with probe **17**. Several bands below 30 kDa were observed in lanes 3 and 4, indicating nonspecific binding of the probe. However, one difference between probe **17** and **18** was evident around 38kDa, indicating that this protein showed stereospecific recognition of the aglycone of the probe. Additionally, the binding of probe **7** with this protein was competitively inhibited by the photoaffinity labeling experiment when an excess amount (1×10^{-3}M) of **1** was present. Our enantiodifferential approach clearly discriminated specific from nonspecific binding of the probe. The observation that only the biologically active stereoisomer was recognized by this protein strongly suggested that this membrane protein is the true target protein of **1** involved in the control of nyctinasty in *A. saman*.

3.3.3 Double Fluorescence Labeling of Plant Pulvini Using Fluorescence-Labeled Leaf-Closing and Leaf-Opening Factors

There are two types of motor cells in pulvini of nyctinastic plants: extensors and flexors. Since leaflets move upward during closure and downward during opening, extensors are located on the upper (adaxial) side of the leaf and flexors on the lower (abaxial) side. To examine whether closing and opening factors differentially target

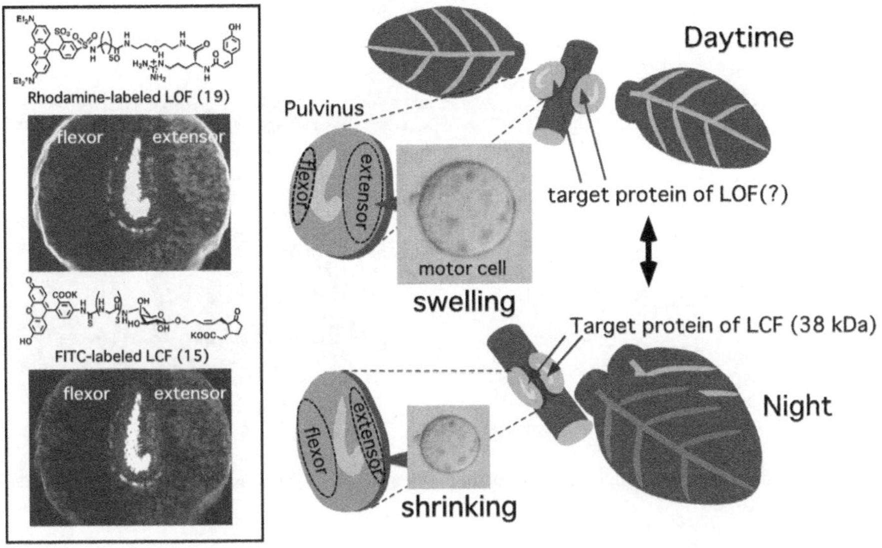

Fig. 11 Double fluorescence labeling of *Albizzia* pulvinus using fluorescence-labeled leaf-closing and -opening factors

extensors and flexors, we performed a double fluorescence labeling study using FITC-labeled leaf-closing factor **15** and rhodamine-labeled leaf-opening factor **19** (Fig. 11) (Nakamura et al. 2006b). Figure11 shows a photograph of the fluorescence image of a plant section that was cut perpendicular to the vessel. Somewhat unexpectedly, both of the probes bound to the extensor cells but not the flexor cells in the pulvini. Therefore, the motor cell with a set of target proteins for leaf-movement factors is located in the extensor side of the pulvini in *A. saman*. As extensor cells are defined as cells that increase turgor during opening and decrease turgor during closing, the leaf-movement factors may regulate potassium channels, which in turn change potassium salt levels and thus turgor pressure.

As described, leaf-closing and -opening factors act in a genus-specific manner. Therefore, we investigated whether the labeled factors bind to the target cells in a genus-specific manner. As expected, the fluorescence-labeled probes **15** and **19** bound to motor cells of *A. saman* and *A. juribrissin*, whereas they did not bind to the cells of *Cassia mimosoides* L., *Phyllanthus urinaria*, and *Leucaena leucocephara* (Nakamura et al. 2006a; Nagano et al. 2003).

4 The Chemical Mechanism of Rhythm in Nyctinasty

If a pair of leaf-movement factors regulate nyctinasty, there should be some relationship between their levels in plants and the circadian clock. The changes in the contents of leaf-closing and -opening factors in the plant *P. urinaria* over time are highlighted in Fig. 12 (Ueda et al. 1999c). HPLC was used to determine the levels of these factors

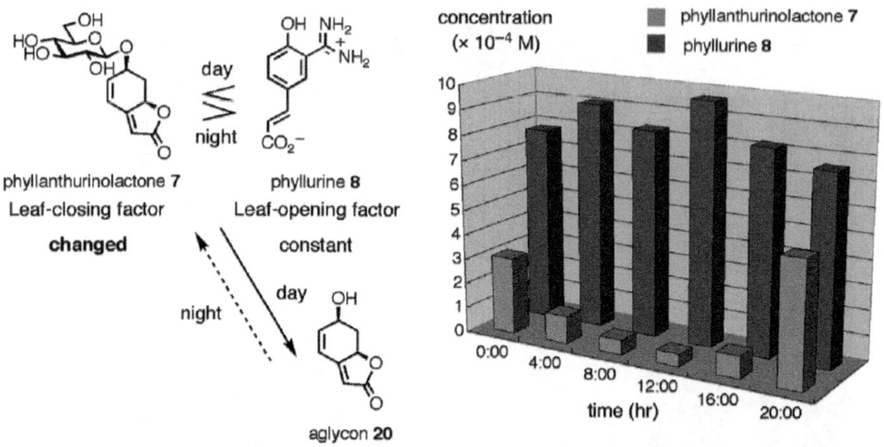

Fig. 12 Changes in the concentrations of leaf-opening and leaf-closing factors in *Phyllanthus urinaria* over time

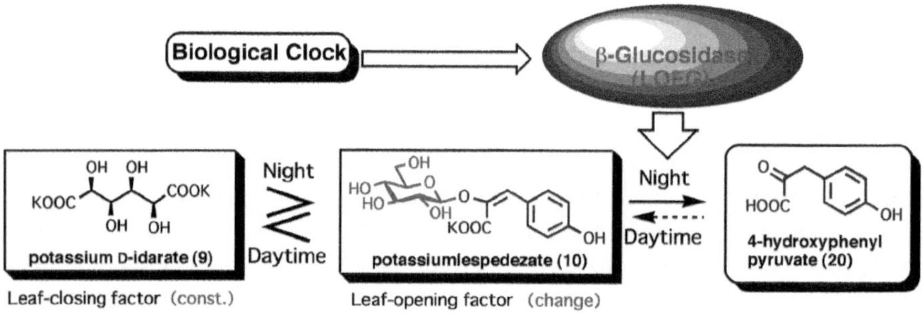

Fig. 13 Chemical mechanism of nyctinasty in *Lespedeza cuniata*

every 4h over a daily cycle. It was found that the content of the leaf-opening factor **8** remains nearly constant during the day, whereas that of the leaf-closing factor **3** changes by as much as 20-fold. This behavior could be accounted for by the conversion of the leaf-closing factor to its corresponding aglycon **20** in a hydrolytic reaction. It follows from this type of analysis that significant changes in the ratio of the concentrations of the leaf-closing and leaf-opening factors in the plant are responsible for leaf movement.

In *Lespedeza cuneata*, the concentration of potassium lespedezate **10** (a glucoside-type leaf-opening factor, Shigemori et al. 1989, 1990) decreases in the evening, whereas the concentration of the leaf-closing factor **9** remains constant during the day (Ohnuki et al. 1998). Leaf-opening factor **10** is metabolized to the biologically inactive aglycon **21** in the evening (Fig. 13). These findings are consistent with the changes in β-glucosidase activity in the plant body that occur during the day, where significant activity is only observed in plants collected in the evening. This suggests that there is a temporal mechanism that regulates β-glucosidase activity and influences these factors during the diurnal cycle. Recently, the β-glucosidase associated with the hydrolysis of **10** was purified and named LOFG (leaf-opening factor β-glucosidase), and was revealed to be a family III type glucosidase from partial sequence analysis (Kato et al. 2008).

In all of the five pairs of leaf-closing and -opening factors **1–10** from the five nyctinastic plants discovered so far, one from each pair of factors is a glycoside, and in all cases the concentrations of these glycoside-type leaf-movement factors change during the day in a similar manner to that described for *L. cuneata*.

This suggests that all nyctinastic leaf movement can be explained by a single mechanism involving two leaf movement factors, of which one is a glucoside. β-Glucosidase activity is then regulated by some mechanism that deactivates the glucoside and controls the relative concentrations of leaf-closing and -opening factors. Thus, nyctinastic leaf movement is controlled by regulated β-glucosidase activity with a daily cycle.

References

Asamitsu Y, Nakamura Y, Ueda M, Kuwahara S, Kiyota H (2006) Synthesis and Odor Descriptions of Both Enantiomers of Methyl 4,5-Didehydrojasmonate, a Component of Jasmin Absolute, Chem. Biodivers. 3: 654–659

Darwin C (1875) Insectivorous plants. John Murray, London

Gorton H, Satter RL (1984) Extensor and flexor protoplasts from *Samanea pulvini.* 2. X-ray analysis of potassium, chlorine, sulfur, phosphorus, and calcium. Plant Physiol 76:680–684

Kato N, Inada M, Sato H, Miyatake R, Kumagai T, Ueda M (2006) Enantio-differential approach using fluorescence-labeled phyllanthurinolactone, a leaf-closing factor of *Phyllanthus urinaria* L. Tetrahedron 62:7307–7318

Kato E, Sasaki T, Ueda M (2008) Affinity purification and characterization of a key enzyme concerning circadian rhythmic control of nyctinasty in *Lespedeza cuneata* L. Bioorg Med Chem 16:4600–4616

Kirchner O (1874) Die botanischen Schriften des Theophrast von Eresos. Jahrb Class Philol 7:449–539

Kotzyba-Hibert F, Kapfer I, Goeldner M (1995) Recent trends in photoaffinity labeling. Angew Chem Int Ed 34:1296–1312

Lee Y (1990) In: Satter RL, Gorton HL, Vogelmann TC (eds) The pulvinus: motor organ for leaf movement. American Society of Plant Physiologists, Rockville, MD, pp 130–141

Miyoshi E, Shizuri Y, Yamamura S (1987) Isolation of potassium chelidonate as a bioactive substance concerning with circadian rhythm in nyctinastic plants. Chem Lett 16:511–514

Moran N (2007a) Osmoregulation of leaf motor cell. FEBS Lett 581:2337–2347

Moran N (2007b) Rhythmic leaf movements: physiological and molecular aspects. In: Mancuso S, Shabala S (eds) Rhythms in plants. Springer, Berlin, pp 3–37

Moshelion M, Moran N (2000) Potassium-efflux channels in extensor and flexor cells of the motor organ of *Samanea saman* are not identical. Effects of cytosolic calcium. Plant Physiol 124:911–919

Moshelion M, Becker D, Czempinski K, Mueller-Roeber B, Attali, B, Hedrich R, Moran N (2002) Diurnal and circadian regulation of putative potassium channels in a leaf moving organ. Plant Physiol 128:634–642

Nagano H, Kato E, Yamamura S, Ueda M (2003) Fluorescence studies on nyctinasty using fluorescence labeled *cis-p*-coumaroylagmatine, a leaf-opening substance of *Albizzia* plants: Existence of genus-specific receptor for leaf-movement factor. Org Biomol Chem 1:3186–3192

Nakamura Y, Miyatake R, Matsubara A, Kiyota H, Ueda M (2006a) Enantio-differential approach to identify the target cell for glucosyl jasmonate-type leaf-closing factor, by using fluorescence-labeled probe compounds. Tetrahedron 62: 8805–8813

Nakamura Y, Matsubara A, Okada M, Kumaga T, Ueda M (2006b) Double fluorescence-labeling study on genus *Albizzia* using a set of fluorecence-labeled leaf-movement factors to identify the spatial distribution of their receptors. Chem Lett 35:744–745

Nakamura Y, Miyatake R, Ueda M (2008) Enantiodifferential approach for the detection of the target membrane protein of the jasmonate glycoside that controls the leaf movement of *Albizzia saman*. Angew Chem Int Ed 47:7289–7292

Ohnuki T, Ueda M, Yamamura S (1998) Molecular mechanism of the control of nyctinastic leaf-movement by biological clock in *Lespedeza cuneata* G. Don. Tetrahedron 54:12173–12184

Satter RL, Galston AW(1981) Mechanisms of control of leaf movements.Annu Rev Plant Physiol 32:83–110

Satter RL, Gorton HL, Vogelmann TC (1990) The pulvinus: motor organ for leaf movement (Current Topics in Plant Physiology, vol 3). American Society of Plant Physiologists, Rockville, MD

Schildcknecht H (1983) Turgorins, hormones of the endogeneous daily rhythms of higher organized plants—detection, isolation, structure, synthesis, and activity. Angew Chem Int Ed 22:695–710

Shigemori H, Sakai N, Miyoshi E, Shizuri Y, Yamamura S (1989) Potassium lespedezate and potassium isolespedezate, bioactive substances concerned with the circadian rhythm in nyctinastic plants. Tetrahedron Lett 30:3991–3994

Shigemori H, Sakai N, Miyoshi E, Shizuri Y, Yamamura S (1990) Bioactive substances from *Lespedeza cuneata* L.G. Don and their biological activities. Tetrahedron 46:383–394

Tamura T, Terada T, Tanaka A (2003) A quantitative analysis and chemical approach for the reduction of nonspecific binding protein on affinity resin. Bioconjugate Chem 14:1222–1230

Ueda M, Nakamura Y (2006) Metabolites involved in plant movement and "memory:" nyctinasty of legumes and trap movement in the Venus flytrap, Nat Prod Rep 23:548–557

Ueda M, Yamamura S (1999a) Leaf-opening substance of *Mimosa pudica* L.; chemical studies on the other leaf-movement of *Mimosa*. Tetrahedron Lett 40:353–356

Ueda M, Yamamura S (1999b) Leaf-closing substance of *Mimosa pudica* L.; chemical studies on another leaf-movement of *Mimosa* II. Tetrahedron Lett 40:2981–2984

Ueda M, Yamamura S (2000) The chemistry and biology of the plant leaf-movements. Angew Chem Int Ed 39:1400–1414

Ueda M, Shigemori-Suzuki T, Yamamura S (1995) Phyllanthurinolactone, a leaf-closing factor of nyctinastic plant, *Phyllanthus urinaria* L. Tetrahedron Lett 36:6267–6270

Ueda M, Ohnuki T, Yamamura S (1997a) The chemical control of leaf-movement in a nyctinastic plant, *Lespedeza cuneata* G. Don. Tetrahedron Lett 38:2497–2500

Ueda M, Tashiro C, Yamamura S (1997b) *cis-p*-Coumaroylagmatine, the genuine leaf-opening substance of a nyctinastic plant, *Albizzia julibrissin* Durazz. Tetrahedron Lett 38:3253–3256

Ueda M, Asano M, Yamamura S (1998a) Leaf-opening substance of a nyctinastic plant, *Phyllanthus urinaria* L. Tetrahedron Lett 39:9731–9734

Ueda M, Ohnuki T, Yamamura S (1998b) Chemical substances controlling the leaf-movement of a nyctinastic plant, *Cassia mimosoides* L. Phytochemistry 49:633–635

Ueda M, Asano M, Sawai Y, Yamamura S (1999c) Leaf-movement factors of nyctinastic plant, *Phyllanthus urinaria* L.; the universal mechanism for the regulation of nyctinastic leaf-movement. Tetrahedron 55: 5781–5792

Ueda M, Okazaki M, Ueda K, Yamamura S (2000a) A leaf-closing substance of *Albizzia julibrissin* Durazz. Tetrahedron 56:8101–8105

Ueda M, Shigemori H, Sata N, Yamamura S (2000b) The diversity of chemical substances controlling the nyctinastic leaf-movement in plants. Phytochemistry 53:39–44

Wilchek M, Miron T, Kohn J (1984) Affinity chromatography. Methods Enzymol 104:3–55

Aposematic (Warning) Coloration in Plants

Simcha Lev-Yadun

Abstract Aposematic (warning) coloration is a common defense in plants, although it was largely ignored before 2001. The fact that many aposematic animals use both plant-based pigments and sequestered poisonous molecules to become aposematic emphasizes the absurdity of neglecting the aposematic nature of so many plants. Similar to the situation in animals, aposematic coloration in plants is commonly yellow, orange, red, brown, black, white, or combinations of these colors. Aposematic coloration is expressed by thorny, spiny, prickly and poisonous plants, and by plants that are unpalatable for various other reasons. Plants that mimic aposematic plants or aposematic animals are also known. Many types of aposematic coloration also serve other functions at the same time, such as physiological, communicative and even other defensive functions. It is therefore difficult in many cases to evaluate the relative functional share of visual aposematism in various color patterns of plants and the specific selective agents involved in their evolution. Aposematic coloration is part of a broader phenomenon of defensive coloration in plants; this topic has also received only limited attention, as is evident from the lack of a regular and systematic description of these color patterns in published floras.

1 Introduction

Most land plants have organs or tissues with colors other than green that should have both a cost and an advantage. The cost to the plant of producing colored organs has three aspects. First, it requires the allocation of resources to synthesize the pigments.

S. Lev-Yadun
Department of Science Education - Biology, Faculty of Science and Science Education, University of Haifa—Oranim, Tivon, 36006 Israel
e-mail: levyadun@research.haifa.ac.il

F. Baluška (ed.), *Plant-Environment Interactions*, Signaling and Communication in Plants,
DOI: 10.1007/978-3-540-89230-4_10,

Second, any color of an organ of a nonwoody aerial plant other than green may in many cases (but certainly not all, see Chalker-Scott 1999; Matile 2000; Hoch et al. 2001, 2003; Lee and Gould 2002; Gould et al. 2002a, b; Close and Beadle 2003; Gould 2004; Ougham et al. 2005; Hatier and Gould 2008) be linked to lower photosynthesis. Third, conspicuousness may attract herbivores. In general, the benefits of coloration should be higher than the costs in order for such character to evolve.

Plant pigments and coloration caused by air spaces or other physical effects serve many physiological and communicative functions, such as photosynthesis, defense from UV light, scavenging of oxygen radicals, pollination, seed dispersal, thermoregulation and defense (e.g., Gould et al. 2002a; Close and Beadle 2003; Lee 2007). Gould et al. (2002b), Lev-Yadun et al. (2002, 2004), Lev-Yadun (2006a), Schaefer and Wilkinson (2004) and Lev-Yadun and Gould (2007, 2008) have already argued that nonphotosynthetic plant pigments have the potential to serve more than one function concurrently. I stress that I fully agree with Endler (1981), who proposed in relation to animal coloration that "we must be careful not to assume that because we have found one apparent function to a color pattern, it necessarily means that we have a complete explanation." Thus, various hypotheses concerning the coloration of leaves and other plant parts need not contrast with or exclude any other functional explanation of specific types of plant coloration, and traits such as coloration that may have more than one type of benefit may be selected for by several agents. Consistent with Grubb's (1992) view that defense systems are not simple, I consider that the evolution of plant coloration reflects an adaptation to both physiological pressures and to relations with other organisms.

Here I will describe and discuss the facts and questions related to aposematic coloration in plants in an attempt to outline this phenomenon and compare it with the broad knowledge of visual aposematism in animals. I will refer to aposematic coloration in the broadest sense, considering any visual warning phenomenon associated with unpalatability that may deter herbivores. The goal of this chapter is to stimulate further research into this generally overlooked phenomenon in plant biology.

1.1 Partial Descriptions of Color Patterns in Floras

One major obstacle to studies of defensive plant coloration in general and aposematic coloration in particular is provided by the fragmentary and inconsistent descriptions of plant coloration, especially of vegetative organs. Taxonomists have usually referred only to flower colors, and even this character has not always been fully described. Thorn, spine and prickle color, unripe and ripe fruit color, leaf colors, bark color and color changes in all of these organs have usually not been systematically described. In his seminal book on plant demography, Harper (1977) commented as follows on the possibility that defensive plant coloration operates: "botanists were reluctant to accept things that are commonplace for zoologists." The surprisingly small number of papers in botany related to defensive coloration as compared to zoology is clearly reflected in the annotated bibliography by Komárek (1998), which has thousands of related publications on animals and only a few about plants. The significant progress made in understanding the defensive role of pigmentation in zoology and the basics of the genetic mechanisms involved took over a century to achieve (e.g., Majerus 1998; Ruxton et al. 2004; Hoekstra 2006), and the effort needed to reach the same level of progress in botany is probably not any smaller. Lev-Yadun and Gould (2008) emphasized that in spite of all of the current difficulties involved in accepting, understanding and proving defensive plant coloration, there is no reason to continue with the long tradition of botanists (or, to give them their current popular name, "plant scientists") of neglecting the study of defensive plant coloration including aposematism. Moreover, even zoologists studying animal aposematism who studied plant–animal interactions related to herbivory overlooked this issue. An intermediate stage of imperfect explanations, which in any case are common in many areas of biology and other sciences, will still allow progress to be made in the issue of aposematic coloration and may stimulate thinking by other scientists who may develop even better theoretical or experimental ideas than the ones that exist today.

2 Aposematism

Aposematic (warning) coloration is a biological phenomenon in which poisonous, dangerous or otherwise unpalatable organisms visually advertise these qualities to other animals (Cott 1940; Edmunds 1974; Gittleman and Harvey 1980; Ruxton et al. 2004). The evolution of aposematic coloration is based on the ability of target enemies to associate the visual signal with risk, damage, or nonprofitable handling, and thus to avoid such organisms as prey (Edmunds 1974; Gittleman and Harvey 1980; Ruxton et al. 2004). Typical colors of aposematic animals are yellow, orange, red, purple, black, white and brown, or combinations of these (Cott 1940; Edmunds 1974; Wickler 1968; Savage and Slowinski 1992; Ruxton et al. 2004). The common defense achieved by aposematic coloration has resulted in the evolution of many mimicking animals. The mimics belong to two general categories, although there are intermediate situations. One is Müllerian mimics: here, defended animals mimic each other, sharing the cost of predator learning among more participants. The other is Batesian mimics, which are undefended animals that benefit from the existence of common defended aposematic models (Cott 1940; Edmunds 1974; Wickler 1968; Savage and Slowinski 1992; Ruxton et al. 2004).

2.1 *Olfactory Aposematism*

While this chapter is dedicated to visual aposematism in plants, olfactory aposematism, whereby poisonous plants deter mammalian or insect herbivores, has also been proposed (Eisner 1964; Rothschild 1972, 1973, 1986; Levin 1973; Atsatt and O'Dowd 1976; Wiens 1978; Eisner and Grant 1981; Harborne 1982; Rothschild et al. 1984; Guilford et al. 1987; Rothschild and Moore 1987; Kaye et al. 1989; Moore et al. 1990; Woolfson and Rothschild 1990; Launchbaugh and Provenza 1993; Provenza et al. 2000; Massei et al. 2007). It is probable that—similar to pollination (Faegri and van der Pijl 1979; Dafni 1984; Jersáková et al. 2006) and seed dispersal (Pijl 1982), where certain plants use both visual and olfactory signals simultaneously for animal attraction—double signaling also holds for plant aposematism. In the case of the very spiny zebra-like rosette annual *Silybum marianum* (Asteraceae), which was proposed to use visual aposematic markings—white stripes (Lev-Yadun 2003a), Rothschild and Moore (1987) proposed that it uses olfactory aposematism via pyrazine. It is likely that both types of aposematism operate simultaneously in the case of *Silybum*, possibly towards different herbivores. The possibility that thorny, spiny and prickly plants use visual and olfactory aposematism simultaneously should be studied systematically. I should stress that olfactory aposematism is especially important as a defense against nocturnal herbivores, as has been shown for many fungi (Sherratt et al. 2005).

2.2 The Anecdotal History of Discussions of Aposematic Coloration in Plants

A database search of "aposematism in plants" does not yield anything earlier than the year 2001. After it became clear to me in January 1996, following compelling evidence in the field, that aposematic coloration probably exists in many thorny, spiny and prickly plants, 12 years of thorough library study resulted in a very short pre-2000 list of authors who discussed it (usually very briefly) in poisonous plants (Cook et al. 1971; Hinton 1973; Harper 1977; Wiens 1978; Rothschild 1980, 1986; Harborne 1982; Williamson 1982; Knight and Siegfried 1983; Smith 1986; Lee et al. 1987; Givnish 1990; Tuomi and Augner 1993). Moreover, several of these references (Knight and Siegfried 1983; Smith 1986; Lee et al. 1987) dismissed the existence of aposematic coloration in the plants they studied. These few early mentions of visual aposematism in plants referred to poisonous ones, while papers published since 2001 have given more attention to thorny, spiny and prickly ones (Lev-Yadun 2001, 2003a, b, 2006b; Midgley et al. 2001; Gould 2004; Midgley 2004; Lev-Yadun and Ne'eman 2004, 2006; Rubino and McCarthy 2004; Ruxton et al. 2004; Speed and Ruxton 2005; Halpern et al. 2007a, b; Lev-Yadun and Gould 2008; Lev-Yadun and Halpern 2008) and less attention to poisonous ones (Lev-Yadun and Ne'eman 2004; Hill 2006; Lev-Yadun 2006b; Lev-Yadun and Gould 2007, 2008).

Cook et al. (1971) is the earliest reference I managed to find that briefly proposed that aposematic coloration occurs in poisonous seeds of the plant *Eremocarpus setigerus* (Euphorbiaceae). Seeds of this plant are either camouflaged by mottling when less poisonous, or are much more poisonous and have a plain gray color that does not camouflage them much. The mountain dove (*Zenaidura macroura*) rejects the gray seeds and eagerly eats the mottled ones, leading Cook et al. (1971) to propose that they are aposematic. Hinton (1973), who was a zoologist, gave the first detailed hypothesis for a possible defense from herbivory of yellow, red and other types of vivid flower coloration. Hinton proposed that colorful poisonous flowers should be considered to be aposematic and that they probably have mimics. His review of deception in nature was published in a book about illusion which was not a biology book, but rather dealt with art. This hypothesis, which was also briefly referred to by the very influential Miriam Rothschild (1980) when she discussed the roles of carotenoids, did not cause botanists or zoology-oriented ecologists to pursue this issue. It seems that the relevant community was willing to consider plant coloration only for physiological issues, or when plant–animal relations were considered, mostly to attract pollinators and seed dispersers. Later, Rothschild (1986) proposed that red may serve as an aposematic color in poisonous plants, without giving examples. Harper, who wrote his comment about botanists that were reluctant to accept things that are commonplace for zoologists around the same time (1977), did not explain why zoologists who dealt with animal aposematism and were also involved in research on plant–animal interactions had not recognized how common these phenomena are in many plant habitats. The fact that many aposematic animals use both plant-based pigments and sequestered poisonous molecules to become aposematic highlights the absurdity of neglecting the aposematic nature of so many plants. Wiens (1978), in his review on mimicry in plants, mentioned that many examples of striking, contrasting, and often variegated or mottled patterns of coloration characterize plants, particularly leaves, and asked if they serve aposematic functions. He proposed that if herbivores primarily orient their feeding selection visually, these patterns should function aposematically. Wiens (1978) gave several examples of poisonous plants with patterned leaf coloration, such as *Caladium* and *Dieffenbachia*, and colorful poisonous seeds of various plants including *Ricinus*. Wiens (1978) also mentioned personal communication with C. Dodson that suggested that young red leaves in tropical plants may also be aposematic. Eisner (1981), without using the term aposematic, actually described visual aposematism in the thorny plant *Schrankia microphylla*, which folds its leaves when touched, further exposing its thorns. Harborne (1982), in his book on chemical ecology, proposed that the brightly colored, purple-black berries of the deadly *Atropa belladonna* warn grazing mammals of the dangers of consuming them. Williamson (1982) also briefly proposed that brightly colored (red or red and black) seeds lacking an arillate or fleshy reward (e.g., *Erythrina*, *Ormosia*, and *Abrus*) might be aposematically colored to warn seed eaters of their toxicity. Knight and Siegfried (1983) raised the question of whether green fruits signal unpalatability, and concluded that green does not provide enough contrast to be aposematic in the forest canopy. Smith (1986) hypothesized that leaf variegation may be aposematic in theory, but concluded that for the vine species (*Byttneria aculeata*) he studied, the variegation was actually related to defense from herbivory, mimicking leaf mining damage but not actually aposematic. Although Smith rejected the operation of aposematism in the plant species he studied, he gave a clear and detailed formulation of the aposematic hypothesis for poisonous plants:

"The benefits to the plant of chemical defense against herbivores would be greater if herbivores avoided such plants altogether, rather than testing leaves for palatability, and so causing some damage. A distinct leaf color pattern linked with chemical defense might function in this way. Polymorphism for leaf color should then coincide with polymorphisms for chemical defense. Müllerian and Batesian mimicry could result in evolution of similar patterns of variegation, with or without associated toxicity, among other species which have herbivore species in common with the model species" (Smith 1986). Lee et al. (1987) concluded that anthocyanins in developing leaves of mango and cacao are not aposematic. Givnish (1990) noted that Smith's (1986) rejected hypothesis regarding the aposematic value of leaf variegation should be considered, but did not elaborate on this issue when he proposed that the understory herbs he studied use leaf variegation as camouflage. Tuomi and Augner (1993) mentioned a possible association between bright colors in plants and toxicity. Augner (1994) modeled and discussed the conditions needed for the operation of aposematism in plants, focusing on chemical-based aposematism with no direct reference to a visual one, although it can be understood from the text that visual aposematism was not opposed. Augner and Bernays (1998) modeled the possibilities of plant defense signals and their mimics, and although they did not refer directly to visual aposematism, it is again clear from the text that they concluded that Batesian mimics of plant defense signals may be common (see proposed Müllerian and Batesian mimics in Lev-Yadun 2003a, 2006b; Lev-Yadun and Gould 2007, 2008). Archetti (2000), in his discussion of red and yellow autumn leaves that were proposed to signal aphids about the defensive qualities of trees, rejected the possibility that these leaves are aposematic.

Another issue of importance concerning poison-related aposematism is the relativity of aposematism. Deciding that a certain branch, root, leaf, flower, fruit or seed is poisonous or unpalatable is a relative issue. Certain frugivores can consume fruits that are poisonous to other animals (Janzen 1979), and the same is true of any plant organ or tissue. Therefore, a chemically defended plant that is aposematic for certain animal taxa may be edible and nonaposematic for other taxa.

2.3 *Aposematic Coloration in Thorny, Spiny, and Prickly Plants*

There are three terms for sharp defensive plant appendages: thorns, when they are made of branches; spines, when they are made of leaves; and prickles, when they are made of cortical tissues (e.g., in roses). Thorns, spines and prickles provide mechanical protection against herbivory (Janzen and Martin 1982; Janzen 1986; Tomlinson 1990; Myers and Bazely 1991; Grubb 1992; Rebollo et al. 2002) because they can wound mouths, digestive systems (Janzen and Martin 1982; Cooper and Owen-Smith 1986; Janzen 1986), and other body parts of herbivores. Thus, theoretically, once herbivores learn to identify thorns, spines and prickles (and their bright colors or associated markings should help in their recognition), they can avoid the harmful plants advertising them. The fact that thousands of thorny, spiny and prickly species have colorful and sharp defensive structures or that they are otherwise conspicuous due to their white or colorful markings somehow escaped the notice of botanists and zoologists, although cacti and other spiny taxa are found in the majority of botanical gardens.

Since what is toxic to one animal might be harmless to another (Laycock 1978; Janzen 1979; Gleadow and Woodrow 2002), chemical-based aposematism may not operate for all herbivores. For sharp defensive organs, the situation is somewhat different. There are differences in the sensitivity of herbivores to sharp objects, but even specialized mammalian herbivores like woodrats and collared peccaries, which are well adapted to deal with and exploit very spiny *Opuntia* plants, tend to choose the less spiny ones (Brown et al. 1972; Theimer and Bateman 1992). The need to touch and ingest sharp objects makes all large vertebrate herbivores sensitive to such plants. Thorns, spines and prickles may therefore be more universal than poisons in relation to aposematism.

The recent proposals that thorny, spiny and prickly plants may be visually aposematic (Lev-Yadun 2001, 2003a, b, 2006b; Midgley et al. 2001; Gould 2004; Midgley 2004; Lev-Yadun and Ne'eman 2004, 2006; Rubino and McCarthy 2004; Ruxton et al. 2004; Speed and Ruxton 2005; Halpern et al. 2007a, b; Lev-Yadun and Gould 2008; Lev-Yadun and Halpern 2008) were based on the fact that thorns, spines and prickles are usually colorful or are conspicuous because they are marked by various types of associated coloration in the tissues that form them, including white markings. Similarly, it has also recently been proposed that many spiny animals have colorful spines and so they are aposematic (Ruxton et al. 2004; Inbar and Lev-Yadun 2005; Speed and Ruxton 2005), a fact that was discussed only briefly in the classic monograph by Cott (1940).

After realizing that the thorns, spines and prickles of many wild plants in Israel are usually colorful or are associated with conspicuous white or colorful markings, I decided to examine whether this principle is true in four very spiny taxa (cacti, *Agave*, *Aloe*, *Euphorbia*. When the examination of many species of these taxa clearly indicated that the sharp defensive appendages are usually conspicuous, I proposed that these plants are visually aposematic (Lev-Yadun 2001).

Lev-Yadun (2001) showed that two types of thorn conspicuousness are typical of many plant species: (1) colorful thorns and spines, and (2) white and colorful spots and stripes associated with thorns and spines in leaves, stems, and fruits. Both types of aposematic coloration dominate the spine systems of taxa rich in spiny species: cacti and the genera *Agave*, *Aloe*, and *Euphorbia*. It has been recorded in over a thousand species originating in America and Africa. The colorful spine systems are commonly multicolored (spines are brown, yellow, red, white, gray, pink, black, and tan). For instance, in cacti (the spiniest taxon), in more than 50% of the species for which there are detailed data (e.g., Benson 1982), the spines are pigmented with 3–7 colors, and 88.6% of the 973 cacti species described in Preston-Mafham and Preston-Mafham (1994) have white markings associated with their spines (Lev-Yadun 2001). It has been proposed that conspicuous spines are beneficial for plants, since herbivorous vertebrates remember the signal and thus tend to avoid sampling these conspicuous spiny plants subsequently. Furthermore, herbivores may pass over the aposematic individuals and eat their nonaposematic neighbors, thus reducing competition between aposematic and their neighboring plants (Lev-Yadun 2001). Rubino and McCarthy (2004) tested Lev-Yadun's (2001) aposematic hypothesis by examining the presence of aposematic coloration in thorny, spiny, and prickly vascular plants of southeastern Ohio, and because of their similar field results, reached the same conclusions.

This phenomenon of aposematism in thorny, spiny and prickly plants, which seems to be very common, has been described and discussed at three levels: (1) the floristic approach, where it is studied across large taxa (Lev-Yadun 2001) or floras or ecologies (Lev-Yadun and Ne'eman 2004; Rubino and McCarthy 2004); (2) the individual species level (Lev-Yadun 2003a; Lev-Yadun and Ne'eman 2006; Halpern et al. 2007a, b), and; (3) mimicry of the phenomenon (Lev-Yadun 2003a, b, 2006b; Lev-Yadun and Gould 2008). Although Midgley et al. (2001) and Midgley (2004) did not use the word aposematic, they described the typical conspicuous white thorns of many African *Acacia* trees as visually deterring large herbivores, supporting the aposematic hypothesis. Ruxton et al. (2004) and Speed and Ruxton (2005) elaborated on the principle that, unlike poisons, aposematic thorns advertise their own dangerous quality (self-advertisement).

Lev-Yadun (2003a) showed that the rosette and cauline leaves of the highly thorny winter annual plant species of the Asteraceae in Israel (*S. marianum*) resemble green zebras. The widths of typical variegation bands were measured and found to be highly correlated with leaf length, length of the longest spine at leaf margins, and the number of spines along the leaf circumference. Thus, there was a significant correlation between the spininess and strength of variegation. Lev-Yadun (2003a) proposed that this was a special case of aposematic (warning) coloration. However, additional defensive and physiological roles of the variegation, such as mimicry of the tunnels of flies belonging to the Agromyzidae, reducing the number of insects landing on the leaves in general, just as zebra stripes defend against tsetse flies (Lev-Yadun 2003a and citations therein), were also proposed.

2.4 Pathogenic Bacteria and Fungi and Thorns

Three recent publications showed that spines harbor an array of pathogenic bacteria and fungi (Halpern et al. 2007a, b; Lev-Yadun and Halpern 2008). Spines from date palm (*Phoenix dactylifera*) trees, thorns from common hawthorn (*Crataegus aronia*) trees and two thorny shrub species, thorny burnet (*Sarcopoterium spinosum*) and manna tree (*Alhagi graecorum*), were sampled in Israel. Every typical mature individual of these trees and shrubs carries hundreds or even thousands of conspicuous and therefore potentially aposematic spines or thorns. The severity and frequency of infections among orchard workers in Israel following date-palm spine wounding has necessitated the costly practice of removing all of the millions of spines from many of the orchards using mechanical saws. Even the small number of spines and thorns studied resulted in a list of aerobic and anaerobic bacteria species including *Clostridium perfringens*, *Bacillus anthracis* and *Pantoea agglomerans* (Halpern et al. 2007a, b).

C. perfringens is known to be a flesh-eater in that it can produce a necrotizing infection of the skeletal muscle called gas gangrene (Shimizu et al. 2002). *Clostridium tetani*, the etiological agent of tetanus, a serious disease in humans and animals, can be fatal when left untreated. Thorn injuries have been known to cause tetanus in the USA, Ethiopia, and Turkey (Hodes and Teferedegne 1990; Ergonul et al. 2003; Pascual et al. 2003). *B. anthracis* is the etiological agent of anthrax, a notoriously acute fatal disease in both domesticated and wild animals, particularly herbivorous ones, and humans (Jensen et al. 2003). The cutaneous form of the disease is usually acquired through injured skin or mucous membranes, a typical thorn injury. None of the published medical data discussed ecological or evolutionary issues or aposematism, but were instead only published in the interests of medical practice. However, these data showed that plant thorns, spines and prickles may regularly harbor various toxic or pathogenic bacteria (Halpern et al. 2007a, b).

In their review of the medical literature, Halpern et al. (2007b) found that septic inflammation caused by plant thorn injury can result from not only bacteria but also pathogenic fungi. Dermatophytes that cause subcutaneous mycoses are unable to penetrate the skin and must be introduced into the subcutaneous tissue by a puncture wound (Willey et al. 2008).

2.5 *Do Spiny Plants Harbor Microbial Pathogens on their Spines, Unlike Nonspiny Plants?*

Given that microorganisms are generally ubiquitous, there is no reason to assume that only specific plants or specific plant organs will be rich in microorganisms. Despite this ubiquitous occurrence, however, certain plants or plant organs may have specific chemical components or structures on their surfaces that either reduce or increase the possibility that microorganism taxa will survive. Microorganisms can grow on plant surfaces in biofilms, which are assemblages of bacterial cells that are attached to a surface and enclosed in adhesive polysaccharides excreted by the cells. Within the biofilm matrix, several different microenvironments can exist, including anoxic conditions that facilitate the existence of anaerobic bacteria. Considering the findings of Halpern et al. (2007a, b) in regard to spines and thorns, it is clear that anaerobic bacteria can survive on these defensive structures. Although it is assumed that an array of biofilm types is formed on plant surfaces, this issue should be studied systematically in relation to defense from herbivory in order to gain a better understanding of the antiherbivory role of microorganisms.

2.6 *Silica Needles and Raphids Made of Calcium Oxalate*

An obvious question concerning the potential defensive role of pathogenic microorganisms on plant surfaces concerns those not found on thorns, spines and prickles.

The positive answer in many cases is simple. Thousands of plant species have a sharp microscopic alternative to insert the pathogens into the tissues of the herbivores.

Lev-Yadun and Halpern (2008) proposed that many plant species without thorns, spines, or prickles possess an alternative: one of two types of usually internal (but sometimes external), sharp, microscopic defensive structures: silica needles and raphids (which are needles made of calcium oxalate). Silica bodies in plants are formed by the ordered biological deposition of silicon that enters the plant via the roots (Richmond and Sussman 2003). Silica bodies have several known functions: structural, serving as cofactors in the detoxification of heavy metals, and defense from herbivory (e.g., Richmond and Sussman 2003; Wang et al. 2004). Lev-Yadun and Halpern (2008) discussed their specific potential defensive function: enabling the penetration of microorganisms into the bodies of herbivores. Thousands of plant species belonging to many families produce raphids (Franceschi and Horner 1980). Usually, raphids are formed in specific parenchymal cells that differ from their neighboring cells and are called idioblasts (Fahn 1990). The raphids are formed in idioblasts in large numbers and are packed compactly (aligned parallel to each other), but spread when the tissue is wounded. Raphids are always elongated, needle-shaped, and have two sharp, pointed ends. This, however, is not the whole structural story. Studies conducted with a scanning and transmission electron microscope have revealed that, in many cases, the raphids may be barbed or may have deep grooves along them. The grooves serve as channels through which plant toxins are introduced into the tissues of the herbivores (Sakai et al. 1972; Franceschi and Horner 1980). Like silica bodies in plants, calcium oxalate bodies have several functions, including tissue calcium regulation, defense from herbivory, metal detoxification, and structural functions (Franceschi and Horner 1980; Ruiz et al. 2002; Nakata 2003; Franceschi and Nakata 2005).

In addition to the ability of both types of internal microscopic spines (raphids and silica needles) to introduce plant toxins into the wounded tissues of the herbivore by causing mechanical irritation, Lev-Yadun and Halpern (2008) proposed that they are also able to introduce pathogenic microorganisms. Because of their small size, raphids and silica needles can internally wound the mouth and digestive systems of not only large vertebrates but also insects and other small herbivores that manage to avoid thorns, spines and prickles by passing between them. Through the wounds inflicted by the silica needles and raphids, microorganisms found on the plant surfaces themselves as well as in the mouth and digestive tract of the herbivore may cause infection. Like thorns, spines and prickles, the raphids and silica needles actually inject the pathogenic microorganisms into the sensitive mouth and digestive tract of the herbivore.

The use of pathogenic microorganisms to harm animals by wounding is already known from zoology. For instance, the huge predaceous lizard known as the Komodo dragon (*Varanus komodoensis*) seems to use the pathogenic bacteria found in its saliva as an additional advantage in hunting, like snake's venom. Animals wounded by a bite from the Komodo dragon commonly suffer from bacteremia and thus can be caught later after being incapacitated if not killed by the primary attack (Montegomery et al. 2002).

2.7 *Plant Biological Warfare: Thorns Inject Pathogenic Bacteria into Herbivores, Enhancing the Evolution of Aposematism*

The physical defense provided by thorns, spines, prickles, silica needles, and raphids against herbivores might be only the tip of the iceberg in a much more complicated story. All of these sharp plant structures may inject bacteria into herbivores by wounding, enabling the microorganisms to pass the animal's first line of defense (the skin), and in so doing may cause severe infections that are much more dangerous and painful than the mechanical wounding itself (Halpern et al. 2007a, b; Lev-Yadun and Halpern 2008).

Another theoretical aspect is the delay between the thorn's contact and wounding and the microorganism's action. While the pain induced by contact with thorns is immediate, the microorganism's action is delayed. However, the same is true for the delayed action of poisons in aposematic poisonous organisms, and yet there is general agreement that colorful poisonous organisms are aposematic (e.g., Cott 1940; Edmunds 1974; Gittleman and Harvey 1980; Harvey and Paxton 1981; Ruxton et al. 2004). Therefore, there is no reason to view a microorganism's contamination and its delayed action any differently.

Lev-Yadun and Halpern (2008) proposed that thorns, spines, prickles, silica needles and raphid-injected microorganisms play a considerable potential role in antiherbivory, actually serving as a biological warfare agent, and they may have uniquely contributed to the common evolution of aposematism (warning coloration) in thorny plants or on the surfaces of plants that have internal microscopic spines (Halpern et al. 2007a, b; Lev-Yadun and Halpern 2008). While it now seems clear that thorny plants are aposematic, the issue of potential aposematism in plants with microscopic internal spines in the form of raphids and silica needles has not yet been systematically addressed.

2.8 *Color Changes in Old Aposematic Thorns, Spines, and Prickles*

Among the various colorful plant/animal communication systems, adaptive color changes are known to take part in the two extensively studied gene dispersal systems: pollination and frugivory. Young and unrewarding animal-pollinated flowers and young and unripe fleshy fruits are usually green and cryptic. Flowers usually become colorful and conspicuous only towards anthesis, when they open and offer nectar and pollen as rewards to pollinators. Many flowers retain their conspicuous advertising colors until they wilt. However, many others change color after pollination (Weiss 1991, 1995; Weiss and Lamont 1997). A change in flower color that occurs during an inflorescence may reduce the flower's advertising intensity, and thus its detectability by pollinators. On the other hand, retaining the coloration after pollination, or after such flowers turn unreceptive, may reduce pollinator visits to unpollinated flowers, thus diminishing the plant's reproductive success. By simultaneously reducing the reward after pollination and their attractiveness by changing their color, plants direct pollinators to unpollinated flowers within the same inflorescence or plant. Floral color change is a well-documented phenomenon in various taxa and life forms on all continents except Antarctica (Weiss 1991, 1995; Weiss and Lamont 1997; Bradshaw and Schemske 2003). Fleshy fruits usually become colorful (yellow, pink, orange, red, brown, blue, purple and black) only toward ripening, when they become edible by lowering the content of protective, poisonous, and otherwise harmful secondary metabolites, and by increasing their sugar, protein and fat contents as well as their flavor and softness (Ridley 1930; van der Pijl 1982; Snow and Snow 1988; Willson and Whelan 1990; Schaefer and Schaefer 2007), a phenomenon that is also considered to be at least partly adaptive (Willson and Whelan 1990).

While the adaptive significance and the broad occurrence of color change in flowers (Weiss 1991, 1995), fruits (van der Pijl 1982; Willson and Whelan 1990) and leaves (Matile 2000; Archetti 2000; Hamilton and Brown 2001; Hoch et al. 2001; Lee et al. 2003; Schaefer and Wilkinson 2004; Lev-Yadun and Gould 2007) has been widely discussed, the phenomenon of color change in thorns, spines and prickles has only recently been described as being a widespread phenomenon and discussed as such (Lev-Yadun and Ne'eman 2006).

Patterns of color changes of senescent colorful aposematic thorns, spines and prickles were described in Lev-Yadun and Ne'eman (2006). Color changes make them less conspicuous, and they lose most or even all their aposematic character. The scale of this phenomenon on a taxon, flora, continent or global scale is still unknown. Lev-Yadun and Ne'eman (2006) emphasized that color changes in thorns, spines and prickles are not mandatory. Color changes and the aposematic character losses occur when the defended organs become less edible to large herbivores because of their increased size, mechanical rigidity or chemical defense, or when there is no need for defense. Reducing the cost of defense seems to be the reason for the ephemeral nature of the conspicuousness of plant thorns, spines and prickles (Lev-Yadun and Ne'eman 2006). The adaptive value may lie in reducing the investment in coloration, since a thin ephemeral coloration layer demands fewer resources. Keeping a thorn, spine, or prickle colorful for a long time is more costly, and the benefit of being aposematic is smaller in older, larger, or otherwise better protected organs. The tendency of plants to lower the cost of defense by thorns, spines and prickles is a well-known phenomenon. For instance, African acacias and other woody plants have longer thorns on the lower branches than on the higher ones (Cooper and Owen-Smith 1986; White 1988; Milewski et al. 1991; Brooks and Owen-Smith 1994; Young and Okello 1998; Gowda and Palo 2003). Certain trees (e.g., various citruses and palms) have large thorns or spines only when juvenile and none or fewer when mature (e.g., Kozlowski 1971; Cooper and Owen-Smith 1986; Cornett 1986; Clement and Manshardt 2000). Moreover, like several other types of induced defenses, thorns and spines are known to increase in size and number following herbivory (e.g., Milewski et al. 1991; Perevolotsky and Haimov 1991; Young et al. 2003). There is no theoretical difficulty in proposing that color changes in thorns, spines and prickles also reflect conservation of resources (Lev-Yadun and Ne'eman 2006). However, a simple alternative explanation exists: the thorns, spines, and prickles are colorful simply because the hard polymers composing them are colorful by nature. Lev-Yadun and Ne'eman (2006) dismissed this possibility because the thorns, spines and prickles that lose or change color remain hard and functional. The layer of coloration does not seem to have a significant, or even any, role in producing their sharpness. The broad taxonomic distribution of color changes in thorns, spines and prickles indicates that this character has evolved repeatedly and independently (convergent character) in both gymnosperms and angiosperms, probably in response to selection by visually oriented herbivores.

2.9 *Biochemical Evidence of Convergent Evolution of Aposematic Coloration in Thorny, Spiny and Prickly Plants*

There is very strong indirect evidence for the operation of aposematic coloration in thorny and spiny plants and its convergent evolution in the fact that conspicuous thorn and spine coloration is found in angiosperm taxa that have mutually exclusive biochemical pathways of pigmentation. For instance, taxa belonging to the Caryophyllales (e.g., Cactaceae, Caryophyllaceae, Chenopodiaceae) produce yellow and red pigments via the betalain pathway (Stafford 1994). Most other angiosperm families use anthocyanins for similar patterns of coloration. The fact that spines of cacti are usually conspicuous because of their coloration (Lev-Yadun 2001), commonly including yellow, orange and red coloration resulting from betalain derivatives, indicates that this group of pigments may, among their various functions, be involved in aposematic coloration. By contrast, in Rosaceae, Asteraceae and Fabaceae as well as in many other angiosperm families that use anthocyanins for yellow, orange, pink, red, blue and black coloration of thorns, spines and prickles, the chemical origin of the aposematic coloration is different (Lev-Yadun 2001, 2006b; Lev-Yadun and Gould 2008). It seems therefore that the aposematic coloration of thorny, spiny and prickly plants is a good case of convergent evolution.

2.10 *Mimicry of Aposematic Thorns, Spines, and Prickles*

Mimicry of aposematic animals is very common (Cott 1940; Edmunds 1974; Wickler 1968; Ruxton et al. 2004), and several authors have already proposed that mimicry also operates in plants as an antiherbivore mechanism. Wiens (1978) estimated that about 5% of land plants are mimetic, listing several types of defensive plant mimicry. For instance, mimicry of host leaf morphology is common in mistletoes and was proposed to give rise to crypsis and thus to reduce herbivory (Ehleringer et al. 1986). Since there are so many colorful (aposematic) thorns, spines and prickles, mimics of them are expected. Indeed, various plant taxa from several continents mimic thorns, spines and prickles. Lev-Yadun (2003b) described two types of thorn mimicry: (1) a unique type of weapon (spine) automimicry (within the same spiny or prickly individual), a phenomenon previously known only in animals (e.g., Guthrie and Petocz 1970), and (2) mimicry of aposematic colorful thorns, spines and prickles by colorful elongated and pointed plant organs (buds, leaves and fruit), which, despite their appearance, are not sharp. The discussion of mimicry of thorny, spiny, and prickly plants may be addressed at different taxonomic levels: (1) Müllerian mimicry among thorny, spiny and prickly plant taxa, (2) weapon (spine and prickle) automimicry (within the same individual), and (3) Batesian mimicry, when nonspiny plants mimic thorny, spiny and prickly ones. Interestingly, some insects mimic colorful aposematic plant thorns to escape predation (Purser 2003).

When the proportion of aposematic spiny plants in a given habitat increases for a period that is long enough for an evolutionary change, Müllerian mimicry may lead to the establishment of defense guilds (see Waldbauer 1988). Müllerian mimicry does indeed seem to occur within the group of spiny plants; for instance, there are three very spiny zebra-like annual rosette plant species in the eastern Mediterranean region (*S. marianum*; *Notobasis syriaca*; *Scolymus maculatus*, all of the Asteraceae), and it has been proposed that a defense guild has evolved in these plants (Lev-Yadun 2003a). Similarly, the white spines of many African acacias (Midgley et al. 2001; Midgley 2004) and the yellow, orange, red, brown and black spines of cacti (Lev-Yadun 2001) can all be considered Müllerian mimicry rings of aposematically and physically defended plants.

Weapon (spine) automimicry (within the same individual) occurs when impressions (with or without color printing) of the real spines form on leaves during their development. The developmental mechanism that allows the weapon automimicry to appear is simple. In most if not all *Agave* species, the developing leaves press hard against each other, the spines found along the margins press into the surface of the same or adjacent leaf, and their pattern is copied as a sunken negative and retained along the nonspiny parts of the leaves. For instance, in *Agave americana*, a common ornamental in Israel, the spine copies are seen in many leaves. The species showing the most remarkable spine mimicry is *A. impressa*, in which it is very conspicuous because of a white material that is printed on the false spines. The same type of colorful spines along the margins of the petiole and their mimicry by impression is obvious in the American palm *Washingtonia filifera* (Palmaceae), a common ornamental as well as a feral tree in Israel, and in *Aloe* sp. (Liliaceae) (Lev-Yadun 2003b). This spine automimicry is a vegetal parallel to the "weapon automimicry" of horns or canine teeth known in several mammalian species (Guthrie and Petocz 1970). Weapon (spine) automimicry was found in dozens of species of *Agave*, one species of *Aloe*, and a palm species, which all have spine-like imprints or colorations on their leaves, giving the impression of more extensive spininess (Lev-Yadun 2003b). The mimicry of aposematic colorful thorns, spines and prickles by nonspiny plants is a simple and typical case of Batesian mimicry, but the spine automimicry is a special intraorganismic Batesian mimicry. Lev-Yadun

(2003b, 2006b) proposed that both types of mimicry serve as antiherbivore mechanisms.

When nonthorny plants mimic thorny ones with colorful elongated and pointed plant organs, which despite their appearance and conspicuous coloration are not sharp at all, Batesian mimicry occurs (Lev-Yadun 2003b). Simple mimicry by colorful thorn-like structures was found in several wild species growing in Israel. For example, in several *Erodium* sp. (Geraniaceae), the elongated fruits, which are several centimeters long, beak-like, pointed, and self-dispersing (by drilling into the soil), are red. In *Sinapis alba*, an annual of the Brassicaceae, the elongated and pointed distal part of the fruit, when fully developed but not yet ripe, looks like a spine and is colorful (yellow, red, purple, or various combinations of these). In *Limonium angustifolium*, a wild and domesticated perennial of the Plumbaginaceae, the distal part of its large leaves is red and looks like a spine, although it is soft (Lev-Yadun 2003b).

Lev-Yadun (2006b) and Lev-Yadun and Gould (2008) proposed that there are two possible evolutionary routes towards the mimicry of colorful thorns, spines, or prickles. In the first, an aposematic thorny plant may have lost its thorny character but retained the shape and aposematic signal. In the second, a nonaposematic and nonthorny plant can acquire the signal, becoming a primary mimic. Alternatively, the thorn or spine-like structure and its coloration may have a different, unknown function. There are no field, developmental, or genetic data that may help in distinguishing between these options for any plant species. Concerning aposematism, Ruxton and Sherratt (2006) proposed that defense preceded signaling, which supports both proposed evolutionary routes. In general, the evolution of aposematism in plants is a neglected subject that needs considerable research effort for even a basic level of understanding.

3 Aposematic Coloration in Poisonous Flowers, Fruits, and Seeds

Flower and fruit colors and their chemical defenses were commonly discussed as mechanisms for filtering pollinators and seed dispersers rather than concerning aposematism (Ridley 1930; Faegri and van der Pijl 1979; Herrera 1982; Willson and Whelan 1990; Weiss 1995; Clegg and Durbin 2003; Schaefer et al. 2004, 2007). However, in many cases, the combination of visual signaling and chemical defense and the unpalatability of flowers and fruits should have led to the view that they are aposematic. I will describe the meager information concerning aposematic reproductive structures in plants.

As described above, poisonous seeds were probably the first plant parts that were proposed to be visually aposematic because they are both poisonous and colorful (Cook et al. 1971; Wiens 1978; Williamson 1982). However, aposematism was mentioned only briefly in each of these three papers and further research was not done. The second plant part proposed to be poisonous and colorful and therefore aposematic was the flower (Hinton 1973; Rothschild 1980), but again, this hypothesis was not pursued further. Concerning fruits, Harborne (1982) proposed that the brightly colored, purple-black berries of the deadly *A. belladonna* warn grazing mammals of the dangers of consuming them. Aposematism in fruits mimicking thorns (Lev-Yadun 2003b) or aposematic caterpillars (Lev-Yadun and Inbar 2002) are discussed in other sections of this chapter. Schaefer and Schmidt (2004), without using the term aposematic, actually described visual aposematism in chemically defended fruits, like Eisner concerning the thorny plant *S. microphylla* (1981), and like Midgley et al. (2001) and Midgley (2004) concerning the conspicuous white thorns of many African *Acacia* trees, who described aposematism without mentioning it. Only Hill (2006) experimentally examined the aposematic function of poisonous and colorful fruits and gave good indications for the warning function of the coloration.

There is a large body of evidence for the operation of olfactory and visual aposematism in both flowers and fruits, although the authors of these studies referred to filtering of pollinating and dispersing animals rather than to aposematism. For instance, Pellmyr and Thien (1986), in a broad theoretical study on the origin of angiosperms, proposed that floral fragrances originated from chemicals serving as deterrents against herbivore feeding. In a much more focused study of flower defense in the genus *Dalechampia*, Armbruster (1997) and Armbruster et al. (1997) proposed that defensive resins have evolved into a pollinator-reward system, and that several defense systems have evolved from such advertisement systems. However, the possibility of dual signaling systems that serve to simultaneously attract some animals and repel others has not received much research attention. Pollen odors in certain wind-pollinated plants that do not attract pollinators are rich in defensive molecules such as α-methyl alcohols and ketones (Dobson and Bergström 2000). The dearomatized isoprenylated phloroglucinols may visually attract pollinators of *Hypericum calycinum* by their UV pigmentation properties, but at the same time the plant may use this pigmentation as a toxic substance against caterpillars, defending the flowers from herbivory (Gronquist et al. 2001). The dual action of attracting pollinators while deterring other animals was also found in other taxa, e.g., *Catalpa speciosa* and *Aloe vryheidensis* (Stephenson 1981; Johnson et al. 2006; Hansen et al. 2007). Thus, floral scents may have a defensive role (Knudsen et al. 2006; Junker et al. 2007) in addition to their known attracting function. A similar double strategy of using signals to attract certain animals and repel others occurs in fruits (Cipollini and Levey 1997; Tewksbury and Nabhan 2001; Izhaki 2002). Altogether, in spite of the huge body of research conducted to characterize visual and chemical signaling by plants to animals in flower and fruit biology, the aposematic hypothesis for these very important plant organs, which are commonly visually and chemically conspicuous, has received very little attention.

4 Undermining Insect Camouflage: A Case of Habitat Aposematism

It has recently been suggested that many patterns of plant coloration may undermine the camouflage of small invertebrate herbivores (Lev-Yadun et al. 2004). This hypothesis attempted to provide a unifying general explanation for many of the vegetal coloration types found in nature. The essence of the hypothesis is based on a simple principle that many types of plant coloration undermine the camouflage of small invertebrate herbivores, especially insects, thus exposing them to predation, and in addition causing them to avoid plant organs with unsuitable coloration, to the benefit of the plants. Undermining camouflage is a special case of "the enemy of my enemy is my friend," and a visual parallel of the chemical signals that plants emit to call wasps when attacked by caterpillars (Kessler and Baldwin 2001; Kappers et al. 2005). Moreover, this is a common natural parallel to the well-known phenomenon of industrial melanism (e.g., Kettlewell 1973; Majerus 1998), which illustrates the great importance of plant-based camouflage for herbivorous insect survival and can serve as an independent test for the insect camouflage undermining hypothesis. It was proposed that the enormous variations in coloration of leaves, petioles and stems, as well as of flowers and fruits, undermine the camouflage of invertebrate herbivores, especially insects (Lev-Yadun et al. 2004). For instance, if a given leaf has two different colors—green on its upper (adaxial) side and blue, brown, pink, red, white, yellow or just a different shade of green on its lower (abaxial) side—a green insect (or one of any color) that is camouflaged on one of the sides will not be camouflaged on the other. The same is true for vein, petiole, branch, stem, flower, or fruit coloration. These differences in color are common across diverse plant forms, from short annuals to tall trees, and in various habitats, from deserts to rain forests and from the tropics to the temperate region. Furthermore, leaf color frequently changes with age, season, or physiological condition. Young leaves of many tropical trees and shrubs (Richards 1996; Dominy et al. 2002; Lee 2007)—as well as of many nontropical plants—are red, and later become green, whereas leaves of many woody species in the temperate zones change to yellow and red in autumn (Matile 2000; Hoch et al. 2001).

In heterogeneous habitats, optimal camouflage should maximize the degree of crypsis in the microhabitats used by the prey, and so herbivores may enjoy better crypsis in heterogeneous habitats (Endler 1984; Edmunds and Grayson 1991; Merilaita et al. 1999). Therefore, a plant with many colors may under certain conditions provide better crypsis than a monocolored one. However, the ratio between the size of the herbivore and the size of the color patches on the plants determines whether a certain coloration pattern will promote or undermine crypsis of the herbivore (Lev-Yadun 2006b). Since insects are in general smaller than many of the color patches of leaves, flowers, fruits or branches, they will often be exposed to predators and parasites and will not become more cryptic and better defended. Indeed, certain types of variegation that form small-scale mosaics are not considered to operate to undermine insect camouflage, as has been partly addressed by Schaefer and Rolshausen (2006). The relative colored areas of plant organs (especially leaves) and the sizes of relevant herbivorous invertebrates should be documented and analyzed under natural and experimental conditions to allow a better understanding of the camouflage issue.

Plants provide habitat and food for many animals, so it is logical to assume that visual perception of animals (both herbivores and predators) coevolved with plants. Intuitively, the common optimal camouflage for herbivorous insects should be green, and many insects, e.g., aphids, caterpillars and grasshoppers, have indeed evolved green coloration (Cott 1940; Purser 2003). The effectiveness of green camouflage or gray colors that match bark is impaired by diverse nongreen backgrounds, or even by a variety of green shades of plant background, as was evident with industrial melanism (Kettlewell 1973; Majerus 1998). It has therefore been suggested (Lev-Yadun et al. 2004) that all herbivores that move, feed or rest during the day on plant parts that have different colorations from their own immediately become more conspicuous to their predators. The same is true for insect egg color, which should match the background color for defense. Many plants are simply too colorful to enable a universal camouflage of herbivorous insects and other invertebrates to operate successfully, and so they force small herbivores to cross "killing zones" with colors that do not match their camouflage. Since the variable coloration is usually either ephemeral (red young leaves or red or yellow autumn leaves) or occupies only a small part of the canopy (young leaves, petioles, flowers, and fruits), the gains for insects that have evolved to match such ephemeral or less common coloration are low (Lev-Yadun et al. 2004), and with low gains it is difficult to overcome this type of plant defense by evolution. The excellent color vision possessed by many predators of insects, in particular insectivorous birds (the most common and significant predators of herbivorous invertebrates) (Van Bael et al. 2003), probably makes undermining herbivores' camouflage highly rewarding for plants (Lev-Yadun et al. 2004).

I conclude that since insects, like many other animals, tend to avoid surfaces that don't match their coloration (e.g., Cott 1940; Kettlewell 1973; Endler 1984; Stamp and Wilkens 1993; Carrascal et al. 2001; Ruxton et al. 2004), plant coloration that undermines camouflage can be viewed as habitat aposematism.

5 Delayed Greening as Unpalatability-Based Aposematism

Delayed greening of young leaves of various conspicuous colors (white, pink, very light green) is a common phenomenon in the tropics (Richards 1996). The hypothesis that delayed greening is associated with low nutritive value in young leaves of tropical plants, and that this property defends them from herbivory (Kursar and Coley 1991, 1992, 2003; Coley and Barone 1996), is a special case of a more general hypothesis that low nutritive value acts as a defense (Feeny 1976; Moran and Hamilton 1980; Augner 1995). A similar principle is known to operate well as a defense in many leaves, stems, and young fruit that produce high levels of tannins and other protease inhibitors that decrease protein availability during digestion (Robbins et al. 1987; Bernays et al. 1989; Ryan 1990). Numata et al. (2004) showed that seedlings of various species of the genus *Shorea* (Dipterocarpaceae) that express delayed greening suffer less damage from insect herbivory than species with regular greening. A similar phenomenon occurs in the intraspecific polymorphic *Conocarpus erectus* (buttonwood) leaf color. Some individual plants are silvery and some are green, but some change from green when young to silvery later.

Silvery leaves in buttonwood suffer less insect herbivory (Schoener 1987, 1988; Agrawal and Spiller 2004). Yet, despite the high likelihood that delayed greening is effective and probably also operates outside the tropics, this hypothesis has not received the attention it merits. I propose that the association of being unpalatable with conspicuous colors (delayed greening) may act as a signal to herbivores regarding the lower nutritive value, a typical aposematism. At the same time, such coloration may undermine herbivorous insect camouflage (Lev-Yadun et al. 2004; Lev-Yadun 2006b).

6 Colorful Autumn Leaves

The liveliest recent discussion on defensive plant coloration has centered on the phenomenon of red and yellow autumn leaves. For many decades most people believed that these colors simply appear after the degradation of chlorophyll, which masked these pigments, and that they have no function. However, physiological benefits of autumn leaf coloration, such as protection from photoinhibition and photooxidation, are well indicated (e.g., Chalker-Scott 1999; Matile 2000; Hoch et al. 2001, 2003; Lee and Gould 2002; Gould et al. 2002ab; Close and Beadle 2003; Gould 2004; Ougham et al. 2005; Hatier and Gould 2008). So far, six defensive roles of this coloration against insect herbivory have been proposed. (1) The first, and most discussed, is that the bright colors of autumn leaves signal that the trees are well defended and that this is a case of Zahavi's handicap principle (Zahavi 1975, 1977, 1991; Grafen 1990; Zahavi and Zahavi 1997) operating in plants (Archetti 2000; Hamilton and Brown 2001; Archetti and Brown 2004). (2) Schaefer and Rolshausen (2006) formulated the "defense indication hypothesis." (3) Lev-Yadun and Gould (2007) proposed that the function of the bright autumn leaf coloration may in some cases represent aposematism or its mimicry. (4) Lev-Yadun and Gould (2007) also proposed that the colorful autumn leaves signal that they are going to be shed soon. (5) Yamazaki (2008) proposed that autumn leaf coloration employs plant–ant mutualism via aphids. (6) The last hypothesis concerning the defensive role of bright autumn coloration addresses the undermining of herbivorous insect camouflage (Lev-Yadun et al. 2004), which was discussed above. There are several additional subhypotheses of the defensive role of red and yellow autumn leaves that will not be discussed here because they are less relevant to the discussion on aposematic coloration.

Concerning Zahavi's handicap principle operating in colorful autumn leaves (Archetti 2000, 2007a, b; Hamilton and Brown 2001; Hagen et al. 2003, 2004; Archetti and Brown 2004, 2006; Archetti and Leather 2005; Brown 2005), the idea was partly (Lev-Yadun and Gould 2007, 2008; Ougham et al. 2008; Ramirez et al. 2008) or wholly (Holopainen and Peltonen 2002; Wilkinson et al. 2002; Schaefer and Wilkinson 2004; Ougham et al. 2005; Schaefer and Rolshausen 2006, 2007; Sinkkonen 2006a, b; Chittka and Döring 2007; Rolshausen and Schaefer 2007; Schaefer and Gould 2007; Hatier and Gould 2008; Yamazaki 2008) discounted.

Lev-Yadun and Gould (2007, 2008) emphasized that the operation of aposematism does not exclude the possible simultaneous operation of any other types of visual or nonvisual defense in autumn leaves (see also Hatier and Gould 2008).

The opposition to the handicap hypothesis is based on the complicated biological facts involved (which are also not yet well understood), and on the simultaneous operation of various and sometimes contrasting physiological and defensive functions of autumn leaf coloration. The various functions probably differ in their importance over time, even in a single leaf, let alone in a flora or a broad geographical region (see Lev-Yadun and Gould 2007; Ougham et al. 2008). Holopainen and Peltonen (2002) suggested that leaves that have just turned yellow are a good indication to aphids of the nitrogen available in them in the form of amino acids, an attracting rather than a repelling signal. Wilkinson et al. (2002) held that rather than signaling defensive qualities to aphids, especially since these are drawn to yellow leaves, this coloration serves as a sunscreen (a physiological role), and red colors help to warm leaves, and also function as antioxidants. Ougham et al. (2005) stressed the importance and good documentation of the physiological role of autumn leaf coloration. They argued that the signal is not costly, which, according to the most common view (but not all views, see Lachmann et al. 2001), is a basic feature of signals involved in the operation of Zahavi's handicap principle (Zahavi 1991; Zahavi and Zahavi 1997).

Elaborating on a previous idea by Fineblum and Rausher (1997) about the shared biochemical pathways for flower color and defensive molecules, Schaefer and Rolshausen (2006) formulated the "defense indication hypothesis," a hypothesis of defensive plant coloration, focusing on anthocyanins. It posits that fewer herbivorous insects will feed on plants with strong anthocyanin coloration because it correlates with defensive strength. The biochemical basis for this correlation is that anthocyanins and a number of defense chemicals such as tannins stem from the same biosynthetic pathways. Schaefer and Rolshausen (2006) clearly state that since, according to their understanding, autumn leaf coloration has evolved primarily because of physiological roles, and not as a defense against herbivores, this coloration is not a signal (it is not aposematic), and may be used only as a cue by the insects.

Contrasting views on the phenomenon of bright autumn leaf coloration were presented concerning the hypothesis that these bright leaves may be aposematic. Archetti (2000) specifically rejected the possibility that autumn leaf coloration is aposematic, and aposematism was not discussed in other studies that favored the signaling hypothesis (Hamilton and Brown 2001; Hagen et al. 2003, 2004; Archetti and Brown 2004, 2006; Archetti and Leather 2005; Brown 2005; Archetti 2007a, b). Lee and Gould (2002), Lee (2002), Gould (2004), Sherratt et al. (2005) and Schaefer and Rolshausen (2007) interpreted the hypothesis of handicap-related autumn coloration presented in the papers by Archetti (2000) and Hamilton and Brown (2001) as a case of aposematism, notwithstanding the different views of the authors. Lev-Yadun and Gould (2007), however, proposed that the function of the bright autumn leaf coloration may in many cases represent aposematism or its mimicry. Lev-Yadun and Gould (2007, 2008) proposed that if the defense indication hypothesis is accepted, it directly follows that plant parts rich in anthocyanins may serve in many cases as aposematic (warning) coloration for chemical-based unpalatability. If the red-colored autumn leaves are well defended by various chemicals, as proposed by Schaefer and Rolshausen (2006), or even if red and old yellow autumn leaves are just of low nutritive value (two cases of unpalatability), many bright autumn leaves should be considered aposematic (Lev-Yadun and Gould 2007).

The reason why yellow or red autumn leaves in species that are chemically well defended or unpalatable should be considered aposematic is obvious. Moreover, as in other cases of aposematism (Cott 1940; Wickler 1968; Lev-Yadun 2003b), it is tempting to postulate that mimics of true aposematic autumn leaves also exist. Lev-Yadun and Gould (2007) proposed that the widespread phenomenon of red autumn leaves in some areas may be partly the result of Müllerian and Batesian mimicry. When toxic or unpalatable red leaves of different species mimic each other, they should be considered Müllerian mimics, and when nontoxic and palatable leaves mimic toxic ones, they should be considered Batesian mimics. The question of the potential role of mimicry in the evolution of red (or yellow) autumn coloration is still an enigma. If old yellow leaves are unpalatable, while leaves that have just turned yellow are rich with free amino acids (e.g., Holopainen and Peltonen 2002), then Batesian mimicry by the newly formed yellow leaves seems to operate with the yellow leaves formed earlier on the same tree, or among various trees of the same species that differ in yellowing time, or even among different species. The potential involvement of olfactory cues in autumn leaf aposematism should be studied. Again, Lev-Yadun and Gould (2007) emphasized that the lack of strong attacks on red or yellow autumn leaves does not necessarily prove that there is no risk of herbivory. The possibility of olfactory aposematism of yellow and red autumn leaves operating simultaneously with visual aposematism in unpalatable leaves was not discussed in depth. The fact that there are good physiological indications of significant volatile release from such leaves (Keskitalo et al. 2005) supports such a possibility.

7 Animal and Herbivore Damage Mimicry May Also Serve as Aposematic Coloration or Aposematic Visual Signals

It is probable that various types of defensive mimicry by plants may trick animals into behaving according to the plant's interests, just as they are tricked by bee mimicry of orchid flowers during pollination (e.g., Dafni 1984; Jersáková et al. 2006). Defensive animal mimicry by plants exists in several forms: (1) egg-laying mimicry, (2) ant mimicry, (3) aphid mimicry, (4) caterpillar mimicry, and (5) animal chewing or tunneling damage mimicry.

Butterfly egg mimicry in plants was proposed as a way to reduce egg laying by *Heliconius* butterflies (Benson et al. 1975; Gilbert 1980; Shapiro 1981a, b; Williams and Gilbert 1981).

Three types of visual defensive insect mimicry have been described. In the first type, plants have dark spots and flecks in the epidermis of stems, branches, and petioles that resemble ant swarms in size, shape, and pattern (Lev-Yadun and Inbar 2002). In the second type, dark anthers are the size, shape, and color of aphids, and they sway in the wind like swiveling aphids (Lev-Yadun and Inbar 2002). Finally, stipules along the branches of *Passiflora caerulae* look like caterpillars, slugs or snails climbing along the stems (Rothschild 1974, 1984), and immature pods of several annual legumes have conspicuous reddish spots, arranged along the pods, causing them to look like aposematic lepidopteran caterpillars (Lev-Yadun and Inbar 2002).

It is well known that ants defend plants from insect or mammalian herbivory, and in certain cases their relations with their hosts have been recognized as being mutualistic (e.g., Madden and Young 1992; Jolivet 1998). The potential benefit of ant-attendance mimicry is obvious. Ants bite and sting and are aggressive, and so many animals, including herbivores, will avoid them. Thus, ants have become models for a variety of arthropods that have evolved to mimic them (Wickler 1968; Edmunds 1974). The importance of ants in defending plants was demonstrated in a field experiment in which ant and aphid removal resulted in a 76% increase in the abundance of other herbivores on narrow-leaf cottonwoods (Wimp and Whitham 2001). Many plant species invest resources in attracting ants, providing them with shelter, food bodies and extrafloral nectaries (Huxley and Cutler 1991). Certain plants tolerate aphid infestation to gain antiherbivore protection from aphid-attending ants (Bristow 1991; Dixon 1998). Thus, it is not surprising that ant mimicry is found in plants. Ant mimicry has been found so far on the stems and petioles of *Xanthium trumarium* (Asteraceae) and *Arisarum vulgare* (Araceae) growing in Israel. The ant mimicry was in the form of conspicuous, dark-colored spots and flecks, usually 2–10 mm in size on the epidermis, resembling ants in size, shape and in the direction of their spatial patterns, which resemble a column of ants. Dots predominate in some individual plants; flecks in others (Lev-Yadun and Inbar 2002). Ant swarms are typically composed of many moving dark flecks, each varying in size from several millimeters to over a centimeter. The swaying of leaves, stems or branches in the wind in combination with the dark spots and flecks, many of which are arranged in lines, may give the illusion that the "ants" move. Again, the possibility of the involvement of olfactory mimicry of ants has not been studied yet. In any case, the aggressive and efficient antiherbivore activities of ants seem to make it beneficial for plants to mimic ant attendance in order to deter herbivores (both insects and vertebrates) without paying the cost of feeding or housing them (Lev-Yadun and Inbar 2002).

A mimicry phenomenon similar to ant mimicry is aphid mimicry. Lev-Yadun and Inbar (2002) described aphid mimicry in *Paspalum paspaloides* (= *P. distichum*), where the dark anthers are the size, shape and color of aphids and the anthers sway in the wind like swiveling aphids. Similarly, the stems of *Alcea setosa* are also covered with dark flecks that look like aphids. It has been proposed by Lev-Yadun and Inbar (2002) that plants that look infested may be left untouched by both mammalian grazers and aphids or various other insects. Several studies have shown that early infestation by aphids and other homopterans has a negative impact on host plant preferences and larval performance of other insect herbivores. Finch and Jones (1989) reported that large colonies of the cabbage aphid *Brevicorynebrassicae* and the peach aphid *Myzus presicae* deter ovipositioning by the root fly *Delia radicum*. Inbar et al. (1999) demonstrated that homopterans (whiteflies) not only alter adult cabbage looper (*Trichoplusia ni*) host selection, but also actually reduce the feeding efficiency of their offspring. Aphids respond to crowding by enhanced dispersal (Dixon 1998), and so it is also probable that they may avoid already infested or infestation-mimicking hosts. This clear zoological data set supports the hypotheses about the potential defensive value of aphid mimicry, but experimental data is needed to fully accept this hypothesis. Again, the possible involvement of olfactory cues should not be ruled out.

The third case of conspicuous coloration that mimics insects for defense is that of caterpillar mimicry. It was proposed to operate in two types of mimicry: (1) stipules along the branches of *P. caerulae* look like caterpillars, slugs or snails climbing along the stems, and were proposed to deter butterflies searching for laying sites (Rothschild 1974, 1984); (2) immature legume pods of several wild annual legumes (*Lathyrus ochrus*; *Pisum elatius*; *P. humile*; *Vicia peregrina*) look like aposematic poisonous lepidopteran caterpillars ornamented with spiracles or other spots on their sides due to the presence of conspicuous spots in various shades of red and purple arranged along the pods (Lev-Yadun and Inbar 2002), which may serve as herbivore-repellent cues and form part of the defense system of the plants. Caterpillars employ a large array of defenses that reduce predation. Unpalatable caterpillars with stinging and irritating hairs, functional osmeteria or body-fluid toxins often advertise their presence by aposematic coloration and aggregation (Cott 1940; Bowers 1993; Eisner et al. 2005). The usual warning colors of caterpillars are yellow, orange, red, black and white with stripes along the body and/or arranged in spots, especially around the abdominal spiracles. To conclude the cases of defensive insect mimicry by plants, Lev-Yadun and Inbar (2002) suggested that the cases of ant, aphid and caterpillar mimicry may signal unpalatability (aposematism) to more than one group of animals in two ways: first, insect mimicry may reduce attacks by insect herbivores that refrain from colonizing or feeding on infested plants (because of competition, cannibalism and/or induced plant defenses); and second, where the insect mimicked is aposematic, this could deter larger herbivores from eating the plants. None of these hypotheses about the various types of defensive insect mimicry was tested directly. It has however been shown that ungulates may actively select leaves in the field by shape and color and avoid spotted ones (e.g., Cahn and Harper 1976), but there seems to be no published data on the response of mammalian herbivores to aposematic (or cryptic) caterpillars. Again, the possible involvement of olfactory deterrence was not studied.

Mimicry of feeding damage by caterpillars as the reason for the formation of leaf lobes in some Moraceae (Niemelä and Tuomi 1987) or mimicry of chewed leaf ends in certain palms (Dirzo 2002) has also been posited. The mimicry of tunneling (Smith 1986; Lev-Yadun 2003a; Lee 2007; Campitelli et al. 2008; Soltau et al. 2009) was described above.

A related phenomenon, the use of aposematic insects to defend plants from large herbivores, has been proposed by Rothschild (1972, 1986). Various poisonous aposematic insects aggregate on poisonous plants, adding to the plant's aposematic odor and possibly to its coloration. This type of mutualism via aposematism deserves much more descriptive, theoretical and experimental studies.

8 Plant Aposematism Involving Fungi

The possibility that plants have mutualistic relationships with various fungi including pathogenic ones is not new. Most suggestions for such relations were based on the chemical defenses provided by endophytic or parasitic fungi (Clay 1990; Bush et al. 1997; Omacinl et al. 2001; Clay and Schardl 2002). Recently, there were two suggestions that fungal pigmentation, with or without known toxins, is used as a type of aposematic coloration. In the first case, Lev-Yadun and Halpern (2007) proposed that the very poisonous purple–black sclerotia of the infamous fungus *Claviceps purpurea* (ergot) and many other *Claviceps* species are aposematic. Very toxic fungal sclerotia are associated with conspicuous colors (black, yellow, purple, reddish, brown, violet, white and their combinations), and they severely harm herbivores that consume the infected plants, thus meeting the criteria for aposematism. These fungi, which only moderately reduce the reproductive capacity of their hosts, can protect the host plants from herbivory and weaken the evolutionary tendency of their hosts to evolve better resistance to infection. Moreover, by doing so, the fungi defend the host plant that is their habitat. In the second case, Lev-Yadun (2006a) proposed that whitish-colored plants may appear to be infested by fungal disease. Because there are very good indications that plant parts that may be infested by fungi are rejected by animals—frugivores avoid eating damaged fruits (Janzen, 1977; Herrera, 1982; Manzur and Courtney, 1984; Borowicz, 1988; Buchholz and Levey, 1990)—Lev-Yadun (2006a) proposed that white plant surfaces that mimic fungus-infested plants may reduce the tendency of herbivores to consume such plants. This is a type of visual aposematism.

9 Distance of Action of Aposematic Coloration (Crypsis Versus Aposematism)

The difference between a smart person and a wise one is that the wise one will not get into the difficult situation from which the smart one can find his way out. Ruxton et al. (2004) named their book about defensive mechanisms *Avoiding Attack*. Camouflage of various types is a good way to escape attack, but when it fails, aposematism may operate at close range in various animals (e.g., Mappes et al. 2005; Tullberg et al. 2005). The different visual and cognitive abilities of various animal taxa add to the significance of the variability of aposematic signaling (Endler and Mappes 2004); this is probably also true when plants have several colors in their spine system (Lev-Yadun 2001). While this double strategy (camouflage versus aposematism) has not yet been studied in plants, there are indications that it may operate. For instance, certain cacti use their spines for camouflage from a distance (Benson 1982), while they may be colorful and aposematic at close range (e.g., Lev-Yadun 2001). This issue deserves descriptive, theoretical and experimental studies so that it can be better understood.

10 Aposematic Trichomes: Probably an Overlooked Phenomenon

Trichomes, the unicellular and multicellular appendages of the epidermis (Fahn 1990), are well known for their multiple functions in plants. Trichomes may serve in protecting plants from excess sun irradiation of various wavelengths, including UV (Fahn and Cutler 1992; Manetas 2003); secrete toxic ions, especially in saline habitats (Fahn 1988); function in water absorption (Fahn and Cutler 1992); reduce transpiration (Fahn and Cutler 1992; Werker 2000); defend from insect or other herbivorous invertebrates by reducing accessibility or by actually trapping their legs or by chemical means (Levin 1973; Fahn 1979, 1988; Werker 2000); and defend from large herbivores when they sting, as in *Urtica* (Thurston and Lersten 1969; Levin 1973; Fahn 1990; Fu et al. 2006). In addition, in certain carnivorous plants like *Drosera* and *Dionea*, they may take part in the attraction, capture and digestion of insects (Juniper et al. 1989; Fahn 1990). Many plant trichomes are colorful (red, yellow, orange, blue, white) and very conspicuous. In certain cases, such as in cotton plants, pigmented trichomes produce toxins that defend from caterpillars (Agrawal and Karban 2000). In addition, the trichomes have conspicuous red markings at their base in various plants, e.g., *Echium angustifolium* (Boraginaceae) and *Echinopsadenocaulos* (Asteraceae). Thorns, spines and prickles are large and usually spaced, and their ability to defend from insects is limited (e.g., Potter and Kimmerer 1988), whereas trichomes—because of their size, density and chemical composition—may commonly defend plants from insects (e.g., Levin 1973; Fahn 1979, 1988; Werker 2000). I propose that colorful and poisonous or sticky trichomes may deter insects and serve as aposematic coloration. Because many insects see UV (Briscoe and Chittka 2001), the possibility that trichomes may deter insects in the UV channel should be considered and studied. The possibility that trichomes produce olfactory aposematic signals in addition to visual ones should also be considered, in light of the secretive nature of many trichomes (Fahn 1979, 1988).

11 Experimental Evidence

The experimental evidence for the operation of aposematic coloration in plants is meager. Cook et al. (1971) showed that poisonous gray seeds of *Eremocarpus setigerus* are rejected by the mountain dove. Cahn and Harper (1976) showed that sheep avoid *Trifolium repens* plants with leaf marks, but did not discuss aposematism. Lev-Yadun and Ne'eman (2004) showed that sheep, goats, camels, donkeys and cattle reject conspicuous green plants in the yellow desert in the summer. Numata et al. (2004) found that leaves with delayed greening suffer lower levels of insect damage when they are still young. Hill (2006) showed that the Florida scrub jay (*Aphelocoma coerulescens*) rejects poisonous red fruit. Karageorgou and Manetas (2006) showed that young red leaves of the evergreen oak *Quercus coccifera* are attacked less than green ones by insects, but rejected the aposematic coloration hypothesis. Similar results were found for other species in Greece (Karageorgou et al. 2008). Recently, additional data about the defensive operation of white variegation that mimics insect damage in leaves was published (Campitelli et al. 2008; Soltau et al. 2009). The possibility of olfactory aposematism was not tested in any of these cases.

12 Conclusions

Aposematic (warning) coloration seems to be a common defense in plants, even though it was practically ignored by botanists and zoologists until recently. The first papers fully dedicated to visual aposematism in plants were published after the year 2000. Similar to the situation in animals, aposematic (warning) coloration (yellow, orange, red, brown, black and white, and combinations of these colors) seems to be a common defense in plants, although (except for anecdotal mentions) it has been practically ignored by botanists and zoologists until recently. The fact that many aposematic animals use both plant-based pigments and sequestered poisonous molecules to become aposematic highlights the absurdity of neglecting the aposematic nature of so many plants. Aposematic coloration was referred to here in the broadest sense, considering any visual warning phenomenon that may deter herbivores. Aposematic coloration is expressed by thorny, spiny and prickly plants, by poisonous ones, and by plants that are unpalatable for various reasons. Plants that mimic aposematic plants or aposematic animals are probably common, despite the small number described so far. Colorful mimicry of insect infestation or herbivore damage to tissues in order to repel herbivores is also found in plants. Many types of aposematic coloration simultaneously serve other functions, such as physiological, communicative and even other defensive functions. It is therefore difficult in many cases to evaluate the relative functional share of aposematism in various color patterns of plants and the specific selective agents involved in their evolution. Aposematic coloration is part of a broader phenomenon of defensive coloration in plants, which has also received insufficient attention. The fact that botanists have not usually considered the operation of aposematic coloration or other types of defensive coloration is evident from the lack of a regular and systematic description of these color patterns in the majority if not all of the thousands of published floras. The related phenomenon of olfactory aposematism of poisonous plants has also received very little attention. Simultaneous operation of visual and olfactory aposematism in the same plant is also proposed. Many theoretical aspects of aposematism that were and are currently being studied experimentally in animals have almost never been studied in plants. Aposematic coloration in animals has been broadly studied since the nineteenth century and is still not fully understood. The effort needed to understand aposematic coloration in plants is probably not any smaller. This situation provides the opportunity for ambitious scientists to express their capabilities. Thus, there appears to be a colorful future for the study of aposematic coloration in plants.

Acknowledgements I thank Shahal Abbo, Marco Archetti, Amots Dafni, Moshe Flaishman, Kevin Gould, Malka Halpern, Moshik Inbar, Ido Izhaki, Gadi Katzir, Gidi Ne'eman, Martin Schaefer, Ron Sederoff, Uri Shanas, and Pille Urbas for stimulating discussions, important comments, field trips and collaboration.

References

Agrawal AA, Karban R (2000) Specificity of constitutive and induced resistance: pigment glands influence mites and caterpillars on cotton plants. Entomol Exp Appl 96:39–49

Agrawal AA, Spiller DA (2004) Polymorphic buttonwood: effects of disturbance on resistance to herbivores in green and silver morphs of a Bahamian shrub. Am J Bot 91:1990–1997

Armbruster WS (1997) Exaptations link evolution of plant–herbivore and plant–pollinator interactions: a phylogenetic inquiry. Ecology 78:1661–1672

Armbruster WS, Howard JJ, Clausen TP, Debevec EM, Loquvam JC, Matsuki M, Cerendolo B, Andel F (1997) Do biochemical exaptations link evolution of plant defense and pollination systems? Historical hypotheses and experimental tests with *Dalechampia* vines. Am Nat 149:461–484

Archetti M (2000) The origin of autumn colours by coevolution. J Theor Biol 205:625–630

Archetti M (2007a) Autumn colours and the nutritional translocation hypothesis: a theoretical assessment. J Theor Biol 244:714–721

Archetti M (2007b) Colour preference as evidence for the theories on the evolution of autumn colours. J Theor Biol 245:595–596

Archetti M, Brown SP (2004) The coevolution theory of autumn colours. Proc R Soc Lond B 271:1219–1223

Archetti M, Brown SP (2006) Putting 'red alerts' in an ecological and evolutionary context. BioEssays 28:959

Archetti M, Leather SR (2005) A test of the coevolution theory of autumn colours: colour preference of *Rhopalosiphum padi* on *Prunus padus*. Oikos 110:339–343

Atsatt PR, O'Dowd DJ (1976) Plant defense guilds. Science 193:24–29

Augner M (1994) Should a plant always signal its defence against herbivores? Oikos 70:322–332

Augner M (1995) Low nutritive quality as a plant defence: effects of herbivore-mediated interactions. Evol Ecol 9:605–616

Augner M, Bernays EA (1998) Plant defence signals and Batesian mimicry. Evol Ecol 12:667–679

Benson L (1982) The cacti of the United States and Canada. Stanford University Press, Stanford

Benson WW, Brown KS, Gilbert LE (1975) Coevolution of plants and herbivores: passion flower butterflies. Evolution 29:659–680

Bernays EA, Driver GC, Bilgener M (1989) Herbivores and plant tannins. Adv Ecol Res 19:263–302

Borowicz VA (1988) Do vertebrates reject decaying fruit? An experimental test with *Cornus amomum* fruits. Oikos 53:74–78
Bowers DM (1993) Aposematic caterpillars: life-styles of the warningly colored and unpalatable. In: Stamp NE, Casey TM (eds) Caterpillars: ecological and evolutionary constraints on foraging. Chapman and Hall, New York, pp 331–371
Bradshaw HD Jr, Schemske DW (2003) Allele substitution at a flower colour locus produces a pollinator shift in monkeyflowers. Nature 426:176–178
Briscoe AD, Chittka L (2001) The evolution of color vision in insects. Annu Rev Entomol 46:471–510
Bristow CM (1991) Why are so few aphids ant attended? In: HuxleyCR, Cutler DF (eds) Ant–plant interactions. Oxford University Press, London, pp 104–119
Brooks R, Owen-Smith N (1994) Plant defences against mammalian herbivores: are juvenile *Acacia* more heavily defended than mature trees? Bothalia 24:211–215
Brown SP (2005) A view from Mars. In: Ridley M (ed) Narrow roads of gene land—the collected papers of W. D. Hamilton, vol 3—last words. Oxford University Press, Oxford, pp 350–356
Brown JH, Lieberman GA, Dengler WF (1972) Woodrats and cholla: dependence of a small mammal population on the density of cacti. Ecology 53:310–313
Buchholz R, Levey DJ (1990) The evolutionary triad of microbes, fruits, and seed dispersers: an experiment in fruit choice by cedar waxwings, *Bombycilla cedrorum*. Oikos 59:200–204
Bush LP, Wilkinson HH, Schardl CL (1997) Bioprotective alkaloids of grass-fungal endophyte symbioses. Plant Physiol 114:1–7
Cahn MG, Harper JL (1976) The biology of the leaf mark polymorphism in *Trifolium repens* L. 2. Evidence for the selection of leaf marks by rumen fistulated sheep. Heredity 37:327–333
Campitelli BE, Steglik I, Stinchcombe JR (2008) Leaf variegation is associated with reduced herbivore damage in *Hydrophyllum virginianum*. Botany 86:306–313
Carrascal LM, Díaz JA, Huertas DL, Mozetich I (2001) Behavioral thermoregulation by treecreepers: trade-off between saving energy and reducing crypsis. Ecology 82:1642–1654
Chalker-Scott L (1999) Environmental significance of anthocyanins in plant stress responses. Photochem Photobiol 70:1–9
Chittka L, Döring TF (2007) Are autumn foliage colors red signals to aphids? PLoS Biol 5:e187
Cipollini ML, Levey DJ (1997) Secondary metabolites of fleshy vertebrate-dispersed fruits: adaptive hypotheses and implications for seed dispersal. Am Nat 150:346–372
Clay K (1990) Fungal endophytes of grasses. Annu Rev Ecol Syst 21:275–297
Clay K, Schardl C (2002) Evolutionary origins and ecological consequences of endophyte symbiosis with grasses. Am Nat 160(Suppl):S99–S127
Clegg MT, Durbin ML (2003) Tracing floral adaptations from ecology to molecules. Nat Rev Genet 4:206–215
Clement CR, Manshardt RM (2000) A review of the importance of spines for pejibaye heart-of-palm production. Sci Hort 83:11–23
Close DC, Beadle CL (2003) The ecophysiology of foliar anthocyanin. Bot Rev 69:149–161
Coley PD, Barone JA (1996) Herbivory and plant defenses in tropical forests. Annu Rev Ecol Syst 27:305–335
Cook AD, Atsatt PR, Simon CA (1971) Doves and dove weed: multiple defenses against avian predation. BioScience 21:277–281
Cooper SM, Owen-Smith N (1986) Effects of plant spinescence on large mammalian herbivores. Oecologia 68:446–455
Cornett JW (1986) Spineless petioles in *Washingtonia filifera* (Arecaceae). Madroño 33:76–78
Cott HB (1940) Adaptive coloration in animals. Methuen, London
Dafni A (1984) Mimicry and deception in pollination. Annu Rev Ecol Syst 15:259–278
Dirzo R (2002) Lying leaves. Science 297:1119
Dixon AFG (1998) Aphid ecology. An optimization approach. Chapman and Hall, London

Dobson HEM, Bergström G (2000) The ecology and evolution of pollen odors. Plant Syst Evol 222:63–87

Dominy ND, Lucas PW, Ramsden W, Riba-Hernandez P, Stoner KE, Turner IM (2002) Why are young leaves red? Oikos 98:163–176

Edmunds M (1974) Defence in animals. A survey of anti-predator defences. Longman, Harlow

Edmunds M, Grayson J (1991) Camouflage and selective predation in caterpillars of the poplar and eyed hawkmoths (*Laothoe populi* and *Smerinthus ocellata*). Biol J Linn Soc 42:467–480

Ehleringer JR, Ullmann I, Lange OL, Farquhar GD, Cowan IR, Schulze E-D, Ziegler H (1986) Mistletoes: a hypothesis concerning morphological and chemical avoidance of herbivory. Oecologia 70:234–237

Eisner T (1964) Catnip: its Raison d'Être. Science 146:1318–1320

Eisner T (1981) Leaf folding in a sensitive plant: a defensive thorn-exposure mechanism? Proc Natl Acad Sci USA 78:402–404

Eisner T, Grant RP (1981) Toxicity, odor aversion, and "olfactory aposematism". Science 213:476

Eisner T, Eisner M, Siegler M (2005) Secret weapons. Defenses of insects, spiders, scorpions, and other many-legged creatures. Harvard University Press, Cambridge

Endler JA (1981) An overview of the relationships between mimicry and crypsis. Biol J Linn Soc 16:25–31

Endler JA (1984) Progressive background matching in moths, and a quantitative measure of crypsis. Biol J Linn Soc 22:187–231

Endler JA, Mappes J (2004) Predator mixes and the conspicuousness of aposematic signals. Am Nat 163:532–547

Ergonul O, Erbay A, Eren S, Dokuzoguz B (2003) Analysis of the case fatality rate of tetanus among adults in a tertiary hospital in Turkey. Eur J Clin Microbiol Infect Dis 22:188–190

Faegri K, Pijl L van der (1979) The principles of pollination ecology, 3rd edn. Pergamon, Oxford

Fahn A (1979) Secretory tissues in plants. Academic, London

Fahn A (1988) Secretory tissues in vascular plants. New Phytol 108:229–257

Fahn A (1990) Plant anatomy, 4th edn. Pergamon, Oxford

Fahn A, Cutler DF (1992) Xerophytes. Gebrüder Borntraeger, Berlin

Feeny P (1976) Plant apparency and chemical defense. Rec Adv Phytochem 10:1–40

Finch S, Jones TH (1989) An analysis of the deterrent effect of Aphids on cabbage root fly (*Delia radicum*) egg-laying. Ecol Entomol 14:387–391

Fineblum WL, Rausher MD (1997) Do floral pigmentation genes also influence resistance to enemies? The *W* locus in *Ipomoea purpurea*. Ecology 78:1646–1654

Franceschi VR, Horner HT Jr (1980) Calcium oxalate crystals in plants. Bot Rev 46:361–427

Franceschi VR, Nakata PA (2005) Calcium oxalate in plants: formation and function. Annu Rev Plant Biol 56:41–71

Fu HY, Chen SJ, Chen RF, Ding WH, Kuo-Huang LL, Huang RN (2006) Identification of oxalic acid and tartaric acids as major persistent pain-inducing toxins in the stinging hairs of the Nettle, *Urtica thunbergiana*. Ann Bot 98:57–65

Gilbert LE (1980) Ecological consequences of a coevolved mutualism between butterflies and plants. In: Gilbert LE, Raven PH (eds) Coevolution of animals and plants. University of Texas Press, Austin, pp 210–240

Gittleman JL, Harvey PH (1980) Why are distasteful prey not cryptic? Nature 286:149–150

Givnish TJ (1990) Leaf mottling: relation to growth form and leaf phenology and possible role as camouflage. Funct Ecol 4:463–474

Gleadow RM, Woodrow IE (2002) Constraints on effectiveness of cyanogenic glycosides in herbivore defense. J Chem Ecol 28:1301–1313

Gould KS (2004) Nature's Swiss army knife: the diverse protective roles of anthocyanins in leaves. J Biomed Biotechnol 2004:314–320

Gould KS, Neill SO, Vogelmann TC (2002b) A unified explanation for anthocyanins in leaves? Adv Bot Res 37:167–192

Gowda JH, Palo RT (2003) Age-related changes in defensive traits of *Acacia tortilis* Hayne. Afr J Ecol 41:218–223

Grafen A (1990) Biological signals as handicaps. J Theor Biol 144:517–546

Gronquist M, Bezzerides A, Attygalle A, Meinwald J, Eisner M, Eisner T (2001) Attractive and defensive functions of the ultraviolet pigments of a flower (*Hypericum calycinum*). Proc Natl Acad Sci USA 98:13745–13750

Grubb PJ (1992) A positive distrust in simplicity—lessons from plant defences and from competition among plants and among animals. J Ecol 80:585–610

Guilford T, Nicol C, Rothschild M, Moore BP (1987) The biological roles of pyrazines: evidence for a warning odour function. Biol J Linn Soc 31:113–128

Guthrie RD, Petocz RG (1970) Weapon automimicry among animals. Am Nat 104:585–588

Hagen SB, Folstad I, Jakobsen SW (2003) Autumn colouration and herbivore resistance in mountain birch (*Betula pubescens*). Ecol Lett 6:807–811

Hagen SB, Debeausse S, Yoccoz NG, Folstad I (2004) Autumn coloration as a signal of tree condition. Proc R Soc Lond B 271(Suppl):S184-S185

Halpern M, Raats D, Lev-Yadun S (2007a) Plant biological warfare: thorns inject pathogenic bacteria into herbivores. Environ Microbiol 9:584–592

Halpern M, Raats D, Lev-Yadun S (2007b) The potential anti-herbivory role of microorganisms on plant thorns. Plant Signal Behav 2:503–504

Hamilton WD, Brown SP (2001) Autumn tree colours as a handicap signal. Proc R Soc Lond B 268:1489–1493

Hansen DM, Olesen JM, Mione T, Johnson SD, Müller CB (2007) Coloured nectar: distribution, ecology, and evolution of an enigmatic floral trait. Biol Rev 82:83–111

Harborne JB (1982) Introduction to ecological biochemistry. Academic, London

Harper JL (1977) Population biology of plants. Academic, London

Harvey PH, Paxton RJ (1981) The evolution of aposematic coloration. Oikos 37:391–396

Hatier J-HB, Gould KS (2008) Foliar anthocyanins as modulators of stress signals. J Theor Biol 253:625–627

Herrera CM (1982) Defense of ripe fruit from pests: its significance in relation to plant-disperser interactions. Am Nat 120:218–241

Hill ME (2006) The effect of aposematic coloration on the food preference of *Aphelocoma coerulescens*, the Florida scrub jay. Bios 77:97–106

Hinton HE (1973) Natural deception. In: Gregory RL, Gombrich EH (eds) Illusion in nature and art. Duckworth, London, pp 97–159

Hoch WA, Zeldin EL, McCown BH (2001) Physiological significance of anthocyanins during autumnal leaf senescence. Tree Physiol 21:1–8

Hoch WA, Singsaas EL, McCown BH (2003) Resorption protection. Anthocyanins facilitate nutrient recovery in autumn by shielding leaves from potentially damaging light levels. Plant Physiol 133:1296–1305

Hodes RM, Teferedegne B (1990) Tetanus in Ethiopia: analysis of 55 cases from Addis Ababa. E Afr Med J 67:887–893

Hoekstra HE (2006) Genetics, development and evolution of adaptive pigmentation in vertebrates. Heredity 97:222–234

Holopainen JK, Peltonen P (2002) Bright autumn colours of deciduous trees attract aphids: nutrient retranslocation hypothesis. Oikos 99:184–188

Huxley CR, Cutler DF (1991) Ant–plant interactions. Oxford University Press, London

Inbar M, Lev-Yadun S (2005) Conspicuous and aposematic spines in the animal kingdom. Naturwissenschaften 92:170–172

Inbar M, Doostdar H, Mayer RT (1999) Effects of sessile Whitefly nymphs (Homoptera: Aleyrodidae) on leaf-chewing larvae (Lepidoptera: Noctuidae). Environ Entomol 28:353–357

Izhaki I (2002) Emodin—a secondary metabolite with multiple ecological functions in higher plants. New Phytol 155:205–217

Janzen DH (1977) Why fruits rot, seeds mold, and meat spoils. Amer Nat 111:691–713

Janzen DH (1979) New horizons in the biology of plant defenses. In: Rosenthal GA, Janzen DH (eds) Herbivores their interaction with secondary plant metabolites. Academic, Orlando, pp 331–350

Janzen DH (1986) Chihuahuan desert nopaleras: defaunated big mammal vegetation. Annu Rev Ecol Syst 17:595–636

Janzen DH, Martin PS (1982) Neotropical anachronisms: the fruits the gomphotheres ate. Science 215:19–27

Jensen GB, Hansen BM, Eilenberg J, Mahillon J (2003) The hidden lifestyles of *Bacillus cereus* and relatives. Environ Microbiol 5:631–640

Jersáková J, Johnson SD, Kindlmann P (2006) Mechanisms and evolution of deceptive pollination in orchids. Biol Rev 81:219–235

Johnson SD, Hargreaves AL, Brown M (2006) Dark, bitter-tasting nectar functions as a filter of flower visitors in a bird-pollinated plant. Ecology 87:2709–2716

Jolivet P (1998) Interrelationship between insects and plants. CRC, Boca Raton

Juniper BE, Robins RJ, Joel DM (1989) The carnivorous plants. Academic, London

Junker R, Chung AYC, Blüthgen N (2007) Interaction between flowers, ants and pollinators: additional evidence for floral repellence against ants. Ecol Res 22:665–670

Kappers IF, Aharoni A, van Herpen TWJM, Luckerhoff LLP, Dicke M, Bouwmeester HJ (2005) Genetic engineering of terpenoid metabolism attracts bodyguards to *Arabidopsis*. Science 309:2070–2072

Karageorgou P, Manetas Y (2006) The importance of being red when young: anthocyanins and the protection of young leaves of *Quercus coccifera* from insect herbivory and excess light. Tree Physiol 26:613–621

Karageorgou P, Buschmann C, Manetas Y (2008) Red leaf color as a warning signal against insect herbivory: honest or mimetic? Flora 203:648–652

Kaye H, Mackintosh NJ, Rothschild M, Moore BP (1989) Odour of pyrazine potentiates an association between environmental cues and unpalatable taste. Anim Behav 37:563–568

Keskitalo J, Bergquist G, Gardeström P, Jansson S (2005) A cellular timetable of autumn senescence. Plant Physiol 139:1635–1648

Kessler A, Baldwin IT (2001) Defensive function of herbivore induced plant volatile emissions in nature. Science 291:2141–2144

Kettlewell B (1973) The evolution of melanism. Clarendon, Oxford

Knight RS, Siegfried WR (1983) Inter-relationships between type, size and color of fruits and dispersal in Southern African trees. Oecologia 56:405–412

Knudsen JT, Eriksson R, Gershenzon J, Ståhl B (2006) Diversity and distribution of floral scent. Bot Rev 72:1–120

Komárek S (1998) Mimicry, aposematism and related phenomena in animals and plants: bibliography 1800–1990. Vesmir, Prague

Kozlowski TT (1971) Growth and development of trees. vol. I. Academic, New York

Kursar TA, Coley PD (1991) Nitrogen content and expansion rate of young leaves of rainforest species: implications for herbivory. Biotropica 23:141–150

Kursar TA, Coley PD (1992) Delayed greening in tropical leaves: an antiherbivory defense? Biotropica 24:256–262

Kursar TA, Coley PD (2003) Convergence in defense syndromes of young leaves in tropical rainforests. Biochem Syst Ecol 31:929–949

Lachmann ML, Számadó S, Bergstrom CT (2001) Cost and conflict in animal signals and human language. Proc Natl Acad Sci USA 98:13189–13194

Launchbaugh KL, Provenza FD (1993) Can plants practice mimicry to avoid grazing by mammalian herbivores? Oikos 66:501–504

Laycock WA (1978) Coevolution of poisonous plants and large herbivores on rangelands. J Range Manage 31:335–342

Lee DW (2002) Anthocyanins in autumn leaf senescence. Adv Bot Res 37:147–165

Lee D (2007) Nature's palette. The science of plant color. University of Chicago Press, Chicago

Lee DW, Gould KS (2002) Anthocyanins in leaves and other vegetative organs: an introduction. Adv Bot Res 37:1–16
Lee DW, Brammeler S, Smith AP (1987) The selective advantages of anthocyanins in developing leaves of mango and cacao. Biotropica 19:40–49
Lee DW, O`Keefe J, Holbrook NM, Feild TS (2003) Pigment dynamics and autumn leaf senescence in a New England deciduous forest, eastern USA. Ecol Res 18:677–694
Levin DA (1973) The role of trichomes in plant defense. Q Rev Biol 48:3–15
Lev-Yadun S (2001) Aposematic (warning) coloration associated with thorns in higher plants. J Theor Biol 210:385–388
Lev-Yadun S (2003a) Why do some thorny plants resemble green zebras? J Theor Biol 244:483–489
Lev-Yadun S (2003b) Weapon (thorn) automimicry and mimicry of aposematic colorful thorns in plants. J Theor Biol 244:183–188
Lev-Yadun S (2006a) Defensive functions of white coloration in coastal and dune plants. Isr J Plant Sci 54:317–325
Lev-Yadun S (2006b) Defensive coloration in plants: a review of current ideas about anti-herbivore coloration strategies. In: Teixeira da Silva JA (ed) Floriculture, ornamental and plant biotechnology: advances and topical issues. vol. IV. Global Science Books, London, pp 292–299
Lev-Yadun S, Gould KS (2007) What do red and yellow autumn leaves signal? Bot Rev 73:279–289
Lev-Yadun S, Gould KS (2008) Role of anthocyanins in plant defense. In: Gould KS, Davies KM, Winefield C (eds) Life's colorful solutions: the biosynthesis, functions, and applications of anthocyanins. Springer, Berlin, pp 21–48
Lev-Yadun S, Halpern M (2007) Ergot (*Claviceps purpurea*)—an aposematic fungus. Symbiosis J 43:105–108
Lev-Yadun S, Halpern M (2008) External and internal spines in plants insert pathogenic microorganisms into herbivore's tissues for defense. In: Van Dijk T (ed) Microbial ecology research trends. Nova Scientific, New York, pp 155–168
Lev-Yadun S, Inbar M (2002) Defensive ant, aphid and caterpillar mimicry in plants. Biol J Linn Soc 77:393–398
Lev-Yadun S, Ne'eman G (2004) When may green plants be aposematic? Biol J Linn Soc 81:413–416
Lev-Yadun S, Ne'eman G (2006) Color changes in old aposematic thorns, spines, and prickles. Isr J Plant Sci 54:327–333
Lev-Yadun S, Dafni A, Inbar M, Izhaki I, Ne'eman G (2002) Colour patterns in vegetative parts of plants deserve more research attention. Trends Plant Sci 7:59–60
Lev-Yadun S, Dafni A, Flaishman MA, Inbar M, Izhaki I, Katzir G, Ne'eman G (2004) Plant coloration undermines herbivorous insect camouflage. BioEssays 26:1126–1130
Madden D, Young TP (1992) Symbiotic ants as an alternative defense against giraffe herbivory in spinescent *Acacia drepanolobium*. Oecologia 91:235–238
Majerus MEN (1998) Melanism. Evolution in Action. Oxford University Press, Oxford
Manetas Y (2003) The importance of being hairy: the adverse effects of hair removal on stem photosynthesis of *Verbascum speciosum* are due to solar UV-B radiation. New Phytol 158:503–508
Manzur MI, Courtney SP (1984) Influence of insect damage in fruits of hawthorn on bird foraging and seed dispersal. Oikos 43:265–270
Mappes J, Marples N, Endler JA (2005) The complex business of survival by aposematism. Trends Ecol Evol 20:598–603
Massei G, Cotterill JV, Coats JC, Bryning G, Cowan DP (2007) Can Batesian mimicry help plants to deter herbivory? Pest Manage Sci 63:559–563
Matile P (2000) Biochemistry of Indian summer: physiology of autumnal leaf coloration. Exp Gerontol 35:145–158
Merilaita S, Tuomi J, Jormalainen V (1999) Optimization of cryptic coloration in heterogeneous habitat. Biol J Linn Soc 67:151–161

Midgley JJ (2004) Why are spines of African Acacia species white? Afr J Range Forage Sci 21:211–212
Midgley JJ, Botha MA, Balfour D (2001) Patterns of thorn length, density, type and colour in African Acacias. Afr J Range Forage Sci 18:59–61
Milewski AV, Young TP, Madden D (1991) Thorns as induced defenses: experimental evidence. Oecologia 86:70–75
Montegomery JM, Gillespie D, Sastrawan P, Fredeking TM, Stewart GL (2002) Aerobic salivary bacteria in wild and captive Komodo dragons. J Wildl Dis 38:545–551
Moore BP, Brown WV, Rothschild M (1990) Methylalkylpyrazines in aposematic insects, their hostplants and mimics. Chemoecology 1:43–51
Moran N, Hamilton WD (1980) Low nutritive quality as defence against herbivores. J Theor Biol 86:247–254
Myers JH, Bazely D (1991) Thorns, spines, prickles, and hairs: are they stimulated by herbivory and do they deter herbivores? In: Tallamy DW, Raupp MJ (eds) Phytochemical induction by herbivores. Wiley, New York, pp 325–344
Nakata PA (2003) Advances in our understanding of calcium oxalate crystal formation and function in plants. Plant Sci 164:901–909
Niemelä P, Tuomi J (1987) Does the leaf morphology of some plants mimic caterpillar damage? Oikos 50:256–257
Numata S, Kachi N, Okuda T, Manokaran N (2004) Delayed greening, leaf expansion, and damage to sympatric *Shorea* species in a lowland rain forest. J Plant Res 117:19–25
Omacinl M, Chaneton EJ, Ghersa CM, Müller CB (2001) Symbiotic fungal endophytes control insect host–parasite interaction webs. Nature 409:78–81
Ougham HJ, Morris P, Thomas H (2005) The colors of autumn leaves as symptoms of cellular recycling and defenses against environmental stresses. Curr Top Dev Biol 66:135–160
Ougham H, Thomas H, Archetti M (2008) The adaptive value of leaf colour. New Phytol 179:9–13
Pellmyr O, Thien LB (1986) Insect reproduction and floral fragrances: keys to the evolution of the angiosperms? Taxon 35:76–85
Perevolotsky A, Haimov Y (1991) Structural response of Mediterranean woodland species to disturbance: evidence of different defense strategies. Isr J Bot 40:305–313
Pijl L van der (1982) Principles of dispersal in higher plants, 3rd edn. Springer, Berlin
Potter DA, Kimmerer TW (1988) Do holly leaf spines really deter herbivory? Oecologia 75:216–221
Preston-Mafham R, Preston-Mafham K (1994) Cacti: the illustrated dictionary. Timber, Portland
Provenza FD, Kimball BA, Villalba JJ (2000) Roles of odor, taste, and toxicity in the food preferences of lambs: implications for mimicry in plants. Oikos 88:424–432
Purser B (2003) Jungle bugs: masters of camouflage and mimicry. Firefly, Toronto
Ramirez CC, Lavandero B, Archetti M (2008) Coevolution and the adaptive value of autumn tree colours: colour preference and growth rates of a southern beech aphid. J Evol Biol 21:49–56
Rebollo S, Milchunas DG, Noy-Meir I, Chapman PL (2002) The role of spiny plant refuge in structuring grazed shortgrass steppe plant communities. Oikos 98:53–64
Richards PW (1996) The tropical rain forest: an ecological study, 2nd edn. Cambridge University Press, Cambridge
Richmond KE, Sussman M (2003) Got silicon? The non-essential beneficial plant nutrient. Curr Opin Plant Biol 6:268–272
Ridley HN (1930) The dispersal of plants throughout the world. L. Reeve, Ashford
Robbins CT, Hanley TA, Hagerman AE, Hjeljord O, Baker DL, Schwartz CC, Mautz WW (1987) Role of tannins in defending plants against ruminants: reduction in protein availability. Ecology 68:98–107
Rolshausen G, Schaefer HM (2007) Do aphids paint the tree red (or yellow): can herbivore resistance or photoprotection explain colourful leaves in autumn? Plant Ecol 191:77–84
Rothschild M (1972) Some observations on the relationship between plants, toxic insects and birds. In: Harborne JB (ed) Phytochemical ecology. Academic, London, pp 1–12

Rothschild M (1973) Secondary plant substances and warning colouration in insects. In: van Emden HF (ed) Insect/plant relationships (Symposia of the Royal Entomological Society of London, no. 6). Blackwell, Oxford, pp 59–83
Rothschild M (1974) Modified stipules of *Passiflora* which resemble horned caterpillars. Proc R Entomol Soc Lond 39:16
Rothschild M (1980) Remarks on carotenoids in the evolution of signals. In: Gilbert LE, Raven PH (eds) Coevolution of animals and plants. University of Texas Press, Austin, pp 20–51
Rothschild M (1984) Aide mémoire mimicry. Ecol Entomol 9:311–319
Rothschild M (1986) The red smell of danger. New Sci 111:34–36
Rothschild M, Moore B (1987) Pyrazines as alerting signals in toxic plants and insects. In: Labeyrie V, Fabres G, Lachaise D (eds) Insects–plants. Junk, Dordrecht, pp 97–101
Rothschild M, Moore BP, Brown WV (1984) Pyrazines as warning odour components in the Monarch butterfly, *Danaus plexippus*, and in moths of the genera *Zygaena* and *Amata* (Lepidoptera). Biol J Linn Soc 23:375–380
Rubino DL, McCarthy BC (2004) Presence of aposematic (warning) coloration in vascular plants of southeastern Ohio. J Torrey Bot Soc 131:252–256
Ruiz N, Ward D, Saltz D (2002) Calcium oxalate crystals in leaves of *Pancratium sickenbergeri*: constitutive or induced defence? Funct Ecol 16:99–105
Ruxton GD, Sherratt TN (2006) Aggregation, defence and warning signals: the evolutionary relationship. Proc R Soc B 273:2417–2424
Ruxton GD, Sherratt TN, Speed MP (2004) Avoiding attack. The evolutionary ecology of crypsis, warning signals and mimicry. Oxford University Press, Oxford
Ryan CA (1990) Protease inhibitors in plants: genes for improving defenses against insects and pathogens. Annu Rev Phytopathol 28:425–449
Sakai WS, Hanson M, Jones RC (1972) Raphids with barbs and grooves in *Xanthosoma sagittifolium* (Araceae). Science 178:314–315
Savage JM, Slowinski JB (1992) The colouration of the venomous coral snakes (family Elapidae) and their mimics (families Aniliidae and Colubridae). Biol J Linn Soc 45:235–254
Schaefer HM, Gould KS (2007) Modelling the evolution of leaf colouration with binary assumptions is barking up the wrong tree. J Theor Biol 249:638–639
Schaefer HM, Rolshausen G (2006) Plants on red alert: do insects pay attention? BioEssays 28:65–71
Schaefer HM, Rolshausen G (2007) Aphids do not attend to leaf colour as visual signal, but to the handicap of reproductive investment. Biol Lett 3:1–4
Schaefer HM, Schaefer V (2007) The evolution of visual fruit signals: concepts and constraints. In: Dennis AJ Schupp EW, Green R, Wescott DW (eds) Seed dispersal: theory and its application in a changing world. CAB International, Wallingford, pp 59–77
Schaefer HM, Schmidt V (2004) Detectability and content as opposing signal characteristics in fruits. Proc R Soc Lond B(Suppl) 271:S370–S373
Schaefer HM, Wilkinson DM (2004) Red leaves, insects and coevolution: a red herring? Trends Ecol Evol 19:616–618
Schaefer HM, Schaefer V, Levey DJ (2004) How plant–animal interactions signal new insights in communication. Trends Ecol Evol 19:577–584
Schaefer HM, Schaefer V, Vorobyev M (2007) Are fruit colors adapted to consumer vision and birds equally efficient in detecting colorful signals? Am Nat 169(suppl):S159–S169
Schoener TW (1987) Leaf pubescence in buttonwood: community variation in a putative defense against defoliation. Proc Natl Acad Sci USA 84:7992–7995
Schoener TW (1988) Leaf damage in island buttonwood, *Conocarpus erectus*: correlations with pubescence, island area, isolation and the distribution of major carnivores. Oikos 53:253–266
Shapiro AM (1981a) Egg-mimics of *Streptanthus* (Cruciferae) deter oviposition by *Pieris sisymbrii* (Lepidoptera: Pieridae). Oecologia 48:142–143
Shapiro AM (1981b) The pierid red-egg syndrome. Am Nat 117:276–294
Sherratt TN, Wilkinson DM, Bain RS (2005) Explaining Dioscorides' "double difference": why are some mushrooms poisonous, and do they signal their unprofitability? Am Nat 166:767–775

Shimizu T, Ohtani K, Hirakawa H, Ohshima K, Yamashita A, Shiba T, Ogasawara N, Hattori M, Kuhara S, Hayashi H (2002) Complete genome sequence of *Clostridium perfringens*, an anaerobic flesh-eater. Proc Natl Acad Sci USA 99:996–1001

Sinkkonen A (2006a) Sexual reproduction advances autumn leaf colours in mountain birch (*Betula pubescens* ssp. *czerepanovii*). J Evol Biol 19:1722–1724

Sinkkonen A (2006b) Do autumn leaf colours serve as a reproductive insurance against suckng herbivores? Oikos 113:557–562

Smith AP (1986) Ecology of leaf color polymorphism in a tropical forest species: habitat segregation and herbivory. Oecologia 69:283–287

Snow B, Snow D (1988) Birds and berries. A study of an ecological interaction. Poyser, Calton

Soltan U, Dotterl S, Liede-Schumann S (2009) Leaf variegation in caladixm stendenerifoium (Araceae): a case of mimery? Evol ecol (in press)

Speed MP, Ruxton GD (2005) Warning displays in spiny animals: one (more) evolutionary route to aposematism. Evolution 59:2499–2508

Stafford HA (1994) Anthocyanins and betalains: evolution of the mutually exclusive pathways. Plant Sci 101:91–98

Stamp NE, Wilkens RT (1993) On the cryptic side of life: being unapparent to enemies and the consequences for foraging and growth caterpillars. In Stamp NE, Casey TM (eds) Caterpillars: ecological and evolutionary constraints on foraging. Chapman and Hall, New York, pp 283–330

Stephenson AG (1981) Toxic nectar deters nectar thieves of *Catalpa speciosa*. Am Midl Nat 105:381–383

Tewksbury JJ, Nabhan GP (2001) Directed deterrence by capsaicin in chillies. Nature 412:403–404

Theimer TC, Bateman GC (1992) Patterns of prickly-pear herbivory by collared peccaries. J Wildl Manage 56:234–240

Thurston EL, Lersten NR (1969) The morphology and toxicology of plant stinging hairs. Bot Rev 35:393–412

Tomlinson PB (1990) The structural biology of palms. Clarendon, Oxford

Tuomi J, Augner M (1993) Synergistic selection of unpalatability in plants. Evolution 47:668–672

Tullberg BS, Merilaita S, Wiklund C (2005) Aposematism and crypsis combined as a result of distance dependence: functional versatility of the colour pattern in the swallowtail butterfly larva. Proc R Soc Lond B 272:1315–1321

Van Bael SA, Brawn JD, Robinson SK (2003) Birds defend trees from herbivores in a Neotropical forest canopy. Proc Natl Acad Sci USA 100:8304–8307

Waldbauer GP (1988) Aposematism and Batesian mimicry. Measuring mimetic advantage in natural habitats. Evol Biol 22:227–259

Wang Y, Stass A, Horst WJ (2004) Apoplastic binding of aluminum is involved in silicon-induced amelioration of aluminum toxicity in maize. Plant Physiol 136:3762–3770

Weiss MR (1991) Floral colour changes as cues for pollinators. Nature 354:227–229

Weiss MR (1995) Floral colour change: a widespread functional convergence. Am J Bot 82:167–195

Weiss MR, Lamont BB (1997) Floral colour change and insect pollination: a dynamic relationship. Isr J Plant Sci 45:185–199

Werker E (2000) Trichome diversity and development. Adv Bot Res 31:1–35

White PS (1988) Prickle distribution in *Aralia spinosa* (Araliaceae). Am J Bot 75:282–285

Wickler W (1968) Mimicry in plants and animals. Weidenfeld and Nicolson, London

Wiens D (1978) Mimicry in plants. Evol Biol 11:365–403

Wilkinson DM, Sherratt TN, Phillip DM, Wratten SD, Dixon AFG, Young AJ (2002) The adaptive significance of autumn leaf colours. Oikos 99:402–407

Willey JM, Sherwood LM, Woolverton CJ (2008) Prescot, Harley, and Klein's microbiology, 7th edn. McGraw Hill, New York

Williams KS, Gilbert LE (1981) Insects as selective agents on plant vegetative morphology: egg mimicry reduces egg laying by butterflies. Science 212:467–469

Williamson GB (1982) Plant mimicry: evolutionary constraints. Biol J Linn Soc 18:49–58
Willson MF, Whelan CJ (1990) The evolution of fruit color in fleshy-fruited plants. Am Nat 136:790–809
Wimp GM, Whitham TG (2001) Biodiversity consequences of predation and host plant hybridization on an aphid–ant mutualism. Ecology 82:440–452
Woolfson A, Rothschild M (1990) Speculating about pyrazines. Proc R Soc Lond B 242:113–119
Yamazaki K (2008) Autumn leaf colouration: a new hypothesis involving plant–ant mutualism via aphids. Naturwiss 95:671–676
Young TP, Okello BD (1998) Relaxation of an induced defense after exclusion of herbivores: spines on *Acacia drepanolobium*. Oecologia 115:508–513
Young TP, Stanton ML, Christian CE (2003) Effects of natural and simulated herbivory on spine lengths of *Acacia drepanolobium* in Kenya. Oikos 101:171–179
Zahavi A (1975) Mate selection: a selection for a handicap. J Theor Biol 53:205–214
Zahavi A (1977) The cost of honesty (further remarks on the handicap principle). J Theor Biol 67:603–605
Zahavi A (1991) On the definition of sexual selection, Fisher's model, and the evolution of waste and of signals in general. Anim Behav 42:501–503
Zahavi A, Zahavi A (1997) The handicap principle. A missing piece of Darwin's puzzle. Oxford University Press, New York

Deceptive Behavior in Plants. I. Pollination by Sexual Deception in Orchids: A Host–Parasite Perspective

Nicolas J. Vereecken

Abstract Sexually deceptive orchids attract male insects as pollinators by mimicking the reproductive signals emitted by the targeted females. Since this mimicry system involves the imitation of female mating signals of certain insects, and since mating signals, especially sex pheromones, generally act on a species-specific basis, theory holds that each sexually deceptive orchid is usually pollinated by only one or a few male insect species. While these orchids rely exclusively on their specialized pollinators for their own reproduction, the male insects derive no benefit from this interaction. In this chapter, I will argue that incorporating questions relevant to the field of animal-centered host–parasite interactions into investigations on the evolutionary ecology of orchid pollination by deception will provide important insights at both the proximate (or mechanistic) and at the ultimate (or evolutionary) levels.

1 Introduction

Despite the popular belief that plant pollination by insects epitomizes the ideal mutually beneficial partnership, observational evidence indicates that flowering plants sometimes exploit insects in complex and quite devious ways. This is particularly true in the Orchidaceae, where approximately one-third of all orchid species (i.e., ca. 10,000 species worldwide) achieve insect-mediated cross-pollination without providing a floral reward of any kind to their pollen vectors (Dafni 1984, 1987; Ackerman 1986; Nilsson 1992; Schiestl 2005; Jeraskova et al. 2006). Whereas a large proportion of deceptive orchids achieve cross-pollination by emitting generalized pollinator attractants that innately evoke the presence of a reward to a wide taxonomical range of insects,

N.J. Vereecken
Behavioural and Evolutionary Ecology, Free University of Brussels (U.L.B.) CP 160/12, Av. F.D. Roosevelt 50, B-1050 Brussels, Belgium
Institute of Systematic Botany, University of Zürich, Zollikerstrasse 107, CH-8008 Zürich, Switzerland
e-mail: nicolas.vereecken@ulb.ac.be, nicolas.vereecken@systbot.uzh.ch

F. Baluška (ed.), *Plant-Environment Interactions*, Signaling and Communication in Plants, 203
DOI: 10.1007/978-3-540-89230-4_11, © Springer-Verlag Berlin Heidelberg 2009

other species have narrowed their pollinator spectrum by evolving a much more specialized and fine-tuned imitation of specific rewarding models. Such instances fall into the category of Batesian mimicry sensu stricto (Wickler 1968; Wiens 1978; Pasteur 1982), where the orchid (i.e., the *mimic*) impersonates another species (i.e., the *model*) and thereby attracts its pollinator(s) (i.e., the *operator*) (Schiestl 2005; Ayasse 2006).

As research efforts proliferate in different areas of expertise in ecological research, scientists inevitably develop compartimentalized research agendas, and barriers to effective communication between these disciplines unfortunately become more and more apparent (Brooks and McLennan 1991; Bush et al. 1997; Waser and Price 1998; Wyatt 2003). In this chapter, we will argue that incorporating questions relevant to the field of animal-centered host–parasite interactions into investigations on the evolutionary ecology of orchid pollination by deception will provide important insights at both the proximate (or mechanistic) and at the ultimate (or evolutionary) levels. Although several examples of Batesian food-deceptive mimicry have been reported in the literature (Dafni and Ivri 1981; Nilsson 1983; Dafni 1987; Johnson 1994, 2000; Galizia et al. 2005; Johnson and Shelah 2006) (see also the chapter by Jeraskova et al.), here we shall limit ourselves mostly to orchid pollination by sexual deceit, i.e., the imitation of female insects by orchid flowers, since research in this field has been particularly prominent over the past decade. This chapter aims to encourage investigations of mimicry systems from a behavioral perspective, by pointing out specific gaps in our knowledge and possible avenues for future research that will help to fill them.

2 Sexual Deception: Parasitism of Insect Sexual Behavior

2.1 Why Parasitism?

The term *parasite* in the broadest sense refers to organisms that benefit from their interactions with other organisms (i.e., the *hosts*) by deriving advantages (habitat, nutrients, motility, or other services) at the latter's expense (Barnard 1990; Combes 2001; Poulin 2007). But why should sexually deceptive orchids be classified as parasites? The flowers of so-called sexually deceptive orchids do not produce nectar, pollen or any form or edible reward that their pollinators could collect during their floral visits. Pollinator attraction is almost exclusively mediated by the emission of specific signals emitted by these flowers, in particular female sex pheromone compounds in their odor bouquet (Schiestl et al. 1999). The male insects are therefore drawn to the orchid flowers as they patrol for mates, and subsequently attempt copulation or a precopulatory routine with the female decoys on the flowers, unwittingly taking up the orchid's pollen masses (the *pollinia*) in the process. Cross-pollination is then ensured as the insect transfers the pollen grains contained in the pollinia onto the flower of a nearby orchid during another *pseudocopulation* (Correvon and Pouyanne 1916, 1923; Pouyanne 1917; Coleman 1928; Kullenberg

1961). In short, these orchids lure male bees with the false promise of sex, and exploit the males' drive for specific reproductive signals to ensure their pollination. Since there is exploitation of the male bees' sexual behavior at the core of these interactions, pollination by sexual deception should therefore be viewed as asymmetrical or one-sided, with the orchid relying exclusively on its duped pollinator for its own reproduction, while the male insects derive no benefit from this interaction. From the plant's perspective (i.e., the parasite), the pollinator (i.e., the host) can therefore be viewed as an ephemeral resource that is used to facilitate cross-pollination and thereby maximize its overall reproductive success.

2.2 The Cost of Parasitism

A central issue related to host–parasite interactions and epidemiology is the level of harm or virulence caused by a parasite to its host. Although it has become commonplace to view parasites as having a negative impact on their hosts' fitness, investigations into host–parasite interactions have helped to develop an alternative and much more subtle picture of the extent to which parasites may affect their hosts. The available data reveal that host–parasite interactions fall along a continuum, from one end where the parasite has no negative fitness impact on its host(s) whatsoever, all the way to the other extreme, where the parasite causes the death of its host(s) (Poulin 2007). In light of all this, where do orchid–pollinator interactions stand along this host–parasite "harmfulness" continuum? To date, we only have a very limited knowledge of the costs, if any, that are incurred by the host species. However, it seems quite clear that, given the taxonomic range of orchid pollinators, each with their own ecological and behavioral peculiarities, no general statement can be made regarding the overall or optimal host exploitation strategy by these parasites. Since deceptive orchids, and sexually deceptive ones in particular, are known to be limited in their reproductive success by access to their hosts (Darwin 1862; Kullenberg 1961; Neiland and Wilcock 1995, 1998; Ayasse et al. 2000; Tremblay et al. 2005; Vandewoestijne et al. 2008), it is expected that selection should favor host exploitation strategies at a rate that makes these intimate interactions sustainable across generations, each in their own particular way. Like all parasites involved in highly specific interactions, sexually deceptive orchids face a major constraint: a fitness impact on its host that was too strong, e.g., death caused during pseudocopulation in an extreme theoretical case, would severely compromise the orchids' chances of reproducing. Hence, it is expected that the orchids' parasitic strategy should allow the local persistence of their host population, ensuring their own survival, while adapting to the local mating preferences of their associated host(s) for female sex pheromone signals.

Although quantitative measurements on the level of harm incurred by the hosts are lacking at this stage, evidence from field observations and a selection of recent studies may help to sketch trends observed in several orchid genera. At first glance, it can be postulated that the orchids' trickery is likely to cause a waste of time and energy for male insects that are otherwise patrolling for mates during their reproductive period. Pseudocopulations with the orchid flowers typically last between a few seconds and as long as a few tens of minutes (Vereecken, pers. obs.), and can therefore potentially cause a decrease in the number of mating opportunities for males that are visiting the orchid flowers instead of searching for access to freshly emerged females. This situation is notably exemplified in the West Palearctic genus *Ophrys*, where the hosts—male bees, wasps and sometimes even beetles—hatch first and generally outnumber (on a daily operational basis) receptive females during the reproductive period. The female insects usually mate only once after their emergence and before they initiate the construction of their nest and oviposition, and males sometimes compete quite intensively with one another to mate with their freshly emerged partners at the nesting/emergence site or at "rendezvous" spots (Alcock et al. 1978; Paxton 2005). In this climate of male–male competition over mates, it can therefore be assumed that male insects attempting copulation with the female decoys on the orchid flowers might miss out on occasions to engage in competitions over mates and sometimes even fail to reproduce altogether during their relatively short lifetime.

Recent studies carried out in Australia on sexually parasitic orchids have shown for the first time that the orchid parasitism might reduce or inhibit the mating opportunities of their hosts. In a series of experiments, Wong and Schiestl (2002) provided evidence that the hosts (males of the Thynnine wasp *Neozeleboria cryptoides*) learn to avoid their associated orchid parasite (*Chiloglottis trapeziformis*) after subsequent and unsuccessful copulation attempts, and that the hosts' wingless females experience a significant decrease in attractiveness when they are "calling" for mates from inside an orchid patch (Wong and Schiestl 2002). The female wasps are only capable of restoring their original attractiveness towards the hosts as the distance between their calling spot and the parasite colony increases (Wong et al. 2004). In another recent study on the ichneumonid host *Lissopimpla excelsa* and its orchid parasites *Cryptostylis erecta* and *C. leptochila* from Australia, Gaskett et al. (2008) confirmed Coleman's (1928) earlier findings that the hosts regularly ejaculate in the flowers during pseudocopulations. Such a wastage of sperm can potentially lead to transient gamete depletion, which might in turn compromise the opportunities of the hosts to transfer their sperm during subsequent matings with genuine females (Damiens and Boivin 2006; Gaskett et al. 2008).

The mimicry systems described above illustrate how the orchid parasites not only channel both time and energy away from the search for genuine mating partners by patrolling males, but can also impact on their hymenopteran hosts' fitness in much more dramatic ways, notably by causing sperm wastage (see also Blanco and Barbosa 2005). As we have hypothesized above, each host–parasite association might be characterized by a balance between the host exploitation strategy that maximizes the parasites' overall fitness and how it affects the hosts' reproductive output (Fig. 1). In a meta-analysis of literature records on the mean reproductive success in sexually deceptive orchids and the associated vigor of the behavioral responses of the male insects to the orchid flowers, Gaskett et al. (2008) have suggested that orchids that trigger more intense pollinator behavior (high sexual arousal, e.g., by causing ejaculation) have a higher reproductive output than other species where pollinator visits are brief (Fig. 1). Studies are now needed to test this hypothesis by including more species within each sexually deceptive orchid genus, but they represent an interesting parallel to experiments across many types of host–parasite interaction where parasite virulence correlates positively with parasite reproduction or dissemination rates (Turner et al. 1995; Ebert 1998, 2000).

Research on the reproductive biology of sexually deceptive orchids is a very promising field in many respects (see Schiestl 2005; Peakall 2007; Waterman and Bidartondo 2008), and efforts should now be made to determine the level of harm/costs incurred by the hosts, e.g., by investigating the flowers for the presence of host sperm in other genera of sexually deceptive orchids (see the method used by Gaskett et al. 2008), or by identifying other ways in which pseudocopulations can affect the hosts' reproductive output. By quantifying the costs, it will be possible to pinpoint the selection pressures at play on both sides of these host–parasite interactions, and we will gain important insights into the maintenance and the evolution of orchid mimicry.

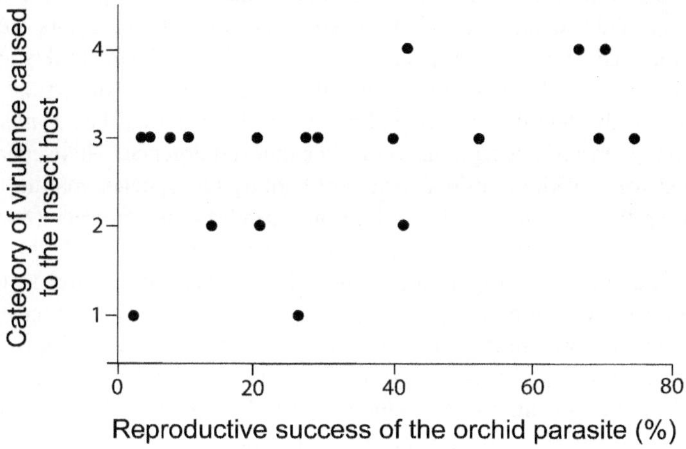

Fig. 1a–d Classes of parasite virulence (measured by the intensity of host behavioral response) and the orchid parasites' overall associated reproductive success (in mean % of annual seed set within populations) in different genera of sexually deceptive mimicry systems. **a** The orchid parasite attracts its male insect host briefly without an attempt at copulation (*Pterostylis*; Taylor 1999; Lehnebach et al. 2005); **b** The orchid parasite attracts its male insect host and triggers inspection behavior or a precopulatory routine (*Caladenia*, *Chiloglottis*, *Drakaea*; Peakall 1990; Peakall and Handel 1993; Schiestl 2004; Dickson and Petit 2006); **c** The orchid parasite triggers a host copulation attempt only (*Ophrys*: Darwin 1862; Correvon and Pouyanne 1916; Kullenberg 1961; Ayasse et al. 1997; Neiland and Wilcock 1998; Ayasse et al. 2000; Vandewoestijne et al. 2008; Vereecken, unpublished data; *Geoblasta*: Ciotek et al. 2006); **d** The orchid parasite triggers host copulation attempt and ejaculation (*Cryptostylis:* Schiestl et al. 2004; Gaskett and Herberstein 2006; Gaskett et al. 2008). (Modified from Gaskett et al. 2008)

3 The Evolution of Color Versus Odor in Orchid Mimicry

In all cases of mimicry, the "ménage à trois" is subjected to specific selection pressures stemming from the nature of the interactions involved. From recent studies on the mating behavior of solitary bees and wasps, we now know that female insects attract patrolling males by releasing specific chemical compounds (either in blends or specific compounds) that are capable of triggering genuine copulation attempts by the male insects when tested for their attractiveness on dummies. Besides, comparative analyses of the orchids' floral scent and the female insect sex pheromone have shown that the parasitic orchids use the same odor compounds as the females to deceive the males into pollinating the flowers (Schiestl et al. 1999, 2003; Ayasse et al. 2003; Schiestl 2004; Mant et al. 2005a; Vereecken and Schiestl 2008). In short, this interaction seems to constitute an illustrative case of mimicry that is primarily mediated by chemical signals.

At first glance, it might seem surprising that the showy, colorful flowers of a wide range of sexually deceptive orchid species do not use visual cues to lure their pollinators. What might this kaleidoscope of floral colors be used for? In a recent study on the Cretan species *O. heldreichii,* Spaethe et al. (2007) reported on the synergistic effect of scent and floral perianth color in pollinator attraction, which illustrates that although the mimicry is primarily based on sex pheromone mimicry, visual cues can also enhance the flowers' attractiveness. These authors proposed that selection may have favored the spectral resemblance between the pinkish perianth of the flowers of *O. heldreichii* and the overall reflectance of co-occurring, nonorchid species which the females of the pollinator, the long-horned bee *Synhalonia rufa* (=*Tetralonia berlandi*), forage on right after their emergence. This scenario makes sense when we consider that in a variety of solitary bee species, mating takes place shortly after the emergence of neighboring "rendezvous" flowers (Alcock et al. 1978; Westrich 1990; Paxton 2005), which, as Spaethe et al. (2007) showed, sometimes include species whose inflorescences have a similar spectral reflectance. It seems therefore that some orchid species within the genus *Ophrys* have evolved a multicomponent floral mimicry based on the chemical mimicry of virgin female bees and the visual mimicry of the spectral reflectance of "rendezvous" flowers where mating takes place during the reproductive period of the bees' life cycle (Fig. 2).

To date, the extent to which these results may be applicable to other species within the 250+ species-rich genus *Ophrys* and in other orchid genera remains poorly understood. Ongoing investigations on the adaptive significance of perianth color polymorphism and its influence on pollinator visitation rates indicate that even within the orchid genus *Ophrys*, certain species might incorporate visual cues in pollinator attractiveness, while others clearly don't. In the Mediterranean species *Ophrys arachnitiformis* for instance, preliminary results suggest that (i) there is no differentiation in either relative or absolute amounts of behaviorally active compounds produced by the flowers between different color morphs of the orchid, and (ii) that neither the presence/absence nor the color of the perianth influences visitation rates of the pollinators to either odorless controls or dummies scented with odor extracts of *Ophrys* flowers (Vereecken and Schiestl, unpublished manuscript).

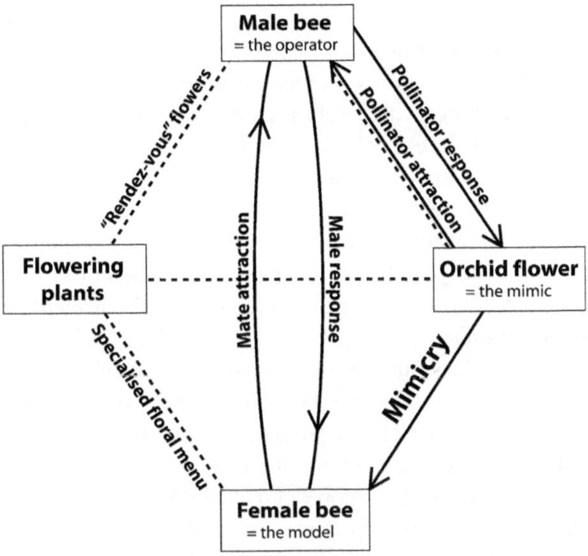

Fig. 2 A schematic view of the selection pressures at play and the nature of the signals involved in *Ophrys* mimicry systems. The orchid flowers hijack the communication channel used by males and females during their courtship, and some *Ophrys* taxa have even evolved a multicomponent mimicry that also incorporates signals involved in the localization of females by males on so-called "rendezvous" flowers. The *dashed lines* represent the involvement of signals of nonorchid flowering plants in the species interactions: the female bees use these signals to locate their foraging resources, the male bees use them to locate the "rendezvous" flowers, and some orchids mimic these signals along with the chemical signals of female bees to attract their pollinators

Hence, it can be postulated that floral color polymorphism in *O. arachnitiformis* is barely subjected to selection imposed by its pollinator, contrary to the results found by Spaethe et al. (2007), and that mate-searching flights of the hosts are primarily driven by odor signals, with no intervention of visual cues of any kind (Vereecken and Schiestl, unpublished manuscript).

More in-depth investigations into the chemical basis of *Ophrys*-pollinator investigations have been performed recently with the orchid *O. exaltata sensu lato*. An initial series of experiments showed that the females of the solitary bee *Colletes cunicularius* use population-specific ratios of specific chemical compounds, mostly alkenes (monounsaturated, straight-chained hydrocarbons with a carbon chain length of 21–25). These experiments have also demonstrated that patrolling males in this bee species are more attracted by sex pheromone mimicking compound mixtures of females from other populations (i.e., allopatric) over local ones (i.e., sympatric) (Vereecken et al. 2007b). Following on from this study, Vereecken and Schiestl (2008b) have undertaken to investigate whether the parasitic orchids released patterns of the key odor compounds matching those emitted by the sympatric females of their male hosts. Contrary to theoretical expectations that the parasitic orchids should imitate the chemical signals of the sympatric female bees as closely as possible (i.e., a "perfect match"), Vereecken and Schiestl (2008) found that the scent composition of the orchid flowers was consistently slightly different from the local female bees' sex pheromone in any given population. The parasitic orchid scent, being different from the "model" signal, was even found to be actively preferred by the male hosts when both the orchid scent and the bee sex pheromone were assayed for their attractiveness. Vereecken and Schiestl (2008) have interpreted their results as a case of imperfect chemical mimicry driven by the predilection of hosts for odor blends released by "exotic" (pseudo)females. It remains to be tested whether the male host preferences can potentially change during the reproductive season, or even from one year to the next (see e.g. Kasumovic et al. 2008), and how this phenomenon might translate into selection for floral scent evolution in their associated parasitic orchids.

Collectively, these results alone demonstrate that we can enhance our understanding of the roles of different floral traits and the factors driving their evolution if investigations are performed from a behavioral perspective, by attempting to approach interspecific interactions from a different angle of view, in this case by deciphering the relevant signals in the insects' reproductive biology. Such background data are often decisive in subsequent investigations aimed at assessing whether the parasitic orchids have succeeded in impersonating the female insects by hijacking chemical communication channels only, or if pollinator attraction is mediated by a combination of other signals.

4 Host Specificity in Sexually Deceptive Orchids

4.1 Defining Host Specificity

The term host specificity relates to "the extent to which a parasite taxon is restricted in the number of host species used at a given stage in the life cycle" (Poulin 2007). For orchids that have co-opted male insects as pollinators, host specificity can therefore be viewed as the taxonomical spectrum of insects that can act as pollinators during their flowering season. Since sexually deceptive orchids have evolved the imitation of female mating signals of certain insects, and since mating signals, especially sex pheromones, generally act on a species-specific basis (Thornhill and Alcock 2000; Wyatt 2003), theory holds that each parasitic orchid is usually pollinated by only one or a few male insect species (Kullenberg 1961; Paulus and Gack 1990).

4.2 The Determinants of Host Specificity

Several important issues need to be discussed before making statements on host specificity in sexually deceptive orchids. The only data available in the literature are lists of insects that have been observed pseudocopulating (Fig. 3) with the

Fig. 3 Pseudocopulating male of *Andrena flavipes* (Kirby) (Hymenoptera, Andrenidae) on flower of the sexually deceptive orchid *Ophrys bilunulata* Rossi. (Photo: NJ Vereecken)

orchid flowers. Based on these pollinator records, interactions between the orchid parasites and their male insect hosts may appear to be relatively species-specific, but such data should be analyzed with caution (see Poulin 2007).

First, it should be considered that high levels of host specificity can be the direct outcome of inadequate sampling effort (Poulin 2007). Pollination events or flower visits are relatively rare under natural conditions, since most sexually deceptive orchids are severely limited in their reproductive success by access to pollinators (Tremblay et al. 2005). Hence, the likelihood of observing a pseudocopulation event under natural conditions during a quick visit in a population of these parasitic orchids is certainly much lower than one would expect. Not only that, but a great number of these orchids also have wide geographic range across which no pollinator record is available. Consequently, two parasitic orchids might each have a single pollinator host, but detailed investigations across their geographic ranges and over the years sometimes reveal that these orchids are instead exploiting a broader range of insect hosts (see, e.g., Lorella et al. 2002; Tyteca et al. 2006; Vereecken and Patiny 2006; Bower 2006; Vereecken et al. 2007a). The real range of hosts that can be exploited by a given parasite species is illustrated by Combes' (2001) "filters" concept (Fig. 4), which shows the mechanisms restricting the number of potential hosts. Unquestionably, the first mechanism that determines host–parasite interactions is the "encounter" filter: parasites must live in the ecosystem of their hosts and have contacts with them at specific stages in their life cycle. Any orchid parasite that does not satisfy this condition, e.g., by being deprived of access to its male insect hosts, will not reproduce. Experiments performed with picked inflorescences of *Ophrys* species transferred from southern France outside their home range provide evidence that the fresh, unpollinated flowers can be attractive to

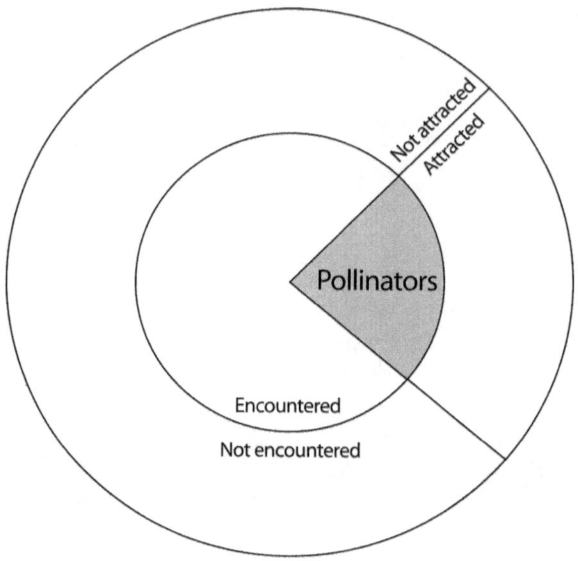

Fig. 4 A schematic view of "encounter" and "attraction" filters that determine the spectrum of pollinators attracted by a sexually deceptive orchid species. Not all male insects respond adequately to the signals released by the orchid flowers: only those that perform stereotyped pseudocopulations on the flower labellum and withdraw pollinia are considered "potential pollinators." The diversity of potential pollinators is further restricted by the encounter filter: only those insect species that co-occur with the orchids can act as pollinators. (Modified from Combes 2001)

novel pollinator taxa such as males of *Andrena flavipes* and *A. bicolor* in England for *O. lupercalis*, a Mediterranean orchid previously reported to be pollinated only by males of *A. nigroaenea*. Similar observations were made in Australian parasitic orchids by Bower (2006), who, by performing pollinator choice experiments under natural conditions, showed that novel pollinators might be attracted when picked inflorescences of a single orchid are translocated to allopatric populations. The second factor that determines the formation of the host spectrum is the "attraction" (or "compatibility") filter, which, in the present case, relates to the ability of the parasitic orchids to successfully (1) attract male insects, and (2) trigger copulation attempts on the flower labellum, (3) resulting in pollinia removal or deposition. This succession of events is not systematically performed by all male insects initially attracted by the orchid flowers; some insects can be observed pseudocopulating (Fig. 3) on the flowers, but the mismatch between their body size/corpulence and the floral architecture hinders pollinia removal or deposition (Fig. 5).

Second, the apparent species-specificity of orchid–pollinator interactions can be an artefact of incorrect species identification. Besides situations where the insect hosts have been assigned to another species by mistake (see the discussion in Schiestl and Vereecken 2008), the problem lies in the taxonomical interpretation of host–parasite associations. Consider the situation depicted in Fig. 6, where a parasitic orchid is reported to have a low host specificity compared to its congeners,

Fig. 5 The sexually deceptive orchid *Ophrys lupercalis* is pollinated by males of *Andrena nigroaenea* (Hym. Andrenidae) (*left*). The flowers of this parasitic orchid are also attractive to males of other wild bee species, such as *A. minutula* (*right*). The *right photograph* illustrates Combes' (2001) "compatibility" filter concept: the male *A. minutula* is attracted by the flowers and makes an attempt at copulation on the flower labellum, but there is a mismatch between its body size and the floral architecture of *O. lupercalis*, preventing these small-sized male bees from withdrawing the orchid's pollen masses. (Photos: NJ Vereecken)

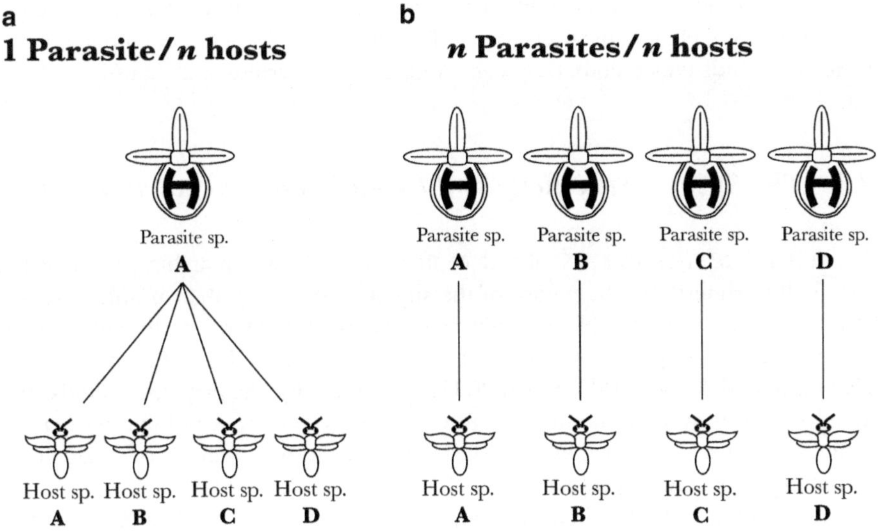

Fig. 6a–b A schematic view of taxonomic considerations that might affect estimates of host specificity in sexually deceptive orchids. **a** The orchid parasite species *a* has a low host specificity and exploits the range of host species *a–d* as pollinators across its home range; **b** the orchid parasite species *a* may also prove to be a complex of *n* cryptic parasite taxa, each highly host-specific and associated with *n* different insect taxa

e.g., because its pollinator records encompass *n* host taxa. With the growing incorporation of modern analytical techniques into studies on species interactions, different schools of thought have emerged: while some scientists might still consider this orchid parasite to be a single species with a lower host specificity, others, helped by analytical tools that are sometimes capable of identifying cryptic taxa, would tend to regard this species as a group of host-specific cryptic taxa, each associated with a single host (see Bower 2006). These contrasting standpoints can potentially lead to extreme levels of taxonomical confusion and affect our estimates and understanding of host specificity in these groups of parasitic orchids. As is observed in different areas of the field of host–parasite interactions, some authors would even go as far as to suggest that new species should be proposed every time an alternative host species is found to be interacting with a parasitic orchid for the first time (see Paulus and Gack 1990; Delforge 2005). Preliminary field experiments with fresh orchid inflorescences tested for their attractiveness to different pollinator taxa in sympatry and allopatry indicate that individual flowers of a single orchid species can successfully attract male bees of different species as hosts (Vereecken, unpublished data), which suggests that lower levels of host specificity may indeed occur.

More experimental investigations of host specificity are now needed to help to sketch a more detailed picture of host spectra in each group of sexually deceptive orchids. From these and parallel studies on the behavioral and chemical ecology of the (more or less species-specific) hosts, it will be possible to investigate the selection pressures exerted on the floral traits, and possibly to assess the extent to which the attraction of different insects in different regions can lead to the evolution of divergent combinations of phenotypic traits (e.g., different ratios of key odor compounds) in the parasitic orchids under study (see, e.g., Aigner 2006; Herrera et al. 2006).

4.3 The Species Specificity and Evolution of Chemical Signals

As we have seen above, sexually deceptive orchids owe their high level of host specificity primarily to the nature of the signals involved in their mimicry system. Mating signals, and female sex pheromones in particular, are thought to rank among the most species-specific communication channels in the insect world (Thornhill and Alcock 2000; Wyatt 2003). Theoretically, the parasitic orchids' fine-tuned mimicry system should therefore not allow for cross-pollination between different orchid taxa that have evolved towards the exploitation of male insects belonging to different species.

Investigations into the chemical communication of orchid–pollinator interactions have indeed reported that orchids attracting different male insects as pollinators have odor bouquets consisting of different ratios of identical or structurally similar odor compounds (Borg-Karlson et al. 1993; Schiestl and Ayasse 2002; Ayasse et al. 2003; Stökl et al. 2005; Mant et al. 2005a; Véla et al. 2007). An exception to this premise is found in the Australian orchid pair *Chiloglottis trapeziformis* and *C. valida*, which are pollinated in a highly specific manner by the male thynnine wasps *Neozeleboria cryptoides* and *N. monticola*, respectively. In their recent study, Schiestl and Peakall (2005) have demonstrated that these two parasitic orchids attract their specialized host via the emission of a single odor compound, 2-ethyl-5-propyl-1,3-cyclohexandione ("chiloglottone"). Assortative pollinator attraction between these two orchids is nevertheless maintained to a large extent by the mating preferences of the male hosts for different heights and the corresponding differences in floral heights between *C. trapeziformis* and *C. valida* (Schiestl and Peakall 2005).

Despite the evidence that different sexually deceptive orchid species associated with different male insect hosts have distinct odor bouquets, observations from the field, particularly in the Mediterranean genus *Ophrys*, indicate that hybridization in this group of orchids does occur in natural populations where two or more species grow in sympatry and bloom at the same period of the year (see, e.g., Stebbins and Ferlan 1956; Danesch and Danesch 1972). This phenomenon suggests that cross-attraction and interspecific gene flow may indeed occur. Furthermore, behavioral bioassays performed with the male insects have conclusively demonstrated that although pollinators are more attracted by the floral scent of the parasitic orchid they are associated with, heterotaxic visits occur and might under certain circumstances lead to successful pollination and the formation of natural hybrids (Stokl 2007; Cortis et al. 2008; Vereecken et al., unpublished manuscripts). Reproductive isolation in *Ophrys* is thought to be primarily mediated by host specificity, and post-mating barriers have been reported to be relatively weak compared to other European orchid genera where pollinator specificity is low (Cozzolino et al. 2005) and post-mating barriers generally keep co-occurring species reproductively isolated (Moccia et al. 2007; Scopece et al. 2007).

The investigations into the ecology and the evolutionary consequences of hybridization in sexually deceptive orchids challenge the commonly held view that these host–parasite interactions are strictly species-specific, and that "each [sexually deceptive orchid] species [...] attracts different sets of [...] pollinator species" (Grant 1994). From the recent studies on the topic, it has become apparent that, despite the evolution of a highly specific host exploitation mechanism, parasitic orchids have retained the genetic and signal variation required to adapt to fluctuating pollinator assemblages and selection pressures, and have evolved a higher flexibility in host specificity than previously thought. These observations make sense when we consider that the reproductive success of sexually deceptive orchids is typically limited by access to suitable hosts (see Tremblay et al. 2005 for a review). As a consequence, any mechanism, such as the ability to co-opt alternative host species, that contributes to conferring a higher reproductive output on its bearer should be favored by selection across generations and should lead to a more adaptive and flexible host exploitation strategy overall. This, in turn, could help these parasitic orchids when they face situations where their "official" host is temporarily unavailable or locally extinct.

4.4 Signal Evolution Above the Species Level

Most sexually deceptive orchid species usually display conservative patterns of pollinator attraction (i.e., they attract closely related pollinator taxa) (Mant et al. 2002; Vereecken and Patiny 2005), along with a weak differentiation in pollinator-attracting scents (see above; Mant et al. 2005b). Consequently, it has been suggested that pollinator shifts through minor changes in floral odor bouquets could be the driving force for speciation (here, prezygotic reproductive isolation) in these parasitic orchids (Schiestl and Ayasse 2002). Furthermore, it is postulated that reproductive isolation between sympatric species can be strengthened when pollinator shifts involve sister species of insects that are reproductively isolated by using of specific ratios of similar compounds for their female sex pheromone, as is observed in mining bees of the genus *Andrena* (Schiestl and Ayasse 2002; Stökl et al. 2005).

In their study on the Australian *Chiloglottis–Neozeleboria* interactions, Mant et al. (2002) have found that sister orchid species tend to be pollinated by insects related taxonomically, which they interpreted as a case of phylogenetic conservatism in orchid–pollinator interactions. However, by incorporating an array of nonpollinating *Neozeleboria* species in their phylogeny of *Chiloglottis* host wasp species, Mant et al. (2005b) found that the host species exploited by the orchids did not form a monophyletic group, and that nonhosts clustered within the host species. Hence, they hypothesized that the parasitic *Chiloglottis* orchids species have diversified through repeated switches to congeners and alternative wasp species with similar traits to the ancestral pollinator. The host records in the three species of the *Ophrys insectifera* species group suggest that another scenario might be applicable here, namely that the parasitic orchids have not adapted to sister insect host species with similar mating signals to pollinators, but rather to unrelated insects (a sawfly, a mining bee and a digger wasp in this case) that presumably use similar chemical communication channels during their courtship (Vereecken et al., unpublished manuscript). Future studies should investigate whether the male insect hosts in groups of closely related sexually deceptive orchids are attracted by overlapping patterns of identical odor compounds in the floral odor, or if different odor compounds mediate the specific interactions between orchids and their pollinators.

5 Transitions to Parasitism by Sexual Deception in Orchids

To date, sexual deceit is thought to be exclusive to the family Orchidaceae, where it has evolved independently on multiple occasions and on different continents, with representatives found across Australia (ten genera: Coleman 1928; Stoutamire 1975; Peakall et al. 1987; Jones 1988; Peakall 1990; Bower 1996), Central and South America (seven genera: Van der Pijl and Dodson 1966; Dod 1976; Singer 2002; Singer et al. 2004; Blanco and Barbosa 2005; Ciotek et al. 2006), South Africa (genus *Disa*; Steiner et al. 1994), and the West Palaearctic (Delforge 2005; Schiestl 2005; Ayasse 2006; Jeraskova et al. 2006). Pollination by sexual deception is thought to be a derived pollination strategy within the family Orchidaceae (Van der Pij and Dodson 1966; Van der Cingel 1995; Alcock 2005). Based on recent advances in molecular phylogenetic analyses and investigations into the chemical ecology and behavioral ecology of orchid–pollinator interactions, it is now hypothesized that this most unusual mode of pollinator attraction in the plant world is derived from food deception (see e.g. Bateman et al. 2003), another common pollination strategy in orchids (see also the chapter by Jeraskova et al.).

A recurrent idea in studies on the evolution of host–parasite interactions is that pre-adaptations must precede the emergence of an alternative kind of species interaction (Rothshild and Clay 1952). As we have seen above the signals involved in the attraction of male insects by the parasitic orchids in the genus *Ophrys* include the emission of patterns of alkenes (monounsaturated alkanes) (reviewed by Schiestl 2005), a specific class of chemical compounds that are otherwise present in the cuticles of a wide array of insect and plant species, where they avoid dehydration (Hadley 1981). An important first step towards an understanding of the evolution of these compounds has been made by Schiestl and Cozzolino (2008), who mapped the alkene production of flowers onto the phylogeny of a selection of European orchid species in the subtribe Orchidinae. Their results show that the emission of these compounds is detectable in the floral odors of most orchid species investigated, which suggests that the production of alkenes is a pre-adaptation to sexual deception, as defined in the case of *Ophrys* (Schiestl and Cozzolino 2008). Furthermore, their results show that the flowers in a group of non-*Ophrys* orchids that are primarily pollinated by male bees, notably in the genus *Serapias* and in *Anacamptis papilionacea*, also emit high amounts of alkenes. Collectively, these data suggest that the emission of these compounds in important amounts may have initiated the attraction of male insects as specialized flower visitors and facilitated the evolution of pollination by pseudocopulation in their sister (and presumably more derived, see Bateman et al. 2003) genus *Ophrys* (Schiestl and Cozzolino 2008), and that selection for pollination by males has resulted in increased production of alkenes.

If alkenes are considered pre-adaptations to the evolution of parasitism by sexual deceit as we know it in *Ophrys* (Schiestl and Cozzolino 2008), then they must have provided fitness gains to the parasite precursors before being favored by natural selection (Poulin 2007). Undoubtedly, male insects are not regarded as being the most effective pollinators of flowering plants *a priori*, since they are usually short-lived and they only occasionally interrupt their patrolling bouts for mates to collect nectar on the flowers, sometimes inadvertently collecting small amounts of pollen in the process. In contrast, female insects, and female bees in particular, spend a considerable proportion of their life cycles visiting flowers and collecting pollen, thereby contributing to pollen transfer from one flower to the next as they forage on different flower patches (Proctor and Yeo 1972; Michener 2007). So what would be the advantage, if any, from the plants' perspective, of relying on male insects for their pollination? First, male insects, and again bees in particular, focus almost exclusively on looking for mates in a highly specialized manner (Alcock et al. 1978; Paxton 2005), which provides the parasites-to-be with an opportunity to exploit hosts with a high degree of specificity in pollen transfer. Second, male insects are thought to promote outcrossing through their long-distance patrolling flights for mating partners (Williams and Dodson 1972; Peakall 1990; Peakall and Beattie 1996). These are potentially important aspects for the reproductive biology of orchids, since such patterns in pollen flow both within and among populations could considerably improve seed quality and decrease pollen loss (see, e.g., Johnson et al. 2004). It is therefore expected that there are indeed fitness advantages associated with the stepwise evolution of floral signals that could ensure the specific attraction of male insects as pollinators by orchid parasite precursors.

Although significant advances have been made in our understanding of the evolution of signals and orchid–pollinator interactions over the past decade, we still have a very fragmentary view of several important aspects of the behavioral ecology and the host specificity of these parasitic orchids. For example, there is a dramatic lack of data on the chemical ecology and the evolution of many groups of sexually deceptive orchids, including those in the genus *Ophrys* which have been the focus of intense scrutinity over the past few years. The roles of chemical and visual cues in the attraction of the male hosts have received very little attention so far, despite the potential avenues of research provided by the extraordinary kaleidoscope of their floral colors and the associated chemical repertoire of their floral odor bouquets. The status of the current lists of pollinator records calls for repeated observations of orchid pollination under natural conditions, both in different regions across the orchids' home range and outside their natural habitat in order to properly characterize the host spectrum of all known parasitic orchids. Several orchids in different genera are pollinated by male insects without attempted copulations with the flowers, which might represent transitional stages in the evolution of sexual deceit from food deception. Finally, comparisons of the reproductive biology between congener species or genera of orchids that differ in the degree of intensity of host behavioral response (or virulence, see Fig. 1) might provide important insights into the factors that drive the evolution of sexual deception and its ecological consequences, from the most primitive forms to the most elaborate, where copulation attempts with the flowers and male insect ejaculation are observed.

Acknowledgements I am grateful to Florian P. Schiestl (University of Zürich, Switzerland) for his continuous support and encouragements during my research, as well as for his comments on this manuscript. I have received financial support from two Belgian institutions, namely the "Fonds pour la formation à la Recherche dans l'Industrie et l'Agriculture" (F.R.I.A.) and the "Royal Academy of Sciences, Humanities and Fine Arts" (Agathon de Potter Foundation).

References

Ackerman JD (1986) Mechanisms and evolution of food-deceptive pollination systems in orchids. Lindleyana 1:108–113

Aigner PA (2006) The evolution of specialized floral phenotypes in a fine-grained pollination environment. In: NM Waser, J Ollerton (eds) Plant–pollinator interactions: from specialization to generalization. The University of Chicago Press, Chicago, pp 23–46

Alcock J (2005) An enthusiasm for orchids: sex and deception in plant evolution. Oxford University Press, Oxford

Alcock J, Barrows EN, Gordh G, Hubbard J, Kirkendall L, Pyle DW, Ponder TL, Zalom FG (1978) The ecology and evolution of male reproductive behaviour in the bees and wasps. Zool J Linn Soc 64:293–326

Ayasse M (2006) Floral scent and pollinator attraction in sexually deceptive orchids. In: N Dudareva E Pichersky (eds) Biology of floral scent. CRC, Boca Raton, pp 219–241

Ayasse M, Schiestl FP, Paulus HF, Erdmann D, Francke W (1997) Chemical communication in the reproductive biology of *Ophrys sphegodes*. Mitt D Ges Allg Ang Ent 11:473–476

Ayasse M, Schiestl FP, Paulus HF, Löfstedt C, Hansson BS, Ibarra F, Francke W (2000) Evolution of reproductive strategies in the sexually deceptive orchid *Ophrys sphegodes*: how does flower-specific variation of odor signals influence reproductive success? Evolution 54:1995–2006

Ayasse M, Schiestl FP, Paulus HF, Ibarra F, Francke W (2003) Pollinator attraction in a sexually deceptive orchid by means of unconventional chemicals. Proc R Soc Lond B 270:517–522

Barnard CJ, 1990. Parasitic relationships. In: Barnard CJ, Behnke JM (eds) Parasitism and host behaviour. Taylor and Francis, London, pp 1–33

Bateman RM, Hollingsworth PM, Preston J, Yi-Bo L, Pridgeon AM, Chase MW (2003) Molecular phylogenetics and evolution of Orchidinae and selected Habenariinae (Orchidaceae). Bot J Linn Soc 142:1–40

Blanco MA, Barboza G (2005) Pseudocopulatory pollination in *Lepanthes* (Orchidaceae: Pleurothallidinae) by fungus gnats. Ann Bot 95:763–772

Borg-Karlson A-K, Groth I, Agren L, Kullenberg B (1993) Form-specific fragrances from *Ophrys insectifera* (Orchidaceae) attract species of different pollinator genera. Evidence of sympatric speciation? Chemoecology 4:39–45

Bower CC (1996) Demonstration of pollinator-mediated reproductive isolation in sexually deceptive species of *Chiloglottis* (Orchidaceae: Caladeniinae). Austr J Bot 44:15–33

Bower CC (2006) Specific pollinators reveal a cryptic taxon in the bird orchid, *Chiloglottis valida sensu lato* (Orchidaceae) in south-eastern Australia. Austr J Bot 54:53–64

Brooks DR, McLennan DA (1991) Phylogeny, ecology, and behavior: a research program in comparative biology. University of Chicago Press, Chicago

Bush AO, Lafferty KD, Lotz JM, Shostak AW (1997) Parasitology meets ecology on its own terms: Margolis et al. revisited. J Parasitol 70:315–316

Ciotek L, Giorgis P, Benitez-Vieyra S, Cocucci AA (2006) First confirmed case of pseudocopulation in terrestrial orchids of South America: pollination of *Geoblasta pennicillata* (Orchidaceae) by *Campsomeris bistrimacula* (Hymenoptera, Scoliidae). Flora 201:365–369

Coleman E (1928) Pollination of *Cryptostylis leptochila*. Vict Nat 44:333–340

Combes C (2001) Parasitism: the ecology and evolution of intimate interactions. University of Chicago Press, Chicago

Correvon H, Pouyanne M (1916) Un curieux cas de mimétisme chez les Ophrydées. J Soc Nat Hortic France 4:29–47

Correvon H, Pouyanne M (1923) Nouvelles observations sur le mimétisme et la fécondation chez les *Ophrys speculum* et *lutea*. J Soc Nat Hortic France 4:372–377

Cortis P, Vereecken NJ, Schiestl FP, Barone Lumanga MR, Scrugli A, Cozzolino S (2008) Pollinator convergence and the nature of species boundaries in sympatric Sardinian Ophrys (Orchidaceae). Annals of Botany doi:10.1093/aob/mcn219

Cozzolino S, Schiestl FP, Müller A, De Castro O, Nardella AM, Widmer A (2005) Evidence for pollinator sharing in Mediterranean nectar-mimic orchids: absence of pre-mating barriers? Proc R Soc Lond B 272:1271–1278

Dafni A (1984) Mimicry and deception in pollination. Ann Rev Ecol Syst 15:259–278

Dafni A (1987) Pollination in *Orchis* and related genera: evolution from reward to deception. In: Arditti J (ed) Orchid biology: reviews and perspectives IV. Cornell University Press, Ithaca, pp 80–104

Dafni A, Ivri Y (1981) Floral mimicry between *Orchis israelitica* Baumann and Dafni (Orchidaceae) and *Bellevalia flexuosa* BOISS (Liliaceae). Oecologia 49:229–232

Damiens D, Boivin G (2006) Why do sperm-depleted parasitoid males continue to mate? Behav Ecol 17:138–143

Danesch O, Danesch E (1972) Orchideen Europas, Ophrys-hybriden. Hallwag, Bern

Darwin C (1862) On the various contrivances by which British and foreign orchids are fertilised by insects. John Murray, London

Delforge P (2005) Guide des orchidées d'Europe, d'Afrique du Nord et du Proche-Orient. Delachaux and Niestlé, Paris

Dickson CR, Petit S (2006) Effect of individual height and labellum colour on the pollination of *Caladenia* (syn. *Arachnorchis*) *behrii* (Orchidaceae) in the northern Adelaide region, South Australia. Plant Syst Evol 262:65–74

Dod DD (1976) *Oncidium henekenii*: bee orchid pollinated by bee. Am Orch Soc Bull 45:792–794

Ebert D (1998) Experimental evolution of parasites. Science 282:1432–1435

Ebert D (2000) Experimental evidence for rapid parasite adaptation and its consequences for the evolution of virulence. In: Poulin R, Morand S, Skorping A (eds) Evolutionary biology of host–parasite relationships: theory meets reality. Elsevier, Amsterdam, pp 163–184

Galizia CG, Kunze J, Gumbert A, Borg-Karlson A-K, Sachse S, Markl C, Menzel R (2005) Relationship of visual and olfactory signal parameters in a food-deceptive flower mimicry system. Behav Ecol 16:159–168

Gaskett AC, Herberstein ME (2006) Flowering, pollination and fruit set in tongue orchids, *Cryptostylis* spp. Vict Nat 123:128–133

Gaskett AC, Winnick CG, Herberstein ME (2008) Orchid sexual deceit provokes ejaculation. Am Nat 171(6):E206–E212

Grant V (1994) Modes and origins of mechanical and ethological isolation in angiosperms. Proc Natl Acad Sci USA 91:3–10

Hadley NF (1981) Cuticular lipids of terrestrial plants and arthropods: a comparison of their structure, composition and waterproofing function. Biol Rev 56:23–47

Herrera CM, Castellanos MC, Medrano M (2006) Geographical context of floral evolution: towards an improved research programme in floral diversification. In: Harder LD, Barrett SCH (eds) Ecology and evolution of flowers. Oxford University Press, New York, pp 278–294

Jeraskova J, Johnson SD, Kindlmann P (2006) Mechanisms and evolution of deceptive pollination in orchids. Biol Rev 81:219–235

Johnson SD (1994) Evidence for Batesian mimicry in a butterfly pollinated orchid. Biol J Linn Soc 53:91–104

Johnson SD (2000) Batesian mimicry in the nonrewarding orchid *Disa pulchra*, and its consequences for pollinator behaviour. Biol J Linn Soc 71:119–132

Johnson SD, Shelah M (2006) Lying to Pinocchio: floral deception in an orchid pollinated by long-proboscid flies. Bot J Linn Soc 152:271–278

Johnson SD, Peter CI, Agren J (2004) The effects of nectar addition on pollen removal and geitonogamy in the non-rewarding orchid *Anacamptis morio*. Proc R Soc Lond B 271:803–809

Jones DL (1988) Native orchids of Australia. American Orchid Society, Delray Beach

Kasumovic MM, Bruce MJ, Andrade MCB, Herberstein ME (2008) Spatial and temporal demographic variation drives within-season fluctuations in sexual selection. Evolution 62(9):2316–2325

Kullenberg B (1961) Studies in *Ophrys* pollination. Zool Bid Uppsala 34:1–340

Lehnebach CA, Robertson AW, Hedderley D (2005) Pollination studies of four New Zealand terrestrial orchids and the implication for their conservation. New Zeal J Bot 43:467–477

Lorella B, Mahé G, Séité F (2002) Pollinisateurs d'*Ophrys* en Bretagne. L'Orchidophile 151:91–96

Mant J, Schiestl FP, Peakall R, Weston PH (2002) A phylogenetic study of pollinator conservatism among sexually deceptive orchids. Evolution 56:888–898

Mant JG, Peakall R, Schiestl FP (2005a) Does selection on floral odour promote differentiation among populations and species of the sexually deceptive genus *Ophrys*? Evolution 59:1449–1463

Mant J, Brown GR, Weston PH (2005b) Opportunistic pollinator shifts among sexually deceptive orchids indicated by a phylogeny of pollinating and non-pollinating thynnine wasps (Tiphiidae). Biol J Linn Soc 86:381–395

Michener CD (2007) The bees of the world, 2nd edn. Johns Hopkins University Press, Baltimore

Moccia MD, Widmer A, Cozzolino S (2007) The strength of reproductive isolation in two hybridizing food-deceptive orchid species. Mol Ecol 16:2855–2866

Neiland MRM, Wilcock CC (1995) Maximization of reproductive success by European Orchidaceae under conditions of infrequent pollination. Protoplasma 187:39–48

Neiland MRM, Wilcock CC (1998) Fruit set, nectar reward, and rarity in the Orchidaceae. Am J Bot 85:1657–1671

Nilsson LA (1983) Processes of isolation and introgressive interplay between *Platanthera bifolia* (L.) RICH and *P. chlorantha*. Bot J Linn Soc 87:325–350

Nilsson LA (1992) Orchid pollination biology. Trends Ecol Evol 7:255–259

Pasteur G (1982) A classificatory review of mimicry systems. Ann Rev Ecol Syst 13:169–199

Paulus HF, Gack C (1990) Pollinators as prepollinating isolation factors: evolution and speciation in *Ophrys* (Orchidaceae). Isr J Bot 39:43–79

Paxton RJ (2005) Male mating behaviour and mating systems of bees: an overview. Apidologie 36:145–156

Peakall R (1990) Responses of male *Zaspilothynnus trilobatus* Turner wasps to females and the sexually deceptive orchid it pollinates. Funct Ecol 4:159–167

Peakall R (2007) Speciation in the Orchidaceae: confronting the challenges. Mol Ecol 16:2834–2837

Peakall R, Handel SN (1993) Pollinators discriminate among floral heights of a sexually deceptive orchid: implications for selection. Evolution 47:1681–1687

Peakall R, Beattie AJ (1996) Ecological and genetic consequences of pollination by sexual deception in the orchid *Caladenia tentactulata*. Evolution 50:2207–2220

Peakall R, Beattie AJ, James SH (1987) Pseudocopulation of an orchid by male ants: a test of two hypotheses accounting for the rarity of ant pollination. Oecologia 73:522–524

Poulin R (2007) Evolutionary ecology of parasites, 2nd edn. Princeton University Press, Princeton

Pouyanne M (1917) La fécondation des *Ophrys* par les Insectes. Bull Soc Hist Nat Afr Nord 43:53–62

Proctor M, Yeo P (1972) The pollination of flowers. Taplinger, New York

Rotschild M, Clay T (1952) Fleas, flukes and cuckoos. Collins, London

Schiestl FP (2004) Floral evolution and pollinator mate choice in a sexually deceptive orchid. J Evol Biol 17:67–75

Schiestl FP (2005) On the success of a swindle: pollination by deception in orchids. Naturwissenschaften 92:255–264

Schiestl FP, Ayasse M, Paulus HF, Löfstedt C, Hansson BS, Ibarra F, Francke W (1999) Orchid pollination by sexual swindle. Nature 399:421–422

Schiestl FP, Peakall R, Mant JG, Ibarra F, Schulz C, Franke S, Francke W (2003) The chemistry of sexual deception in an orchid–wasp pollination system. Science 302:437–438

Schiestl FP, Ayasse M (2002) Do changes in floral odor cause speciation in sexually deceptive orchids? Plant Syst Evol 234:111–119

Schiestl FP, Peakall R (2005) Two orchids attract different pollinators with the same floral odour compound: ecological and evolutionary implications. Funct Ecol 19:674–680

Schiestl FP, Cozzolino S (2008) Evolution of sexual mimicry in the Orchidinae: the role of preadaptations in the attraction of male bees as pollinators. BMC Evolutionary Biology 8: 27

Schiestl FP, Vereecken NJ (2008) *Andrena combinata*, ein neuer Bestäuber von Ophrys araneola. J Eur Orchid 40:501–507

Scopece G, Musacchio A, Widmer A, Cozzolino S (2007) Patterns of reproductive isolation in Mediterranean deceptive orchids. Evolution 61:2623–2642

Singer RB (2002) The pollination mechanism in *Trigonidium obtusum* LINDL (Orchidaceae: Maxillariinae): sexual mimicry and trap-flowers. Ann Bot 89:157–163

Singer RB, Flach A, Koehler S, Marsaioli AJ, Amaral ME (2004) Sexual mimicry in *Mormolyca ringens* (LINDL.) SCHLTR. (Orchidaceae: Maxillariinae). Ann Bot 93:755–762

Spaethe J, Moser WH, Paulus HF (2007) Increase of pollinator attraction by means of a visual signal in the sexually deceptive orchid, *Ophrys heldreichii* (Orchidaceae). Plant Syst Evol 264:31–40

Stebbins GL, Ferlan L (1956) Population variability, hybridization, and introgression in some species of *Ophrys*. Evolution 10:32–46

Stökl J, 2007. Pollinator driven radiation in sexually deceptive orchids of the genus *Ophrys*. Dissertation, University of Ulm, Germany

Stökl J, Paulus H, Dafni A, Schulz C, Francke W, Ayasse M (2005) Pollinator attracting odour signals in sexually deceptive orchids of the *Ophrys fusca* group. Plant Syst Evol 254:105–120

Steiner KE, Whitehead VB, Johnson SD (1994) Floral and pollinator divergence in two sexually deceptive Cape orchids. Am J Bot 81:185–194

Stoutamire WP (1975) Pseudocopulation in Australian terrestrial orchids. Am Orch Soc Bull 44:226–233
Taylor I (1999) Monitoring report 1999 for *Pterostylis gibbosa*. National Parks and Wildlife Service, Hurstville
Thornhill R, Alcock J (2000) The evolution of insect mating systems: enlightenment to romanticism. Harvard University Press, Cambridge
Tremblay RL, Ackerman JD, Zimmerman JK, Calvo RN (2005) Variation in sexual reproduction in orchids and its evolutionary consequences: a spasmodic journey to diversification. Biol J Linn Soc 84:1–54
Turner CMR, Aslam N, Dye C (1995) Replication, differentiation, growth and virulence of *Trypanosoma brucei* infections. Parasitology 111:289–300
Tyteca D, Rois AS, Vereecken NJ (2006) Observations on the pollination of *Ophrys fuciflora* by pseudocopulating males of *Phyllopertha horticola*. J Eur Orch 38:203–214
Vandewoestijne S, Rois A-S, Caperta A, Baguette M, Tyteca D (2008) Effects of individual and population parameters on reproductive success in three sexually deceptive orchid species. Plant Biol (in press)
Van der Cingel NA (1995) An atlas of orchid pollination: European orchids. Balkema, Rotterdam
Van der Pijl L, Dodson CH (1966) Orchid flowers—their pollination and evolution. University of Miami Press, Coral Gables
Véla E, Tirard A, Renucci M, Suehs CM, Provost E (2007) Floral chemical signatures in the genus *Ophrys* L. (Orchidaceae): A preliminary test of a new tool for taxonomy and evolution. Plant Mol Biol Rep 25:83–97
Vereecken NJ, Patiny S (2005) On the pollination of *Ophrys catalaunica* (O. Danesch and E. Danesch) by pseudocopulating males of *Chalicodoma pyrenaica* (Lepeletier) (Hymenoptera, Megachilidae). Natural Belges 86 (Orchid 18):57–64
Vereecken NJ, Patiny S (2006) Patrolling males of *Andrena bicolor* F. (Hymenoptera, Andrenidae) as pollinators of *Ophrys massiliensis* Viglione and Véla. Natural Belges 87 (Orchid. 19):63–68
Vereecken NJ, Schiestl FP (2008) The evolution of imperfect floral mimicry. Proc Natl Acad Sci USA 105:7484–7488
Vereecken NJ, Risch S, Cortis P (2007a) Contribution to the pollination biology of *Ophrys scolopax* Cavanilles (Orchidaceae) in southern France. Natural Belges 88(Orchid 20):17–26
Vereecken NJ, Mant J, Schiestl FP (2007b) Population differentiation in female sex pheromone and male preferences in a solitary bee. Behav Ecol Sociobiol 61:811–821
Waser NM, Price MV (1998) What plant ecologists can learn from zoology. Persp Plant Ecol Evol Syst 1/2:137–150
Waterman RJ, Bidartondo MI (2008) Deception above, deception below: linking pollination and mycorhhizal biology of orchids. J Exp Bot 59:1085–1096
Westrich P (1990) Die Wildbienen Baden-Wurttembergs, Teil 1 and 2. Eugen Ulmer, Stuttgart
Wickler W (1968) Mimicry in plants and animals. Weidenfeld and Nicolson, London
Wiens E (1978) Mimicry in plants. Evol Biol 11:365–403
Williams NH, Dodson CH (1972) Selective attraction of male euglossine bees to orchid floral fragrances and its importance in long-distance pollen flow. Evolution 26:84–95
Wong BBM, Schiestl FP (2002) How an orchid harms its pollinator. Proc R Soc Lond B 269:1529–1532
Wong BBM, Salzmann CC, Schiestl FP (2004) Pollinator attractiveness increases with distance from flowering orchids. Biol Lett 171:S212–S214
Wyatt TD (2003) Pheromones and animal behaviour: communication by smell and taste. Cambridge University Press, Cambridge

Deceptive Behavior in Plants. II. Food Deception by Plants: From Generalized Systems to Specialized Floral Mimicry

Jana Jersáková, Steven D. Johnson, and Andreas Jürgens

Abstract Most of the ca. 8,000 angiosperm species that lack floral rewards are orchids pollinated by food-seeking animals. The fitness benefits of floral deception are still being debated, but most of the available evidence suggests that food deception evolves because it strongly promotes cross-pollination. These plants are generally considered to employ either generalized food deception (exploiting innate preferences of pollinators) or Batesian floral mimicry (exploiting conditioned preferences of pollinators). However, we argue that there are also intermediate conditions between these two modes and that exploitation of conditioned preferences is probably more common among food-deceptive orchids than was previously realized. We also review evidence for ecological facilitation of the pollination of deceptive species by rewarding species via the "magnet effect." We then consider the relative importance of visual and olfactory cues in food-deceptive systems. The role of scent in food-deceptive systems, especially generalized ones, is still poorly understood, but the available evidence indicates that floral color matching is a critical component of Batesian floral mimicry.

1 Introduction

The basic animal drive most commonly exploited for pollination by plants is the search for food (Dafni 1984). In order to attract pollinators, the plants use floral signals that advertise more or less concealed rewards such as pollen, nectar, floral

J. Jersáková (✉)
Department of Theoretical Ecology, Institute of Systems Biology and Ecology, AS CR and University of South Bohemia, Braniovská 31, České Budějovice, 37005, Czech Republic
e-mail: jersa@centrum.cz

S.D. Johnson and A. Jürgens
School of Biological and Conservation Sciences, University of KwaZulu-Natal, Private Bag X01, Pietermaritzburg, 3209, South Africa
e-mail: JohnsonSD@ukzn.ac.za
e-mail: ajuergens@hortresearch.co.nz

F. Baluška (ed.), *Plant-Environment Interactions*, Signaling and Communication in Plants, DOI: 10.1007/978-3-540-89230-4_12,

oils, resin or waxes. Some of these plants are frauds that advertise food, yet do not provide it. Such plants can be considered to be "food deceptive"and have evolved in many angiosperm lineages.

Nonrewarding flowers are found in 146 plant genera from 33 plant families (Renner 2005; Jersáková et al. 2006a). In most nonorchid species there are two major forms of deception (Renner 2005): (1) species exploiting carrion and mycophagous flies or carrion beetles seeking food and oviposition sites (e.g., Araceae, Aristolochiaceae, Asclepiadaceae, Hydnoraceae, Rafflesiaceae), and (2) species with unisexual flowers, where male flowers are usually pollen or nectar rewarding, while female flowers lack these rewards (e.g., Arecaceae, Asteraceae, Cucurbitaceae, Myristicaceae; Renner (2005) included also species from Clusiaceae and Euphorbiaceae, which instead offer waxes or resins to reward insects). However, food-deceptive flowers that mimic other rewarding flowers and sexually deceptive flowers are rare outside of the orchid family (e.g., Apocynaceae, Begoniaceae, Berberidaceae, Ranunculaceae).

The Orchidaceae has by far the greatest concentration of nonrewarding species. Approximately 6,000 species (one-third of the family) in 47 genera are estimated to be food deceptive (van der Pijl and Dodson 1966; Ackerman 1986; Renner 2005; Jersáková et al. 2006a), and another 400 species are sexually deceptive (i.e., they attract pollinators by mimicking the mating signals of the female; see the chapter by Vereecken; Dafni and Bernhardt 1990). A few orchid genera employ other deceptive mechanisms, such as shelter or brood-site imitation, or exploit the defensive behavior of territorial bees (Jersáková et al. 2006a). The widespread occurrence of food deception in orchids suggests that this form of pollination by deceit is a highly successful evolutionary strategy (Cozzolino and Widmer 2005).

The selective factors responsible for the high frequency of deception in orchids have received considerable attention in the past decade. Deception would also result in resource savings because plants are spared the cost of producing nectar, but this is unlikely to have been a major factor in selection. The most compelling and empirically well-supported hypothesis for the evolution and maintenance of deception is that it discourages pollinators from visiting many flowers on a plant and thereby promotes cross-pollination (Johnson and Nilsson 1999; Johnson et al. 2003b; Jersáková and Johnson 2006). Because their pollen is usually packaged into pollinia, orchids are especially vulnerable to loss of mating opportunities caused by self-pollination. A plausible complementary explanation for the high frequency of deception in orchids is that the full complement of pollinia in a flower can be removed in a single visit, making orchid reproduction very efficient when pollinator visits to deceptive flowers are infrequent (Harder 2000). In addition, the "lock and key" system whereby pollinia adhere to orchid stigmata means that orchids can easily share pollinators with other plants without losing pollen to their stigmata. Pollinia are also usually combined with accessory structures, such as viscidia that attach pollinia onto specific places on the pollinators' bodies, which diminish losses associated with pollinator grooming, and can often reorient to reduce the incidence of self-pollination (Peter and Johnson 2006). The improved pollination efficiency in orchids is likely to explain both the floral diversity and the widespread occurrence of deceit pollination in this clade (Harder 2000; Johnson and Edwards 2000; Jersáková et al. 2006a; Harder and Johnson 2008).

The mechanisms by which food-deceptive orchids attract pollinators vary from generalized food deception to specific mimicry of other flowers (Jersáková et al. 2006a, Fig. 1). Most orchids with deceptive pollination mechanisms exploit the innate food-foraging behavior of pollinators (Nilsson 1980; Dafni 1983). In order to attract pollinators, orchids advertise general floral signals of rewarding plant species, such as showy inflorescences, spectrally pure flower color, scent, nectar guides, spurs and pollen-like papillae (Gumbert and Kunze 2001; Galizia et al. 2005). Orchids associated with this form of deception typically flower gregariously in early spring and often exhibit color polymorphism (Nilsson 1984; Gigord et al. 2001). The pollinators may be recently emerged insects from hibernation, immigrants, or exploratory pollinators whose food resources are becoming depleted. Here we use the term "generalized food deception" (Steiner 1998) to describe this form of deception. We have not used the alternative term "mimicry based on naïveté" (Little 1983), because several other forms of pollination are based on visits from inexperienced pollinators; nor have we used the term "non-model mimicry" (Dafni 1986), because use of the term "mimicry" in a system lacking a model is a contradiction in terms (Cropper and Calder 1990).

Some food-deceptive orchids are mimics of particular rewarding species, and these have been termed Batesian mimics (Brown and Kodric-Brown 1979;

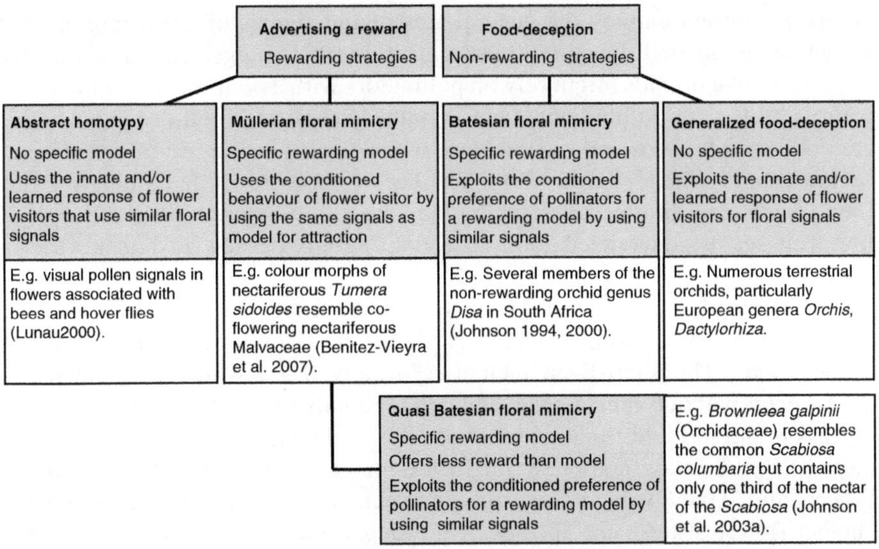

Fig. 1 A scheme with examples of pollination strategies regarding (*1*) presence of a reward, (*2*) presence of a specific model, and (*3*) innate versus learned response

Bierzychudek 1981; Dafni 1984; Johnson 1994; Roy and Widmer 1999). There is increasing evidence that flowers of Batesian mimics bear such a close resemblance to their models that pollinators are literally unable to distinguish between the two kinds of inflorescences (Dafni and Ivri 1981a; Johnson 1994, 2000; Johnson et al. 2003a). The best-studied examples of Batesian floral mimicry in plants involve two species in the genus *Disa*, *D. pulchra* and *D. ferruginea* (Johnson 1994, 2000; Fig. 2). Besides these, there have been several other case studies of nonrewarding plants that may qualify as Batesian mimics (Table 1). Some recent studies show that generalized food-deceptive species may also benefit reproductively from the presence of rewarding species (Johnson et al. 2003b). This raises the interesting possibility of a mimicry continuum between those orchids that exploit instinctive food-seeking behavior of pollinators and those that show an adaptive resemblance to nectar-producing plants.

In this chapter we review the literature on food deception in plants. We focus on the theoretical conditions that have to be met before a particular system qualifies as either generalized food deception or Batesian floral mimicry, and identify plant and pollinator characters that influence the occurrence of these interactions. We then consider the notion of a continuum between these two forms of deception and conclude that pollinator conditioning on "model" plants probably plays a role in the evolution of most food-deceptive systems.

2 Generalized Food Deception

This form of deception is based on a general resemblance of flowers of a nonrewarding plant to those of food plants without any specific mimicry. Such food-deceptive flowers employ the general advertising signals of rewarding plants to attract naïve, unconditioned pollinators. Dafni (1983) suggested that generalized food-deceptive orchids might rely on pollinators with a poor memory or discriminatory abilities for pollination; however, most generalized food-deceptive orchids are pollinated by bees and bumblebees, whose ability to choose a flower is very accurate and reliable (Barth 1991). The bees soon learn to discriminate against food frauds, and diminished visits by individual pollinators lead to the typically low fruit set of generalized food-deceptive orchids (review in Jersáková et al. 2006a).

In a habitat that has several flower species, the foraging of bees is governed by working memory dynamics, which is either short- or long-lived (Chittka and Raine 2006). The short-lived memories usually last few seconds and rapidly decay, even without interference, and can be relatively easily erased by competing information. Field trials with foraging bees showed that in the first few seconds (0–2 s) bees do not easily forget the image of the previously encountered flower, and if newly incoming stimuli match this signal, the bee would visit another flower of the same species. After longer intervals (3–6 s), bees are more likely to switch to visiting different flower species and more likely to those with similar color (Chittka et al. 1997). The particular plant's success will thus depend

Fig. 2 Nonrewarding Batesian mimics and their models: **a** *Disa puchra* (Orchidaceae) mimics **b** *Watsonia lepida* (Iridaceae), a nectar-producing species pollinated by tabanid fly *Philoliche aethiopica* (Johnson 2000); **c** *Disa ferruginea* (Orchidaceae) mimics **d** *Kniphofia uvaria* (Asphodelaceae), which provide nectar rewards for the nymphalid butterfly *Aeropetes* (*Meneris*) *tulbaghia* (Johnson 2004). Photographs by Jana Jersáková

Table 1 List of floral mimicry and magnet species effect studies and conditions used to classify them (or not) as Batesian floral mimicry systems (+: positive; –: negative; 0: no effect; ?: not tested)

Mimic	Model	Pollinator	Pollinator cannot distinguish	Similar shape	Similar spectral reflectance[1]	Similar scent	Overlapping distribution	Overlapping phenology	Mimic is less abundant	Enhanced fitness	Reference
Batesian floral mimicry studies											
Disa cephalotes (Orchidaceae)	*Scabiosa columbaria* (Dipsacaceae), *Brownleea galpinii* (Orchidaceae)	Long-tongued flies (Tabanidae, Nemestrinidae)	+	+	+	? Weak	+	+	+	?	Johnson et al. (2003a)
Disa ferruginea (Orchidaceae)	*Tritoniopsis tritica* (Iridaceae)	Butterflies (Nymphalidae)	+	+	+	? Weak	+	+	+	+	Johnson (1994)
Disa nivea (Orchidaceae)	*Zaluzyanskia microsiphon* (Scophulariaceae)	Long-tongued flies (Nemestrinidae)	+	+	+	? Weak	+	+	+	+	Anderson et al. (2005)
Disa puchra (Orchidaceae)	*Watsonia lepida* (Iridaceae)	Long-tongued flies (Tabanidae)	+	+	+	? Weak	+	+	+	?	Johnson (2000)
Eulophia zeyheriana (Orchidaceae)	*Wahlenbergia cuspidata* (Campanulaceae)	Solitary bees (Halictidae)	+	–	+	?	+	+	–	+	Peter and Johnson (2008)
Potential Batesian floral mimicry or magnet species effect studies											
Cephalanthera longifolia (Orchidaceae)	*Cistus salviifolius* (Cistaceae)	Solitary bees (Halictidae)	?	–	–[2]	?	–	+	?	+	Dafni and Ivri (1981b)
Cephalanthera rubra (Orchidaceae)	*Campanula persicifolia* (Campanulaceae)	Solitary bees (Megachilidae)	?	–	+	– Weak	–	+	+	+	Nilsson (1983)
Cypripedium macranthos var. *rebunense* (Orchidaceae)	*Pedicularis schistostegia* (Scophulariaceae)	Bumblebees (Apidae)	?	–	+	?	+	+	+	?	Sugiura et al. (2002)

Disa draconis (Orchidaceae)	*Pelargonium* sp. (Geraniaceae)	Long-tongued flies (Tabanidae, Nemestrinidae)	?	+	+	?	+	+	+	?	Johnson and Steiner (1997), Van Der Niet, unpublished data
Disa ferruginea (Orchidaceae)	*Kniphofia uvaria* (Asphodelaceae)	Butterflies (Nymphalidae)	?	+	+	? Weak	+	+	+	?	Johnson (1994)
Disa nervosa (Orchidaceae)	*Watsonia densiflora* (Iridaceae)	Long-tongue flies (Tabanidae)	?	+	+	? Weak	+	+	+	0	Johnson and Morita (2006)
Diuris maculata (Orchidaceae)	*Daviesia* sp. (Fabaceae)	Solitary bees (Colletidae)	?	+	(+)	?	?	+	+	?	Beardsell et al. (1986)
Diuris maculata (Orchidaceae)	*Daviesia ulicifolia* subsp. *ulicifolia* (Fabaceae)	Solitary bees (Colletidae)	?	+	+	?	+	+	+	?	Indsto et al. (2006)
Epidendrum ibaguense (Orchidaceae)	*Asclepias curassavica* (Asclepiadaceae), *Lantana camara* (Verbenaceae)	Butterflies (Danaidae)	?	–	(+)	?	+	+	+	?	Boyden (1980)
Epidendrum radicans (Orchidaceae)	*Asclepias curassavica* (Asclepiadaceae), *Lantana camara* (Verbenaceae)	Butterflies	?	–	(+)	?	+	+	+	0	Bierzychudek (1981)
Oncidium cosymbephorum (Orchidaceae)	*Malphigia glabra* (Malphigiaceae)	Solitary bees (Anthophoridae)	?	+	(+)	?	–	+	+	+	Carmona-Diaz (2001)

(continued)

Table 1 (continued)

Mimic	Model	Pollinator	Pollinator cannot distinguish	Similar shape	Similar spectral reflectance[1]	Similar scent	Overlapping distribution	Overlapping phenology	Mimic is less abundant	Enhanced fitness	Reference
Orchis collina	Orchis coriophora	Honeybees, solitary bees (Apidae)	?	+	(+)	?	–	+	?	+	Dafni and Ivri (1979)
Orchis israelitica *Orchis boryi* (Orchidaceae)	Bellevalia flexuosa (Liliaceae)	Bees (Apidae)	?	+	+	–	?	+	?	?	Galizia et al. (2005)
Orchis israelitica (Orchidaceae)	*Bellevalia flexuosa* (Liliaceae)	Solitary bees (Apidae), beeflies (Bombiliidae)	?	+	(+)	?	?	+	?	+	Dafni and Ivri (1981a)
Orchis morio (Orchidaceae)	*Allium schoenoprasum* (Alliaceae)	Bumblebees (Apidae)	?	–	(+)	?	–	+	+	+	Johnson et al. (2003b)
Thelymitra antennifera (Orchidaceae)	*Hibbertia stricta* (Dilleniaceae), Helichrysum (Asteraceae), yellow plants	Solitary bees (Halictidae)	?	–	(+)	?	+	+	?	?	Dafni and Calder (1987)
Non-orchid studies											
Podophyllum peltatum (Berberidaceae)	*Pedicularis canadensis* (Scrophulariaceae)	Bumblebees (Apidae)	?	–	(+)	?	–	+	?	+	Laverty (1992)
Cimicifuga rubifolia (Ranunculaceae)	*Polymnia canadensis* (Asteraceae), *Impatiens pallida* (Balsaminaceae)	Bumblebees (Apidae)	?	–	(–)	?	?	+	+	+	Pellmyr (1986)

[1] Studies in which spectral reflectance was assessed qualitatively by human vision are indicated by parentheses;

[2] The flowers appear similar to humans, but on UV-sensitive photographs, *Cephalanthera longifolia* emits strong UV, unlike *Cistus salviifolius*. The pollinators are probably not deterred by UV differences, but are attracted to pseudopollen

not only on the quality of its own signal but also on the efficiency of the signals of other species in the vicinity, as well as their relative abundance, distribution and degree of spatial intermixing (Chittka and Raine 2006). Therefore generalized food-deceptive orchids often employ strong advertising signals, such as large showy flowers displayed in early spring to outcompete other rewarding plants and benefit from the presence of rewarding species of the similar color (Pellmyr 1986; Laverty 1992; Johnson et al. 2003b).

2.1 Large Floral Displays

This aspect of generalized food-deceptive orchids is considered to be an evolutionary adaptation to increase pollinator attraction in a group of plants with extremely low levels of fruit set. Larger flowers may represent more powerful signals to pollinators than smaller ones and may be favored by selection in a similar manner to flower production (Waite et al. 1991; Murren and Ellison 1996; Tremblay et al. 2005). The recent comparative study of Huda and Wilcock (2008) confirmed that tropical nectarless orchids have displays with larger flowers than their rewarding counterparts, irrespective of their habit.

2.2 Pseudopollen and False Anthers

Besides showy floral displays and the presence of nectar guides, some food-deceptive orchids advertise bright yellow tufts of hairs or pollen-like papillae on the lips to attract pollen-foraging hymenopterans (Davies et al. 2000, 2002). Even though the papillae and trichomes are often rich in protein and starch, and are actively collected by the pollinators (Davies et al. 2003), we have no direct evidence that pseudopollen is used for nutrition by pollinators. Most papillae contain pigments or act as osmophores and probably represent visual or tactile cues that guide pollinators into flowers (Davies and Turner 2004).

2.3 Flowering Early in the Season

This feature is perceived as an adaptation to receive exploratory visits of naïve pollinators which have just emerged from hibernation (Nilsson 1980, 1984) and to lower competition from other rewarding plants in early spring. This idea was supported by Kindlmann and Jersáková (2006), who found the peak flowering for the European deceptive species and genera to be significantly earlier than for the rewarding ones. Many European terrestrial orchids, particularly in the genera *Orchis* and *Dactylorhiza*, flower gregariously in early spring and are thus subjected to increased intraspecific competition for pollinators. Although the number of pollinaria removed and fruit set have a tendency to increase with the number of plants, the proportion of pollinaria removed per plant decreases (Fritz and Nilsson 1994). Therefore individuals flowering early or late in relation to population peak flowering usually have higher pollination success than those flowering in the middle of the blooming period (Fritz 1990; Parra-Tabla and Vargas 2007).

2.4 Magnet Species Effects

Several recent studies have shown that some generalized food-deceptive species have higher reproductive success when growing close to rewarding plants, even though they do not resemble them (Laverty 1992; Pellmyr 1986; Johnson et al. 2003b). In this case, the rewarding plants act as a magnet species that increases the local abundance of pollinators, and the deceptive species may gain a net benefit even though pollinators may show relative constancy to the magnet species and cause reproductive interference (Laverty 1992; Alexandersson and Agren 1996). Interestingly, both deceptive and magnet species often display flowers of similar colors (Table 1).

2.5 High Degree of Variability in Floral Traits

High variation in color (Nilsson 1980; Cropper and Calder 1990; Aragon and Ackerman 2004), shape and size (Ackerman and Galarza-Pérez 1991), and fragrance (Moya and Ackerman 1993; Andersson et al. 2002; Salzmann et al. 2007a, b) is apparent in generalized food-deceptive systems, but relatively few studies have explored the consequences of polymorphism of floral traits on visitation rates and consequently, on reproductive success, in a spatiotemporal framework (Sabat and Ackerman 1996; Ackerman et al. 1997). High levels of variability in floral traits, particularly flower color and floral scent, might be a way to disrupt associative learning of pollinators so that it inhibits the pollinator's ability to recognize nonrewarding flowers (Heinrich 1975; Moya and Ackerman 1993).

3 Batesian Floral Mimicry

According to common convention, a Batesian mimic is a palatable animal that gains protection from predators through its resemblance to an unpalatable model species (Bates 1862). This zoological concept has been further extended to plant pollination systems on the grounds that Batesian mimicry in palatable animals and floral mimicry are outcomes of a common evolutionary process: selection favors a resemblance to an unpalatable animal or rewarding plant model, respectively, because the fitness of the mimic is increased when it is mistaken for the model by a signal receiver (predator or pollinator) (Vane-Wright 1980).

As a consequence of sharing evolutionary properties, the conditions of Batesian floral mimicry (Little 1983; Dafni 1984; Ackermann 1986; Roy and Widmer 1999;

Anderson and Johnson 2006) are essentially the same as those of protective Batesian mimicry in animals (Vane-Wright 1980; Endler 1981):

- The mimic and model should overlap substantially in their flowering phenology.
- The spatial distribution of the mimic should strongly overlap with the distribution of the model and must have done so long enough for evolution to occur.
- The mimic and model share pollinators.
- The mimic should occur at a low frequency relative to the model. In theory, a Batesian mimic always degrades the attractive (protective) value of the model's signal and, because pollinators (predators) can learn about its presence, a Batesian mimic is best off when it is rare and therefore least detectable (Bierzychudek 1981; Joron and Mallet 1998; Anderson and Johnson 2006). Thus according to this theory, new and (by implication) rare forms of mimicking plants should be favored and Batesian species should have some tendency to mimetic polymorphism (Turner 1977; Johnson 1994). In the mimic, the amount of genetic variability for further evolution should thus be greater than in the model.
- The mimic should resemble the model (i.e., in shape, color or scent) to the extent that the conditioned signal receiver is unable to discriminate between mimic and model and moves freely between them (Wiens 1978). The resemblance should be evolutionary adaptive, so the mimic receives more visits and has higher fitness in the presence of the model than in its absence (Dafni 1984). Such adaptive resemblance has evolved or is maintained as a result of selection and should exclude cases of incidental resemblance or convergence due to common adaptive responses to functional requirements or to other selective processes (Starrett 1993).

Several examples of Batesian floral mimicry have been proposed (Table1), and the evidence for these systems varies from observations of sharing pollinators and habitats (Beardsell et al. 1986; Dafni and Calder 1987) to experimental results, which accommodates all or most of the abovementioned predictions (Nilsson 1983; Johnson 2000; Johnson et al. 2003a). Unfortunately, only a few of them yield sufficient evidence demonstrating an evolutionary adaptiveness of the similarity of the mimic to the model (Johnson 1994; Anderson et al. 2005). To establish whether the resemblance is actually adaptive in a Batesian system, several facts can be verified using the following kinds of tests:

- The inability of pollinators to distinguish between model and mimic can be directly tested by an interview method, where the forager is allowed to choose between the model or the mimic mounted to a mobile stick (Thomson 1988; modified by Johnson 2000). The other option is an analysis of foraging bouts in terms of the expected versus observed number of visits and movements between species in natural assemblages or arrays (Johnson 1994).
- Fitness gains based on similarity can be measured in terms of male and female reproductive success. The higher fitness of the mimic in the presence of the model should be maintained throughout time and space (Roy and Widmer 1999; Johnson 2000). This test cannot stand on its own as a way to classify the species as a Batesian mimic, as the increased fitness of the putative mimic due to the presence of a particular rewarding species could simply be a case of the magnet

species effect (Laverty 1992; Johnson et al. 2003b). Some previously published studies on mimicry that are based on increased reproductive success of a species when growing together with its putative model, and which failed to demonstrate other assumptions of floral mimicry, should possibly be reclassified as examples of the magnet species effect.

- Phylogenetic analyses can be used to test if "mimicry traits" represent evolutionary novelties, which are characteristic for adaptations (Johnson et al. 2003a). The supposedly adaptive traits that descended from a common ancestor cannot be accurately described as adaptations to current ecological conditions, even though they may be actively maintained by stabilizing selection (Coddington 1988; Grandcolas and D'Haese 2003). The essential prerequisite for tracing the evolution is an existing species-level phylogeny with complete samplings of taxa. Unfortunately, such phylogenies are not always available.
- The extent to which a mimic resembles its model can be tested as a correlation between among-population variation in attractive traits in the putative mimic and variation in the attractive traits of model species (Johnson 1994; Anderson et al. 2005).
- Since pollinator conditioning should underlie all cases of Batesian floral mimicry, choice experiments can be conducted to establish the importance of conditioned preferences (Gigord et al. 2002). We are not aware of any field-based tests of the importance of pollinator conditioning in a Batesian floral mimicry system (Gumbert and Kunze 2001; Johnson et al. 2003a).

4 The Functional Significance of Floral Traits in Food-Deceptive Pollination Systems

Animal-pollinated flowers advertise themselves using various signals, which are often presented simultaneously. These signals are typically visual, olfactory, and tactile. The relative importance of a signal for attraction depends strongly on the kind of visitor. For example, Roy and Raguso (1997) found in their experiment with the fungal pseudoflowers of a crucifer mimicking a buttercup that halictid bees had greater visual than olfactory response, whereas flies were more dependent on olfactory cues.

4.1 Visual Signals

Color is one of the most important floral signals, and indeed many food-deceptive species possess bright showy floral displays. Although flowers of a mimic may be perceived as being similar to a model by the human eye, the similarity between them should be considered from the pollinators' perceptual point of view. For honeybees, a mathematical model of color vision has been developed that allows the similarity of two colors to be quantified if their reflectance spectra are known (Chittka 1992;

Chittka and Kevan 2005). The model enables the distance between color loci in the color space to be calculated, which increases the frequency of errors in the bee color discrimination (Dyer and Chittka 2004). This model does not however fully account for the intensity of the stimulus (i.e., color saturation), which is also known to drive bees' target detection. Our knowledge of color discrimination at the neural level in other pollinator groups such as Diptera, Lepidoptera and Coleoptera is rather limited, and the spectral sensitivity curves of their color photoreceptors have not yet been studied sufficiently. We might expect that beetles, flies, moths and butterflies would have much in common with bees, based on the monophyly of Insecta and the common dependence of all anthophilous insects on flowers, which should subject them to similar selection pressures (Weiss 2001). Various insect groups, with the exception of ants, have been shown to have color perception and to possess UV, blue and green photoreceptors (Briscoe and Chittka 2001). The color discrimination may however differ in terms of color coding. For example, while color discrimination in bees improves smoothly as the color difference between two stimuli increases, blowflies lump colors into three broad categories each about 100 nm wide, and they treat all colors as either the "same" or "different" to a training stimulus, depending on whether they fall inside the same category (Troje 1993).

4.1.1 Generalized Food Deception

Food-fraud orchids that function without a model often exhibit a high degree of color polymorphism. Variation in flower color is often expressed as continuous gradation in color hue or intensity, as found in *Anacamptis morio* (Nilsson 1984), *Polystachya rosea* (Pettersson and Nilsson 1993) or *Psychilis monensis* (Aragon and Ackerman 2004). A few species exhibit discontinuous, genetically determined variation in flower color resulting in two or several more or less discrete color morphs, such as yellow and purple morphs of *Dactylorhiza sambucina* or *D. romana* (Nilsson 1980). More rarely, petal color is polymorphic, with individuals exhibiting variable-sized spots or nectar guides on darker or lighter backgrounds, as in *Dactylorhiza maculata* (Koivisto et al. 2002).

In general, the variation in flower color within populations has been shown to have an impact on both female and male fitness (Harding 1970; Waser and Price 1981), as well as outcrossing rates (Horovitz and Harding 1972; Brown and Clegg 1984; Schoen and Clegg 1985). Differential fitness can arise when pollinators discriminate among flower color morphs, resulting in one morph being undervisited (Waser and Price 1983; Stanton 1987). Individuals with more deeply colored flowers often receive greater pollinator service than do pale-colored or white (albino)-flowered genotypes (Harding 1970; Waser and Price 1981).

There have been several attempts to grasp the meaning of color polymorphism for fitness in generalized food-deceptive orchids (Aragon and Ackerman 2004; Tremblay and Ackerman 2007), but none of them have confirmed the original idea that color polymorphism slows down the learning processes of naive pollinators and hence increases the number of visits to deceptive plants (Heinrich 1975;

Nilsson 1980; Pettersson and Nilsson 1993). Aragon and Ackerman (2004) found that the reproductive success of *P. monensis*, whose flower color varies from pale pink to deep purple, was mainly related to the flowering phenology of local communities than to color polymorphism. They concluded that natural levels of color variation might be more influenced by genetic drift than selection. Similarly, Koivisto et al. (2002) found no preferences of pollinators for any color variant in *Dactylorhiza maculata*. It seems that not all of the color polymorphism in flowers affects pollinator activity, and that it might be either linked genetically to other adaptive characters, such as flower size or the number of inflorescences, which also impinge on the reproductive success (Wolfe 1993; Galen et al. 1987; Sobrevila et al. 1989), or to the outcome of phenotypic plasticity (Schemske and Bierzychudek 2001; Warren and Mackenzie 2001).

The mechanism that has been proposed to maintain floral color polymorphism in generalized food-deceptive systems is a negative frequency-dependent selection caused by pollinator behavior. As pollinators are expected to learn and avoid floral signals of deceptive plants upon subsequent visits, and because pollinators encounter common morphs more often than rare ones, they may proportionally overvisit rare morphs and hence increase their relative fitness. This hypothesis has not however been thoroughly tested so far, and the only example of the negative frequency-dependent selection affecting plant reproductive success is discrete color polymorphism in *D. sambucina* (Gigord et al. 2001). Other studies found no evidence of negative frequency-dependent selection acting on corolla color polymorphism in orchids (Aragon and Ackerman 2004; Pellegrino et al. 2005; Jersáková et al. 2006b).

4.1.2 Batesian Floral Mimicry

The importance of a close match between the reflectance spectra of the flowers of Batesian mimics and their models has been demonstrated in several studies (Nilsson 1983; Johnson 1994, 2000; Johnson et al. 2003a; Anderson et al. 2005; Peter and Johnson 2008), suggesting that visual signals may be essential for floral mimicry. Peter and Johnson (2008) manipulated the UV component of the reflectance of flowers of the mimic orchid *Eulophia zeyheriana* and found that this led to a strong reduction in visitation by pollinators. Galizia et al. (2005) examined the importance of scent and color in the Batesian floral mimicry system involving rewarding lily *Bellevalia flexuosa* and rewardless orchid *Orchis israelitica*. They found no evidence for scent mimicry in this system and point to visual similarity as being the key to the successful deception performed by *O. israelitica*. Similarly, Gigord et al. (2002) confirmed the potential of a rewardless orchid *D. sambucina* to create mimicry with a rewarding *Mimulus guttatus* through a similarity based on corolla color alone. Both experimental and field observations indicate that the foraging decisions of bees are determined by their previous experience, as bees chose deceptive species more frequently if they foraged on more similarly colored species (Gumbert and Kunze 2001; Gigord et al. 2002).

Other visual stimuli such as flower size and shape seem to play a less important role in mimicry systems. In field experiments with species of different flower shapes and colors, Wilson and Stine (1996) found that bumblebees visited flowers with similar colors but different shapes, but not vice versa. Similarly, Gigord et al. (2002) found that a strong resemblance in floral morphology or traits other than color between the model and the mimic were not necessary for a Batesian mimic to be favored by bumblebees. A low importance of floral shape in the presence of color stimulus was also demonstrated for another pollinator group by Johnson and Morita (2006), who found that long-tongued flies readily probed misshaped and damaged flowers of an orchid *Disa nervosa* mimicking an iris *Watsonia densiflora*. On the other hand, Anderson et al. (2006) found a strong geographical correlation between the flower diameters of the orchid *Disa nivea* and its putative model *Zaluzianskya microsiphon*, which could be interpreted as evidence for selection on flower size in a mimicry system. The entire architecture of the inflorescence may play an important role in the pollinator's decisions, as exemplified by the results of a choice experiment in which flies were not attracted to inflorescences of a mimic that had been artificially reconstructed in the shape of a spike, rather than a flat-topped capitulum typical of the model (Johnson et al. 2003a).

4.2 Olfactory Signals

Odors are important floral signals, and they have been shown to be key signals in sexual deceptive systems. In these systems, the floral scent emitted by the plants contains compounds that are similar to the sex pheromones of female hymenopteran species, and these chemicals alone are sufficient to not only attract the specific pollinator but also to elicit copulatory behavior from the male hymenoptera (e.g., Schiestl et al. 2003; see also the chapter by Vereecken in this volume). An important role has also been suggested for the floral volatiles of food- and brood-site deceptive flowers that imitate the scent of putrification products, decay and fermentation, and attract certain groups of flies and beetles (Wiens 1978; Faegri and Van der Pijl 1979; Jürgens et al. 2006). A meta-analysis of floral scent compositions across different plant groups suggests that different types of odor mimicry are subsumed by the sapromyophilous syndrome (Ollerton and Raguso 2006), such as mimicry of carrion and carnivore feces, herbivore feces, and urine-related odors. In comparison to sexual and brood site deception systems, the role of scent in food-deceptive systems is still poorly understood. It is important to point out that not all species with a sapromyophilous syndrome are deceptive. In a study on the floral odors of 15 sapromyophilous stapeliad species, Jürgens et al. (2006) reported that only *Stapelia asterias* does not produce nectar. The small amounts of nectar found in the other species may be enough for flies to learn the odor source associated with the reward and thus revisit these flowers. However, an interesting question is whether offering the "wrong" food source (carbohydrates instead of protein degradation products) or offering a food source instead of a brood site can be regarded as a deceit or not.

Furthermore, different degrees of deceit (e.g., no reward versus wrong reward) might have different effects on the learning behavior of flies. With respect to the stapeliad study cited above, *Stapelia asterias*, which offers no reward, is the only stapeliad species in the study in which fly eggs have actually been found (Meve, pers. obs., in Jürgens et al. 2006). It is still an open question as to whether the behavior of flies in flowers with nectar changes into food-seeking behavior once the nectar is found, while flies follow their initial behavioral pattern, which is oviposition, in nonrewarding flowers (such as *S. asteris*).

4.2.1 Generalized Food Deception

In contrast to the scents of sexually deceptive orchids, which often comprise fatty acid derivates and various alkenes (Salzmann et al. 2006; Schiestl and Cozzolino 2008), the scents emitted by generalized food-deceptive species usually consist of benzenoids and isoprenoids (Nilsson 1980, 1983, 1984; Andersson et al. 2002; Salzmann et al. 2006, 2007a, b), which are common volatiles present in the floral fragrances of diverse plant families (Knudsen et al. 2006). Although the emission of floral scent has been reported for many food-deceptive orchids (Nilsson 1979, 1983, 1984; Bergstrom et al. 1992; Moya and Ackerman 1993; Ackerman et al 1997; Barkman et al. 1997; Salzmann and Schiestl 2007), there are only a few studies that have investigated the floral scent data for both mimics and their potential models (Galizia et al. 2005), or that have tested the significance of scent variation in generalized food-deceptive systems (Salzmann et al. 2007a, b). In comparison to sexual deception, where the signaling odor is one or a combination of only a few "key active substances," the floral odors advertising food typically consist of tens to hundreds of volatile compounds (Dobson 1994). Salzmann et al. (2006) found that in the food-deceptive orchid *Caladenia londicauda*, not only was the floral scent composition different but the emission was also much higher than in the sexually deceptive *C. arenicola*. Similar to the findings that the flower colors of many generalized food-deceptive species seem to be highly variable between and within populations, a high variability has also been described for the floral scents of some species (Moya and Ackerman 1993; Andersson et al. 2002; Salzmann et al. 2007a, b), and such variability is often attributed either to relaxed selection on this floral trait, or to a strategy aimed at preventing the pollinators from learning to avoid the nonrewarding flowers. Moya and Ackerman (1993) compared the odor compositions of the nonrewarding, moth-pollinated orchid *Epidendrum ciliare* at different levels (inflorescence, plant, population) and at different flowering ages. Their data showed that the fragrances of sample pairs were never identical at any level of analysis. The possible mechanism that could maintain the floral odor polymorphism is negative frequency-dependent selection, where rare floral morphs are expected to experience a selective advantage in comparison with common floral morphs (Smithson and Macnair 1997; Ferdy et al. 1998) because pollinators are not able to learn to avoid them. However, the positive impact of odor polymorphism on plant pollination success was not supported by an odor manipulative experiment in *Anacamptis morio* (Salzmann et al. 2007b).

A recent study by Valterová et al. (2007) showed that the scent of the generalized food-deceptive orchid *Orchis pauciflora* consists of typical floral fragrance compounds (e.g., α-pinene, β-pinene, limonene) as well as sesqui- and diterpenes [(*E,E*)-α-farnesene, (*E*)-β-farnesene, (*S*)-(-)-(*E*)-2,3-dihydrofarnesol, geranylcitronellol] that are frequent constituents of male marking pheromones of bumblebee species. They demonstrated in field experiments that inflorescences enriched with the main compound (*E*)-β-farnesene resulted in significantly increased pollinia export. However, the role of (*E*)-β-farnesene in the pollination of *O. pauciflora* seems to be unclear. Nevertheless, this case might be an interesting example where different behavioral aspects – foraging and bee communication – of the pollinators are exploited at the same time. Although speculative, it seems possible that complex fragrance patterns use multiple communication channels to exploit different behavioral aspects of pollinators to attract and manipulate them. In the case of *O. pauciflora*, further research is needed to test whether the presence of bumblebee male pheromone components in the fragrance increases its fitness.

4.2.2 Batesian Floral Mimicry

In behavioral experiments with bumblebees foraging on artificially scented flower dummies, Gumbert and Kunze (2001) showed that discrimination was poorest if the floral odors of the model and the mimic were identical. Such data point to the importance of floral volatiles, and they reveal a fundamental problem for mimicry systems, where the perfect match of the mimic's odor with its model seems to be essential in order to deceive the pollinator. This raises the question of how mimics could evolve a perfect match to such complex odor patterns. Our field observations and gas chromatography analyses of Batesian mimics suggest that they produce very weak scent signals or avoid odor production altogether. In many studies on floral mimicry, researchers repeatedly reported rather weak or scentless mimics (Johnson 1994, 2000). Floral mimicry appears to be unusually prevalent among southern African plants (cf. Johnson et al. 2003a), in contrast to the situation in Europe, where generalized food-deceptive strategy prevails. Southern African Batesian floral mimicry systems typically involve model plants from the families Iridaceae, Geraniaceae or Scrophulariaceae, which generally have unscented flowers to the human nose (Manning and Goldblatt 1996, 1997; Goldblatt and Manning 2000), and are pollinated by tabanid and nemestrinid flies or butterflies (Boyden 1980; Johnson 1994; Table 1). Both pollinator groups are able to use scent to locate food, but behavioral experiments indicate that vision would be expected to take precedence over any perceived scent (Proctor et al. 1996; Allan et al. 1987). The sole use of visual cues (color and shape) to locate flowers seems to increase the likelihood that mimicry would evolve in such systems (cf. Dafni 1987).

In Europe and South America, the proposed examples of Batesian floral mimicry involve solitary bees and bumblebees (Table 1). The discrimination ability of Hymenoptera is rather complex, and effective mimicry of flowers visited by bees would probably need to involve not only visual resemblance but also similarity in floral fragrances. Of the various signals bees associate with flowers, scent is more rapidly learned than color or shape, and is directly used when making decisions about alighting on flowers (Kunze and Gumbert 2001). Flowery odors are chosen correctly 97–100% of the time after only a single exposure, whereas colors are on average identified reliably (i.e., over 90% of the time) only after 3–5 visits, and shapes only after 20 visits (cf. Barth 1991). Except for the exploratory visits of newly emerged individuals or when switching to new sources, bees display a high degree of constancy in terms of their food sources and effectively avoid nonrewarding plants. The efficient learning and discrimination abilities of bees are probably the reason why few Batesian floral mimicry systems have been described in Europe.

The study of Galizia et al. (2005) suggests that similarity in the floral scents of the mimic and the model may not play a fundamental role in mimicry systems, as they found no evident similarity in the perception of floral scents in the honeybee brain when comparing the odors of the rewardless orchid *Orchis israelitica* and rewarding lily *Bellevalia flexuosa*. Nevertheless, the odor of *O. israelitica* was much weaker than that of its model. As the mimics usually grow intermingled with their models and the odor acts as a diffuse and long-distance signal, mimics growing near scented models might not need to produce odor at all.

5 Conclusions

Data on generalized food-deceptive systems show a threefold continuum in factors that affect avoidance learning of flower visitors. First, we find a continuum of deception ranging from full deception (no reward) to different degrees of deception (low reward, wrong reward). If the offered reward is not sufficient in its quantity and/or quality to satisfy the operator's "expectations," it may affect the behavior of the pollinator directly or after learning in a way that deviates from that associated with the model. An important question is: how do different rewarding "strategies" affect plant fitness by changing the behavior of visitors to the flower?

Second, there is a continuum in the resemblance of the model and the mimic from generalized to specific. However, analyzing the resemblance of the model and the mimic from a flower visitor's perspective implies that we should differentiate between similar and dissimilar aspects of the signaling in order to answer two separate questions. (1) How much *similarity* is needed to deceive a pollinator and to become a mimic? In other words, what are the common floral signals used by the operator to find its host and by the mimic to attract the operator? (2) What are the signals that may be used by the operator to eventually differentiate between model and mimic? Thirdly, there is a continuum in how floral signals exploit purely learned responses to purely innate responses of pollinators.

Finally, studies on the role of scent and color in mimicry systems can give us important insights into the roles of visual and olfactory signals in pollination systems in general. In a more general context, work on mimicry systems laid the conceptual framework for our understanding of how floral signaling deceives flower visitors, and how flower visitors may learn to avoid deceptive systems. However, in natural plant communities, flower visitors are always challenged by the complexity of flower signals and the need to filter the information to find the most rewarding resources. Different plant species often share the same functional pollinator types, because these exhibit similar sensory abilities and preferences, and this is reflected in the concept of pollination syndromes. It is possible that the evolution of features that advertise resources is driven by the most rewarding species of the community, when then serves as the model for the other species that use pollinators with similar sensory abilities and preferences.

Acknowledgments The research into orchid pollination was supported by a grant (No. KJB600870601) from the Academy of Sciences of the Czech Republic to Jana Jersáková and MSM 6007665801 to the Faculty of Science of University of South Bohemia.

References

Ackerman JD (1986) Mechanisms and evolution of food-deceptive pollination systems in orchids. Lindleyana 1:108–113

Ackerman JD, Galarza-Pérez M (1991) Patterns and maintenance of extraordinary variation in the Caribbean orchid, *Tolumnia* (*Oncidium*) *variegata*. Syst Bot 16:182–194

Ackerman JD, Méléndez-Ackerman EJ, Salguero-Faria J (1997) Variation in pollinator abundance and selection on fragrance phenotypes in an epiphytic orchid. Am J Bot 84:1383–1390

Alexandersson R, Agren J (1996) Population size, pollinator visitation and fruit production in the deceptive orchid *Calypso bulbosa*. Oecologia 107:533–540

Allan SA, Day JD, Edman JD (1987) Visual ecology of biting flies. Annu Rev Entomol 32:297–316

Anderson B, Johnson SD (2006) The effects of floral mimics and models on each other's fitness. Proc R Soc Ser B 273:969–974

Andersson S, Nilsson LA, Groth I, Bergström G (2002) Floral scents in butterfly-pollinated plants: possible convergence in chemical composition. Bot J Linn Soc 140:129–153

Anderson B, Johnson SD, Carbutt C (2005) Exploitation of a specialized mutualism by a deceptive orchid. Am J Bot 92:1342–1349

Aragón S, Ackerman JD (2004) Does flower color variation matter in deception pollinated *Psychilis monensis* (Orchidaceae)? Oecologia 138:405–413

Barkman TJ, Beaman JH, Gage DA (1997) Floral fragrance variation in *Cypripedium*: implications for evolutionary and ecological studies. Phytochemistry 44:875–882

Barth FG (1991) Insects and flowers: the biology of a partnership. Princeton University Press, Princeton

Bates HW (1862) Contribution to an insect fauna of the Amazon Valley. Lepidoptera: Heliconidae. Trans Linn Soc Lond Zool 23:495–566

Beardsell D, Clements MA, Hutchinson JF, Williams EG (1986) Pollination of *Diuris maculata* R. Br. (Orchidaceae) by floral mimicry of the native legumes, *Daviesia* spp. and *Pultenaea scabra* R. Br. Aust J Bot 34:165–173

Benitez-Vieyra S, Hempel de Ibarra N, Wertlen AM, Cocucci AA (2007) How to look like a mallow: evidence of floral mimicry between Turneraceae and Malvaceae. Proc R Soc B 274:2239–2248

Bergström GG, Birgersson I, Groth A, Nilsson LA (1992) Floral fragrance disparity between three taxa of lady's slipper *Cypripedium calceolus* L. (Orchidaceae). Phytochemistry 31:2315–2319

Bierzychudek P (1981) *Asclepias*, *Lantana* and *Epidendrum*: a floral mimicry complex? Biotropica 1:54–58

Boyden TC (1980) Floral mimicry by *Epidendrum ibaguense* (Orchidaceae) in Panama. Evolution 34:135–36
Briscoe A, Chittka L (2001) The evolution of colour vision in insects. Annu Rev Entomol 46:471–510
Brown BA, Clegg MT (1984) The influence of flower color polymorphisms on genetic transmission in a natural population of the common morning glory, *Ipomoea purpurea*. Evolution 38:796–803
Brown JH, Kodric-Brown A (1979) Convergence, competition and mimicry in a temperate community of hummingbird pollinated plants. Ecology 60:1022–1035
Carmona-Diaz G (2001) Mimetismo floral entre *Oncidium cosymbephorum* Morren (Orchidaceae) y *Malpighia glabra* L. (Malpighiaceae). Maestro en Neuroetologia. Universidad Veracruzana, Xalapa, Veracruz
Chittka L (1992) The colour hexagon: a chromaticity diagram based on photoreceptor excitations as a general representation of colour opponency. J Comp Physiol A 170:533–543
Chittka L, Kevan PG (2005) Flower colour as advertisement. In: Dafni A, Kevan PG, Husband BC (ed) Practical pollination biology. Enviroquest, Cambridge, ON, pp 157–196
Chittka L, Raine NE (2006) Recognition of flowers by pollinators. Curr Opin Plant Biol 9:428–435
Chittka L, Gumbert A, Kunze J (1997) Foraging dynamics of bumble bees: correlates of movements within and between plant species. Behav Ecol 8:239–249
Coddington JA (1988) Cladistic tests of adaptional hypotheses. Cladistics 4:3–22
Cozzolino S, Widmer A (2005) Orchid diversity: an evolutionary consequence of deception? Trends Ecol Evol 20:487–494
Cropper SC, Calder DM (1990) The floral biology of *Thelymitra epipactoides* (Orchidaceae), and the implications of pollination by deceit on the survival of this rare orchid. Plant Syst Evol 170:11–27
Dafni A (1983) Pollination of *Orchis caspia*: a nectarless plant which deceives the pollinators of nectariferous species from other plant families. J Ecol 71:467–474
Dafni A (1984) Mimicry and deception in pollination. Ann Rev Ecol Syst 15:259–278
Dafni A (1986) Floral mimicry: mutualism and undirectional exploitation of insect by plants. In: Juniper B, Southwood R(eds) Insect and the plant surface. Edward Arnold, London, pp 81–90
Dafni A (1987) Pollination in *Orchis* and related genera: evolution from reward to deception. In: Arditti J (ed) Orchid biology: reviews and perspectives, vol 4. Cornell University Press, Ithaca, pp 79–104
Dafni A, Bernhardt P (1990) Pollination of terrestrial orchids of Southern Australia and the Mediterranean region. Systematics, ecological, and evolutionary implications. In: Hecht MK, Wallace B, Macintyre RJ (eds) Evolutionary biology, vol 24. Plenum, New York, pp 193–252
Dafni A, Calder DM (1987) Pollination by deceit: floral mimesis in *Thelymitra antennifera* Hook f., (Orchidaceae). Plant Syst Evol 158:11–22
Dafni A, Ivri Y (1979) Pollination ecology of and hybridization between *Orchis coriophora* L. and *O. collina*. Sol. ex. Russ (Orchidaceae) in Israel. New Phytol 83:181–187
Dafni A, Ivri Y (1981a) Floral mimicry between *Orchis israelitica* Bauman and Dafni (Orchidaceae) and *Bellevalia flexuosa* Boiss (Liliaceae). Oecologia 49:229–232
Dafni A, Ivri Y (1981b) The flower biology of *Cephalanthera longifolia* (Orchidaceae): pollen imitation and facultative floral mimicry. Plant Syst Evol 137:229–240
Davies KL, Turner MP (2004) Morphology of floral papillae in *Maxillaria* Ruiz and Pav. (Orchidaceae). Ann Bot 93:75–86
Davies KL, Winters C, Turner MP (2000) Pseudopollen: its structure and development in *Maxillaria* (Orchidaceae). Ann Bot 85:887–895
Davies KL, Roberts DL, Turner MP (2002) Pseudopollen and food-hair diversity in *Polystachya* Hook. (Orchidaceae). Ann Bot 90:477–484

Davies KL, Turner MP, Gregg A (2003) Atypical pseudopollen-forming hairs in *Maxillaria* (Orchidaceae). Bot J Linn Soc 143:151–158

Dobson HEM (1994) Floral volatiles in insect biology. In: Bernays EA (ed) Insect-plant interactions. CRC, Boca Raton, pp 47–81

Dyer AG, Chittka L (2004) Bumblebees (*Bombus terrestris*) sacrifice foraging speed to solve difficult colour discrimination tasks. J Comp Physiol A 190:759–763

Endler JA (1981) An overview of the relationship between mimicry and crypsis. Biol J Linn Soc 16:25–31

Faegri K, van der Pijl L (1979) The principles of pollination ecology. Pergamon, Oxford

Ferdy JB, Gouyon PH, Moret J, Godelle B (1998) Pollinator behaviour and deceptive pollination: learning process and floral evolution. Am Nat 152:696–705

Fritz AL (1990) Deceit pollination of *Orchis spitzelii* (Orchidaceae) on the island of Gotland in the Baltic: a suboptimal system. Nord J Bot 9:577–587

Fritz AL, Nilsson LA (1994) How pollinator-mediated mating varies with population size in plants. Oecologia 100:451–462

Galizia CG, Kunze J, Gumbert A, Borg-Karlson AK, Sachse S, Markl C, Menzel R (2005) Relationship of visual and olfactory signal parameters in a food-deceptive flower mimicry system. Behav Ecol 16:159–168

Galen C, Zimmer KA, Newport ME (1987) Pollination in floral scent morphs of *Polemonium viscosum*: a mechanism for disruptive selection on flower size. Evolution 41:599–606

Gigord LDB, Macnair MR, Smithson A (2001) Negative frequency-dependent selection maintains a dramatic flower color polymorphism in the rewardless orchid *Dactylorhiza sambucina* (L.) Soò. Proc Natl Acad Sci USA 98:6253–6255

Gigord LDB, Macnair MR, Stritesky M, Smithson A (2002) The potential for floral mimicry in rewardless orchids: an experimental study. Proc R Soc Lond B 269:1389–1395

Goldblatt P, Manning JC (2000) The long-proboscid fly pollination system in southern Africa. Ann Mo Bot Gard 87:146–170

Grandcolas P, D'Haese C (2003) Testing adaptation with phylogeny: How to account for phylogenetic pattern and selective value together? Zool Scr 32:483–490

Gumbert A, Kunze J (2001) Colour similarity to rewarding model plants affects pollination in a food deceptive orchid, *Orchis boryi*. Biol J Linn Soc 72:419–433

Harder LD (2000) Pollen dispersal and the floral diversity of Monocotyledons. In: Wilson KL Morrison D (ed) Monocots: systematics and evolution. CSIRO, Melbourne, pp 243–257

Harder LD, Johnson SD (2008) Function and evolution of aggregated pollen in angiosperms. Int J Plant Sci 169:59–78

Harding J (1970) Genetics of *Lupinus* II. The selective disadvantage of the pink flower colour mutant in *Lupinus nanus*. Evolution 24:120–127

Heinrich B (1975) Bee Flowers: a hypothesis on flower variety and blooming times. Evolution 29:325–334

Horovitz A, Harding J (1972) Genetics of *Lupinus*. V. Intraspecific variability for reproductive traits in *Lupinus nanus*. Bot Gaz 133:155–165

Huda MK, Wilcock CC (2008) Impact of floral traits on the reproductive success of epiphytic and terrestrial tropical orchids. Oecologia 154:731–41

Indsto JO, Weston PH, Clements MA, Dyer AG, Batley M, Whelan RJ (2006) Pollination of *Diuris maculata* (Orchidaceae) by male *Trichocolletes venustus* bees. Aust J Bot 54:669–679

Jersáková J, Johnson SD (2006) Lack of floral nectar reduces selfpollination in a fly-pollinated orchid. Oecologia 147:60–68

Jersáková J, Johnson SD, Kindlmann P (2006a) Mechanisms and evolution of deceptive pollination in orchids. Biol Rev 81:1–17

Jersáková J, Kindlmann P, Renner SS (2006b) Is the colour dimorphism in *Dactylorhiza sambucina* maintained by differential seed viability instead of frequency-dependent selection? Folia Geob 41:61–76

Johnson SD (1994) Evidence for Batesian mimicry in a butterfly pollinated orchid. Biol J Linn Soc 53:91–104

Johnson SD (2000) Batesian mimicry in the nonrewarding orchid *Disa pulchra*, and its consequences for pollinator behaviour. Biol J Linn Soc 71:119–132

Johnson SD, Edwards TJ (2000) The structure and function of orchid pollinaria. Plant Syst Evol 222:243–269

Johnson SD, Morita S (2006) Lying to Pinocchio: floral deception in an orchid pollinated by long-proboscid flies. Bot J Linn Soc 152:271–278

Johnson SD, Nilsson LA (1999) Pollen carryover, geitonogamy, and the evolution of deceptive pollination system in orchids. Ecology 80:2607–2619

Johnson SD, Steiner K (1997) Long-tongued fly pollination and evolution of floral spur length in the *Disa draconis* complex (Orchidaceae). Evolution 51:45–53

Johnson SD, Alexandersson R, Linder HP (2003a) Experimental and phylogenetic evidence for floral mimicry in a guild of fly-pollinated plants. Biol J Linn Soc 80:289–304

Johnson SD, Peter CI, Nilsson LA, Agren J (2003b) Pollination success in a deceptive orchid is enhanced by co-occurring rewarding magnet plants. Ecology 84:2919–2927

Joron M, Mallet J (1998) Diversity in mimicry: paradox or paradigm? Trends Ecol Evol 13:461–466

Jürgens A, Dötterl S, Meve U (2006) The chemical nature of fetid floral odors in stapeliads (Apocynaceae-Asclepiadoideae-Ceropegieae). New Phytol 172:452–468

Kindlmann P, Jersáková J (2006) Effect of floral display on reproductive success in terrestrial orchids. Folia Geobot 41:47–60

Knudsen JT, Eriksson R, Gershenzon J, Ståhl B (2006) Diversity and distribution of floral scent. Bot Rev 72:1–120

Koivisto AM, Vallius E, Salonen V (2002) Pollination and reproductive success of two colour variants of a deceptive orchid, *Dactylorhiza maculata* (Orchidaceae). Nord J Bot 22:53–58

Laverty TM (1992) Plant interactions for pollinator visits: a test of the magnet species effect. Oecologia 89:502–508

Little RJ (1983) A review of floral food deception mimicries with comments on floral mutualism. In: Jones CE, Little RJ (ed) Handbook of experimental pollination biology. S&E Scientific and Academic, New York, pp 294–309

Manning JC, Goldblatt P (1996) The *Prosoeca peringueyi* (Diptera: Nemestrinidae) pollination guild in southern Africa: long-tongued flies and their tubular flowers. Ann Mo Bot Gard 83:67–86.

Manning JC, Goldblatt P (1997) The *Moegistorhynchus longirostris* (Diptera: Nemestrinidae) pollination guild: long-tubed flowers and a specialized long-proboscid fly pollination system in southern Africa. Plant Syst Evol 206:51–69

Moya SJD, Ackerman JD (1993) Variation in the floral fragrance of *Epidendrum ciliare* (Orchidaceae). Nord J Bot 13:41–47

Murren CJ, Ellison AM (1996) Effects of habitat, plant size, and floral display on male and female reproductive success of the Neotropical orchid *Brassavola nodosa*. Biotropica 28:30–41

Nilsson LA (1979) Anthecological studies on the Lady's Slipper, *Cypripedium calceolus* (Orchidaceae). Bot Not 132:329–347

Nilsson LA (1980) The pollination ecology of *Dactylorhiza sambucina* (Orchidaceae). Bot Not 133:367–385

Nilsson LA (1983) Mimesis of bellflower (*Campanula*) by the red helleborine orchid *Cephalanthera rubra*. Nature 305:799–800

Nilsson LA (1984) Anthecology of *O. morio* (Orchidaceae) and its outpost in the North. Nov Acta Reg Soc Sci Upsal 3:167–180

Ollerton J, Raguso R (2006) The sweet stench of decay. New Phytol 172:382–385

Parra-Tabla V, Vargas CF (2007) Flowering synchrony and floral display size affect pollination success in a deceit-pollinated tropical orchid. Acta Oecol 32:26–35

Pellegrino G, Caimi D, Noce ME, Musacchio A (2005) Effects of local density and flower colour polymorphism on pollination and reproduction in the rewardless orchid *Dactylorhiza sambucina* (L.) Soó. Plant Syst Evol 251:119–129

Pellmyr O (1986) The pollination ecology of two nectarless *Cimifuga* sp. (Ranunculaceae) in North America. Nord J Bot 6:713–723
Peter CI, Johnson SD (2006) Doing the twist: a test of Darwin's cross-pollination hypothesis for pollinium reconfiguration. Biol Lett 2:65–68
Peter CI, Johnson SD (2008) Mimics and magnets: the importance of color and ecological facilitation in floral deception. Ecology 89:1583–1595
Pettersson B, Nilsson LA (1993) Floral variation and deceit pollination in *Polystachya rosea* (Orchidaceae) on an inselberg in Madagascar. Opera Bot 121:237–245
Proctor M, Yeo P, Lack A (1996) The natural history of pollination. HarperCollins, London
Renner SS (2005) Rewardless flowers in the angiosperms and the role of insect cognition in their evolution. In: Waser NM, Olerton J (eds) Plant-pollinator interactions: from specialization to generalization. University of Chicago Press, Chicago, pp 123–144
Roy B, Raguso R (1997) Olfactory vs. visual cues in a floral mimicry system. Oecologia 109:414–426
Roy BA, Widmer A (1999) Floral mimicry: a fascinating yet poorly understood phenomenon. Trends Plant Sci 4:325–330
Sabat AM, Ackerman JD (1996) Fruit set in a deceptive orchid: the effect of flowering phenology, display size, and local floral abundance. Am J Bot 83:1181–1186
Salzmann C, Schiestl F (2007) Odour and colour polymorphism in the food-deceptive orchid *Dactylorhiza romana*. Plant Syst Evol 267:37–45
Salzmann C, Brown A, Schiestl F (2006) Floral scent emission and pollination syndromes: evolutionary changes from food to sexual deception. Int J Plant Sci 167:1197–1204
Salzmann C, Cozzolino S, Schiestl F (2007a) Floral scent in food-deceptive orchids: species specificity and sources of variability. Plant Biol 9:720–729
Salzmann C, Nardella A, Cozzolino S, Schiestl F (2007b) Variability in floral scent in rewarding and deceptive orchids: the signature of pollinator-imposed selection? Ann Bot 100:757–765
Schemske DW, Bierzychudek P (2001) Perspective: evolution of flower color in the desert annual *Linanthus parryae*: Wright revisited. Evolution 55:1269–1282
Schiestl FP, Cozzolino S (2008) Evolution of sexual mimicry in the orchid subtribe orchidinae: the role of preadaptations in the attraction of male bees as pollinators. BMC Evol Biol 8:27
Schiestl FP, Peakall R, Mant JG, Ibarra F, Schulz C, Franke S, Francke W (2003) The chemistry of sexual deception in an orchid-wasp pollination system. Science 302:437–438
Schoen DJ, Clegg MT (1985) The influence of flower color polymorphism on outcrossing rate and male reproductive success in *Ipomoea purpurea*. Evolution 39:1242–1249
Smithson A, Macnair MR (1997) Negative frequency-dependent selection by pollinators on artificial flowers without rewards. Evolution 51:715–723
Sobrevila C, Wolfe LM, Murcia C (1989) Flower polymorphism in the beach plant, *Ipomoea imperati* (Vahl.) Grisebach (Convolvulaceae). Biotropica 21:84–88
Stanton ML (1987) Reproductive biology of petal color variants in wild populations of *Raphanus sativus*: I. pollinator response to color morphs. Am J Bot 74:178–187
Starrett A (1993) Adaptive resemblance: a unifying concept for mimicry and crypsis. Biol J Linn Soc 48:299–317
Steiner KE (1998) The evolution of beetle pollination in a South African orchid. Am J Bot 85:1180–1193
Sugiura N, Goubara M, Kitamura K, Inoue K (2002) Bumblebee pollination of *Cypripedium macranthos* var. *rebunense* (Orchidaceae): a possible case of floral mimicry of *Pedicularis schistostegia* (Orobanchiaceae). Pl Syst Evol 235:189–195
Thomson JD (1988) Effects of variation in inflorescence size and floral rewards on the visitation rates of traplining pollinators of *Aralia hispida*. Evol Ecol 2:65–76
Tremblay RL, Ackerman JD (2007) Floral color patterns in a tropical orchid: Are they associated with reproductive success? Plant Species Biol 22:95–105
Tremblay RL, Ackerman JD, Zimmerman JK, Calvo RN (2005) Variation in sexual reproduction in orchids and its evolutionary consequences: a spasmodic journey to diversification. Biol J Linn Soc 84:1–54

Troje N (1993) Spectral categories in the learning behaviour of blowflies. Z Naturforsch C 48:96–104

Turner JRG (1977) Butterfly mimicry: the genetical evolution of an adaptation. Evol Biol 10:163–206

Valterová I, Kunze J, Gumbert A, Luxová A, Liblikas I, Kalinová B, Borg-Karlson AK (2007) Male bumble bee pheromonal components in the scent of deceit pollinated orchids; unrecognized pollinator cues? Arthropod Plant Interact 1:137–145

van der Pijl L, Dodson CH (1966) Orchid flowers: their pollination and evolution. University of Maimi Press, Coral Gables

Vane-Wright RI (1980) On the definition of mimicry. Biol J Linn Soc 13:1–6

Waite S, Hopkins N, Hitchings S (1991) Levels of pollinia export, import and fruit set among plants of *Anacamptis pyramidalis, Dactylorhiza fuchsii* and *Epipactis helleborine*. In: Wells TCE, Willems JH (eds) Population ecology of terrestrial orchids. SPB Academic, The Hague, pp 103–110

Warren J, Mackenzie S (2001) Why are all colour combinations not equally represented as flower-colour polymorphisms? New Phytol 151:237–241

Waser NM, Price MV (1981) Pollinator choice and stabilizing selection for flower color in *Delphinium nelsonii*. Evolution 35:376–390

Waser NM, Price MV (1983) Pollinator behaviour and natural selection for flower colour in *Delphinium nelsonii*. Nature 302:422–424

Weiss M (2001) Vision and learning in some neglected pollinators: beetles, flies, moths, and butterflies. In: Chittka L, Thomson JD (eds) Cognitive ecology of pollination. Cambridge University Press, New York, pp 171–190

Wiens D (1978) Mimicry in plants. Evol Biol 11:365–403

Wilson P, Stine M (1996) Floral constancy in bumblebees: handling efficiency or perceptual conditioning? Oecologia 106:493–499

Wolfe LM (1993) Reproductive consequences of a flower color polymorphism in *Hydrophyllum appendiculatum*. Am Midl Nat 129:405–408

Cognition in Plants

Paco Calvo and Fred Keijzer

Abstract We discuss the possibility and the meaning of the claim that plants are cognitive from the perspective of *embodied cognition*. In embodied cognition, the notion of cognition can be interpreted in a very broad way and applied to many free-moving creatures. In this chapter, we discuss whether and (if so) how this approach applies to intelligence in plants. Building on work from "plant neurobiology," we discuss the differences in speed between plants and animals, similarities between sensory-driven plant growth and animal memory, and the presence of offline behavior in plants. In our view, these examples show that under a wide, embodied interpretation of cognition, plants may well qualify as being cognitive.

1 Introduction

Plants exhibit much more complex and flexible adaptive behavior than most people would expect (Trewavas 2003). That being said, the very possibility that plant behavior exhibits certain cognitive aspects strikes many people as outrageous, whether they are plant scientists or not. In this chapter, we will discuss the validity and the meaning of the claim that plants may be cognitive by examining current discussions within the field of embodied and embedded cognition ("embodied cognition" hereafter) and its links to biology. In this context, the notion of cognition is drawn around perception–action phenomena, and it appears to encompass a much wider range of phenomena than human-level thought.

P. Calvo
Department of Philosophy, University of Murcia, Spain
e-mail: fjcalvo@um.es

F. Keijzer
Faculty of Philosophy, University of Groningen, The Netherlands
e-mail: F.A.Keijzer@rug.nl

F. Baluška (ed.), *Plant-Environment Interactions*, Signaling and Communication in Plants,
DOI: 10.1007/978-3-540-89230-4_13,

Embodied cognition takes perception–action as its major focus, and within this embodied perspective, most animal and even bacterial behavior can be considered cognitive to a limited extent. Embodied cognition stresses the fact that free-moving creatures are not simple, hard-wired reflex automatons but incorporate flexible and adaptive means to organize their behavior in coherent ways. While it may be going too far to ascribe a mind to such systems, they do deserve recognition of the intelligence involved in the things they do, and for this reason, the notion of cognition seems to be appropriate[1]. So far, the notion of cognition has not been extended to plants. One reason for this is simply that most cognitive scientists, even those involved in embodied cognition, are simply unaware of what plants can do. This is easily remedied. However, a theoretically more important reason for excluding plants from the wider cognitive domain sketched by embodied cognition seems to be the lack of the kind of sensorimotor organization that is the major focus of embodied cognition. We will focus on this issue.

In recent years, the issue of plant intelligence or cognition and even "plant neurobiology" has produced a lively debate among plant scientists (e.g., Alpi et al. 2007; Barlow 2008; Firn 2004; Trewavas 2003, 2005; see also Calvo 2007 and references therein). So far, this debate has gone unnoticed within embodied cognition, and it has not relied on ideas from embodied cognition. We believe that the discussions taking place in the plant sciences can benefit from work in embodied cognition, which has already made significant progress in clarifying the notion of cognition and possible criteria for its use. At the same time, confronting embodied cognition with the "plant question" is sure to raise important new issues in the field of embodied cognition itself.

In this chapter, we will discuss the question of whether an extended reading of cognition, such as that developed within embodied cognition, might apply to plants. Within embodied cognition, the notion of cognition—which is based on perception and action—is used to make sense of a wide range of behaviors exhibited by "simple" animals, like nematodes or flies. The message is clearly that we should avoid generalized dismissive intuitions concerning such animals and attempt a more empirically informed approach. We believe that this open attitude is also beneficial to the study of possible cognitive phenomena in plants.

The chapter has the following structure. In Sect. 3, we introduce the discussion about what cognition is, and sketch the difficulties involved in clarifying this key notion for the cognitive sciences. In Sect. 4, we formulate a wide reading of cognition as derived within embodied cognition, also taking in biological considerations. This wide reading is subsequently condensed into five constraints on cognition, two of which explicitly highlight the need for animal-like sensorimotor organization. In Sect. 5, using the work of Hans Jonas, we discuss *why* having sensorimotor organization is important for cognitive phenomena, and whether plants may fulfill some of these

---

[1]It is thus important to differentiate here between this wide interpretation of *cognition*, which may apply in a meaningful way far beyond the human case, and the notion of *mind*, which may well remain highly restrictive and possibly limited to human beings.

functions in other ways. In the subsequent three sections we discuss plant research findings, and we conclude that—in this light—plants can fulfill these constraints to a significant degree. We introduce the main tenets of plant neurobiology in Sect. 6, and argue that speed and form differences in plant behavior do not exclude a cognitive interpretation. In Sect. 7, we argue for the similarities between plant growth and animal memory. Section 8 deals with plant structures for forms of offline cognition. We conclude that these examples show that plants can be considered to be minimally cognitive and that they constitute an important domain for cognitive studies.

2 What is Cognition?

What is cognition? Although cognition is one of the core concepts in the behavioral and cognitive sciences, there is no generally accepted answer. For example, in his classic book *Cognitive Psychology*, Ulrich Neisser defined cognition as: "all processes by which the sensory input is transformed, reduced, elaborated, stored, recovered, and used" (1967, p 4). However, this definition seems to include many artifacts, like tape recorders, and organisms, like plants, that were not intended to be labeled as cognitive. The classical cognitive sciences that evolved under the influence of people like Neisser used a much more limited interpretation of cognition: not all forms of information processing sufficed. The implicit extra constraint in this definition was that cognition involves the kind of information processing that also occurs in *human intelligence*, where it is described using terms like perception, planning, thinking and action.

Because this field was intrinsically interested in the human mind and not in any other topic, this implicit limitation worked quite well in practice. No one had to provide a reason for why tape recorders and plants were out, as these were not part of the topic studied. The notion of cognition itself became widely interpreted as referring to the information processing mechanisms that describe how the human mind operates. Human cognition became the default interpretation of cognition and also the yardstick by which the cognitive abilities of other animals were to be measured. From that perspective, cognition is tantamount to characteristically human-like capabilities, such as reasoning, problem-solving and symbolization. These processes have to be present to a significant degree before one can speak of a bona fide cognizer (Gould and Gould 1998). Internal, representation-handling processes are considered to be the source of these particular thinking skills. It is a matter of painstaking research to establish whether and (if so) which other creatures also exhibit these refined capabilities (see, for example, Heinrich 2000; Smirnova et al. 2003). In the human-based interpretation of cognition, organisms whose behavior does not unequivocally involve human-style reasoning subsequently remain outside the cognitive domain. Cognition is a scarce commodity in this view.

Having an interest in *human* cognition, and not in the rest, is a way of carving up the domain that leaves little motivation to articulate what falls outside this domain. In the cognitive sciences, the behavior of nonhuman organisms receives comparatively little attention (for a notable exception, see Bekoff et al. 2002). Such behavior is often argued to be predominantly composed of inflexible, hard-wired reactions to environmental stimuli (e.g., Dennett 1984, 1996; Gould and Gould 1998; Sterelny 2001). As a result, such behaviors are not considered to be very interesting from a cognitive perspective, and only require closer attention when they approach human forms of intelligence (see also Godfrey-Smith 1996, 2001; Shettleworth 1998). Dennett (1984) and Hofstadter (1985), for example, talk about "sphexisms" in this context, drawing the term from an anecdote in which a digger wasp of the *Sphex* genus was manipulated so that it remained stuck in an iterative, automatic behavioral loop; endlessly repeating its own inbuilt, behavioral responses[2]. This story is actually a mere anecdote, and not systematically corroborated by evidence (Keijzer 2001). In other words, anthropocentric interpretations of cognition depict a rough dichotomy between intelligent cognizers, and inflexible, mechanic-like organisms that are merely capable of reflexive/instinctive behaviors.

However, there are important practical and theoretical problems with such a general dichotomy between human cognition and "noncognitive" inflexible, mechanistic behavior, as well as with a human-based interpretation of cognition. First, this putative dichotomy does not tally with the underlying intricacies that make so-called noncognitive organisms tick, as it simply fails to provide a realistic account of the behavioral complexities that can be found in nonhuman organisms (Brooks 1999; Keijzer 2001; Roth and Wullimann 2001). When investigated in their own right, the mechanisms and processes required to generate these presumably noncognitive behaviors are found to be very complex and extraordinarily difficult to replicate in robots (Prescott et al. 1999).

Second, the dichotomy cannot cope with any differentiation between the noncognitive organisms. There are huge gaps between the behavioral capabilities of, for instance, nematodes and octopi, or between sharks and squirrel monkeys, all of which are—plausibly—considered noncognitive from an anthropocentric perspective. The assumption that the behavior of these "lower" organisms is entirely composed of reflexes, instincts and/or hardwired reactions does not help to articulate how these very different behavioral capabilities arise.

Third, when one turns to the basic processes of cognition, and leaves aside their anthropocentric interpretation, it is clear that these processes—such as perception, memory, and action—are dispersed extremely widely across and even beyond the animal kingdom. It is now even plausibly maintained that these particular features of cognition are already present in invertebrates (Carruthers 2004; Keijzer 2001), and prokaryotic bacteria (Di Primio et al. 2000; Greenspan and van Swinderen 2004; Lengeler et al. 2000; Müller et al. 2001).

Fourth, a general dichotomy makes it more difficult to develop a gradualistic and diversified evolutionary account of how basic forms of cognition developed into different and more complex ones. As it places the boundaries of cognition very

[2] This story is actually a mere anecdote, and not systematically corroborated by evidence (Keijzer 2001).

high, say at the level of primates or possibly tool users, the relevance of the preceding evolutionary stages of intelligent behavior is blurred. Differences are stressed rather than any possible continuities that are discovered.

All of these considerations have been known for a long time but did not have much impact within the cognitive sciences until the rise of embodied cognition. This new field of cognitive science came to the fore in the late 1980s and the 1990s as a reaction to the strong focus on explicit, reflective human thought. Hurley baptized this view "the sandwich model," as it interpreted cognition as a separate inner filling, wedged between perceptual input and behavioral output (Hurley 1998). Embodied cognition, instead, stressed the ongoing, dynamic interactions between an agent and its environment by means of perception and action. In this view, perception–action processes became the starting point for intelligence, and complex human thought one of its offshoots[3]. For introductions and overviews of embodied cognition, see, e.g., Calvo and Gomila (2008), Clark (1997), Pfeifer and Scheier (1999), or Varela et al. (1991). Perception–action—and the sensorimotor organization it requires—thus becomes the online key feature of cognition, which is evolutionarily expanded in higher-level cognizers to include all sorts of offline processing that exhibit representational characteristics.

One important consequence of the rise of embodied cognition was that the notion of cognition itself started to shift away from its human anchor. Roboticists, for example, started to make insect-like robots that had little to do with human cognition, but a lot to do with investigating a bottom-up interpretation of intelligence, and thus a kind of cognition. In this context, the link to the human case became insufficient, leading for a call to formulate a *mark of the cognitive* (Adams and Aizawa 2001). How such a mark could be given remains a issue, but there seems to be a strong consensus that cognition involves processes such as perception, thinking, memory, and action. As already stressed, it is useful to differentiate between mind and cognition here, even though the two are often used interchangeably within the cognitive sciences. While "mind" may or may not remain closely tied to humans and in particular to consciousness, "cognition" need not be intrinsically connected to either humans or consciousness. Thus, while cognition is a concept that can be stretched beyond its original use, this can be done without requiring additional claims that all cognitive systems should be seen as mindful in a way that is similar to human minds.

We think that the notion of cognition ought and can be developed in new ways to fill the gap between the mindful and the mindless, and turn it into a gradient from human intelligence to inanimate nature. Godfrey-Smith (2001, p 234) argued that cognition initially evolved to enable organisms to control their own behavior, allowing them to cope with environmental complexity. In his view, cognition "shades off" into basic biological processes such as metabolism. We believe that a proper interpretation of cognition should aim to become more specific about this

[3]For introductions and overviews of embodied cognition, see, e.g., Calvo and Gomila (2008), Clark (1997), Pfeifer and Scheier (1999), or Varela et al. (1991).

"shading off," and allow for a better differentiation within the wide array of cognitive capabilities that are found in nature. In Sect. 3, we will develop the outlines of a wide interpretation of cognition, as it can be derived from ongoing theoretical developments within embodied and biological views of cognition.

3 A Biological and Embodied Perspective on Cognition

Since embodied cognition provides the intrinsic connection between humans and cognition, work has begun to provide a more systematic account of cognition. Including artificial systems like robots makes this a very difficult task and possibly one that is impossible. For example, it is hard to make sense of the question of whether a bacterium is as intelligent as a washing machine (Firn 2004), because the aspects of both that one should focus on are utterly unclear. However, restricting oneself to naturally occurring, *biological* systems allows more progress to be made (e.g., Barandiaran 2008; Keijzer 2001, 2003, 2006; Lyon 2006a, b; Moreno and Etxeberria 2005; Moreno et al. 1997; Van Duijn et al. 2006). The main idea here is that cognition is (or that it originated as) a biological phenomenon, and that it is used to manipulate the environment in ways that systematically benefit the living organism that exhibits these cognitive aspects. In this context, the question concerning minimal cognition is important (Beer 2003): what is the minimal biological system to which the notion of cognition applies?

We will disregard claims that make life itself a form of cognition, but it can be argued that bacteria already provide examples of cognition (Di Primio et al. 2000; Lengeler et al. 2000; Lyon 2006a; Van Duijn et al. 2006). Consider chemotaxis in *E. coli*. These free-moving bacteria use flagella to move around and can travel up or down gradients of several substances that they can ingest or that they need to avoid. All of the basic ingredients for a minimal form of cognition are already present here, and discussing them will provide a suitable framework when it comes to judging the possibility of plant cognition.

Since it is a living organism, *E. coli* contains a metabolic biochemical organization, which provides the fundamental energetic and constitutive processes of the bacterium, as it does for all living organisms. To stay alive, the bacterium needs substances that it can incorporate into its structures and it needs energy to drive these biochemical processes. This metabolic organization provides a basic form of normativity (Bickhard 2008)—differentiating "bad for me from good for me"—and it can have adaptive characteristics. For example, a well-known form of metabolic adaptation is the "lac operon" system, which regulates the metabolism of lactose in *E. coli*. This cluster of genes is normally dormant, because the bacterium predominantly metabolizes glucose. However, when the bacterium detects that glucose levels are very low and lactose is abundantly present in the environment, the lac operon system becomes disinhibited, subsequently allowing the transcription and expression of genes that enable lactose metabolism (Todar 2004). This form of metabolic adaptation is induced by environmental conditions but is still part of the organism's metabolic organization. The process consists of a change in the set of chemical reactions that together constitute the bacterium's metabolism.

Chemotaxis, on the other hand, is a different kind of process. It does not comprise chemical reactions but instead physical changes in the position of the bacterium with respect to its environment. In other words, the bacterium interacts with its environment at the larger, physical scale of the *dispersion* of metabolically relevant substances. There are potential metabolic benefits to this interaction, but this manipulation of the environment—moving towards a food source, for example—is not itself a metabolic process. With respect to metabolism, chemotaxis is a second-order process, which is relevant for changing metabolic opportunities and in this way expanding the adaptive opportunities of organisms to a considerable degree.

Manipulating the environment at larger scales to enable or enhance metabolic functioning is a very general biological strategy. Sponges pumping water through their bodies, plants growing leaves oriented toward the light, lions stalking their prey, and even humans discussing which restaurant to go to can be considered as examples. Why would all of these activities be cognitive? In the embodied cognition literature there is a clear, intuitive, cut-off point which limits cognition to systems that show a form of sensorimotor coordination. Cognition applies to free-moving agents, capable of reversible movements and perception. In this view, bacterial chemotaxis is a possible example of minimal cognition, as it uses sensorimotor coordination to expand metabolic forms of adaptation and, in this way, provides a basic example of an organization that is also present in human cognition. Plants, fungi and sessile animals, however, would be left out of the cognitive domain insofar as they (seem to) lack this additional requirement.

Setting up an adequate sensorimotor organization requires a particular physical embodiment of an organism, be it bacterium or monkey. For bacteria, this comes in the form of specific chemical receptors such as methyl-accepting proteins, and actuators such as flagella or pili that enable the bacterium to move about (Berg 2000). It also involves a control system that enables the organism to initiate motion and use the perceptual feedback it generates to guide this motion. Within embodied cognition, it is customary to differentiate between online processing that is under direct perceptual control, and offline processing which is to some extent decoupled from immediate perception–action. Online processing is cast as being more basic, while offline processing is thought to be involved in more complex cognitive tasks, like those relating to memory or planning.

Importantly, *E. coli* already shows some offline capacities. It does not measure a concentration gradient by calculating the simultaneous difference between different sensors at different body parts. Instead, *E. coli*, and many other bacteria, uses a TCST system that acts as a memory and inner connection between sensors and effectors in a way that is functionally similar to the nervous system in multicellular animals. The TCST system allows it to take sequential measurements of the substance concentration. When the bacterium happens to travel down the gradient, according to this memory, it tends to change directions randomly, and when it travels up the gradient, *E. coli* tends to maintain its course. The net result is a systematic form of chemotaxis.

Finally, in this biological and embodied perspective on cognition, the sensorimotor organization does not consist of simple, hard-wired behavior. All animal behavior results from a complex nonlinear dynamical process involving a nervous system, a body and an environment that interact with one another on a continuous basis (Beer 1995, 2000; Keijzer 2001). Even seemingly repetitive and automatic behavior is still the result of complex nonlinear processes, which show up when the larger behavioral repertoire of the organisms is taken into account. This also applies to our example of *E. coli*. We want to stress that the system's chemotaxis is actually much more complex than suggested so far, as it also adapts to the absolute concentration level of the substances, works in conjunction with sensitivity to several other substances, and is itself part of a very flexible organization for behavioral control (Lyon 2006a). Merleau Ponty (1963) provided a clear and detailed analysis of how the seeming simplicity of environmentally controlled stimulus–response behavior very easily results as a methodological artifact when one tries to isolate a specific aspect of behavior from the ordinary behavior of an animal.

To summarize, casting biological forms of sensorimotor coordination as the minimal form of cognition provides a clear and transparent starting point for thinking about cognition. In our view, by framing the issue of the nature of cognition from within the field of embodied cognition, the question of plant intelligence can be suitably addressed. The question that must be answered first, however, is whether plants should not be excluded from the cognitive domain as envisioned within embodied cognition.

4 Embodied Cognition and Plants

An embodied and biological perspective introduces a more open-ended interpretation of cognition and intelligence. The upshot of this enterprise can be summarized in five different constraints that apply to cognition as conceived in this way:

1. Metabolism provides a basic form of biochemical normativity for cognition
2. Cognition proper (initially) consists of exploiting the spatiotemporal dispersal characteristics of metabolically relevant environmental features through the movement of the organism
3. This movement takes the form of various sensorimotor organizations
4. A basic, online, sensorimotor organization can be expanded by using offline control structures that can (but do not need to) involve a nervous system
5. Such a sensorimotor-based cognitive organization is a globally organized cohering unit, not a collection of individual stimulus–response relations.

The question to ask now is whether and (if so) how does this set of constraints apply to plants. To begin with, we must stress that the application of the first two constraints is not disputed. Plants metabolize, of course, giving them a basic motivation for doing things. There is no doubt either that plants manipulate their environment in a second-order way such that their metabolic functions profit from this manipulation. Growing roots downward and light-catching parts upward suffice in this respect (Keijzer 2001).

The real rub for minimal cognition in plants comes from constraints 3 to 5. Constraint 3 imposes being *free-moving*, having a sensorimotor organization, as a requirement for cognition. It is here that the option of plant cognition seems highly problematic within an embodied perspective. However, adhering to being free-moving as an *intuitive* criterion is unsatisfactory. Within the biological domain, free-moving organisms may stand out as potentially intelligent beings, but why should we trust these intuitions? The question should instead be: why should we consider being free-moving to be so important?

Up to now, plants have not received much attention within embodied cognition. Most of those working in the field have employed a default assumption that intelligence is at a minimum an animal thing that was best caught in studies with free-moving agents such as robots, while excluding sessile plants. However, Hans Jonas (1966, 1968), who is now receiving renewed attention as an important thinker on the connections between biology and mind (Barandiaran 2008; Di Paolo 2005; Keijzer 2006), did take plants into consideration. Jonas tried to articulate the differences between plants, animals and humans in a way that highlights the relevance of being free-moving. In his view, the capacity for free movement is a key feature that is required for the development of intelligence as exhibited by animals, and a precondition for the evolution of human thought. Jonas' work can be used to argue that there are fundamental differences between intelligence in plants and animals (e.g., Barandiaran 2008). At the same time, by clarifying why being free-moving is so important, he provides a clearer target for challenges on empirical and theoretical grounds. Plants may very well fulfill these constraints without moving about like animals do.

Jonas provides an analysis of why motility, and in its wake sensing and emotion, are key features when it comes to cognition[4]. For Jonas (1966), motility and perception are also intrinsically linked to emotion and the presence of an inner, phenomenal dimension. We will not discuss these further complexities here. He argues that animal motion is more than an intensified case of vegetative motion, from which it differs in a number of physical respects: "in speed and spatial scale; in being occasional instead of continual; variable instead of predefined; reversible instead of irreversible." (Jonas 1968, p 248). These criteria are important for differentiating between being free-moving—which plants are not, generally speaking—and having self-induced motility, which is present in plants. Jonas uses these physical differences as the foundation for a further argument that animal motion leads to a principled new method of coordinating with the geometry of environmental space (1966; see also Barandiaran 2008). Jonas makes the point as follows:

> Now it is the main characteristic of *animal* evolution as distinct from plant life that *space*, as the dimension of dependence, is progressively transformed into a dimension of freedom by the parallel evolution of these two powers: to move about, and to perceive at a distance. (Jonas 1966, p 100).

[4] For Jonas (1966), motility and perception are also intrinsically linked to emotion and the presence of an inner, phenomenal dimension. We will not discuss these further complexities here.

In his view, only by free-moving and perceiving at a distance, most notably by vision, is "space really disclosed to life." The key issue is that (aspects of) the global spatial structure of the environment must become a feature that is present and accessible for an organism. A fly, for example, is able to orient itself within its environment and home in on places with sweet stuff, while avoiding swapping hands. Animals are sensitive to the spatial layout of the environment, for example in the form of patterns on a sensory surface like the retina or the skin, and their behavior is globally organized as a unit in relation to this layout. Barandiaran uses the nice phrase of being sensitive to the "geometric space where objects can be freely explored" (2008, p 198). Such sensitivity comes in different grades, as the fly will not be sensitive to the highly relevant fact for me that it is trying to land on my child's birthday cake. However, we are both sensitive or "aware" of the environment as a spatially and temporally extended structure in which we can act.

In contrast, plants are presumed to act on local stimuli, which may guide their behavior in globally appropriate ways, but without being directly sensitive to the spatial patterning of the environment. Thus, plants may grow their roots systematically downward strictly based on the locally available perception of gravity in every root. In this way, they can exploit this geometric structure without being sensitive as a single unit to geometric space as induced by free motility. The issue is the extent to which plants are sensitive to and acting on this global spatial structure of the environment. Or do they get by on the basis of a multitude of local interactions or decision-making processes? Thus, the important challenge that Jonas highlights for plant cognition is that only free motility seems to lead to an independent world—a geometric space—in which an agent can act.

Importantly, Jonas shifts the issue from a general unspecific commitment to a sensorimotor organization that plants just do not have, to different, more specific demands that require empirical data to settle. In line with constraint 3, Jonas changes the issue from having an animal-like sensorimotor organization to motility and possible differences in the speed, variability and reversibility of motility. As we will see, recent developments in plant science provide good empirical reasons to downplay the differences between being free-moving and self-induced motility, as imposed by constraint 3. Similarly, sensitivity to the geometric layout of the environment, as stressed by Jonas, may be something that plants are quite capable of without us knowing it. Thus, plants could also very well fulfill constraint 5, also making this an empirical issue rather than a theoretical one.

Constraint 4 stresses the importance of offline control as a way to expand the options of an online operating control structure. Offline control is often considered a very important sign of cognition. Offline control allows an organism to dissociate its behavior from the immediately impinging stimuli and to act in ways that are guided by forms of knowledge. It is ironic then that these more cognitive offline aspects can be comparatively easily established in plants when compared to the motility issue.

In the following three sections, we will turn to plant science and discuss developments and examples that make it plausible that plants can fulfill constraints 1–5 to some degree, and thus can be deemed cognitive in a sense that is regularly used within embodied cognition. In particular, the development of plant neurobiology has been important in establishing many details of plant intelligence.

5 Plant Neurobiology: Intelligence Can Take Different Forms and Speeds

Plant neurobiology (Baluška et al. 2006) has emerged in the last few years based on the integration of results from areas of research such as plant electrophysiology, cell biology, molecular biology, and ecology. The difference between plant neurobiology and other more basic disciplines resides in the target of these interdisciplinary efforts. Plant neurobiology adds up to a scientific understanding of the integration of plant sensation and response. The target is the scientific understanding of how metabolism and growth can be regulated by the endogenous integration and processing of information. More specifically, plant neurobiology stresses the integrated signaling and electrophysiological properties of plant networks of cells. As Baluška et al. (2006) point out:

> Each root apex is proposed to harbor brain-like units of the nervous system of plants. The number of root apices in the plant body is high, and all "brain units" are interconnected via vascular strands (plant neurons) with their polarly-transported auxin (plant neurotransmitter), to form a serial (parallel) neuronal system of plants. (p 28).

The working hypothesis of plant neurobiology is that the integration and transmission of information at the plant level involves neuro-like processes such as action potentials, long-distance electrical signaling, and vesicle-mediated transport of (neurotransmitter-like) auxin (Brenner et al. 2006).

Interestingly, from the point of view of the study of plants as information-processing systems, the issue of plant intelligence is less mysterious, or at least subject to investigation in the very same way that cognitive neuroscience deals with in its respective target domain. As recent research in plant neurobiology shows, plants are not passive systems that build up photosynthates. Plants exhibit sophisticated forms of behavior, and are able to assess current data that can lead to an advantage at a later stage. Roots, for instance, exhibit patterns of growth that depend upon future acquisition of minerals and water. Plants are indeed sensitive to a minimum of 15 biotic and abiotic signals, which include not only water, light, minerals and gravity, but also soil structure, neighbor competition, herbivory, allelopathy, and wind, to name but a few (Trewavas 2003). Likewise, plant roots can, for instance, sense volume, discriminate self from alien nonself roots, and allow for phenotypic root reordering as a function of competition for nutrients. Thus, current evidence easily shows that plant behavior is not predefined at all, as Jonas claimed, but highly variable[5].

[5]For more sophisticated plant competencies, see Trewavas (2005).

The commonality in the variability of the behavioral repertoires of animals and plants can be strengthened when their different functional and architectural setups are taken into account. The first thing to note concerning plants is that the architectural constraints on cognition may be pretty minimal. When Darwin took the Venus flytrap as a clear example of a plant with "animal" features (1875), he was not searching for brains in the plant's root tips, even though they clearly exhibit sophisticated forms of computation. Darwin (1880) concluded that: "…the tip of the root acts like the brain of one of the lower animals, the brain being seated within the anterior end of the body receiving impressions from the sense organs and directing the several movements" (p 573). As Darwin observed, forms of plant behavior can be highly sophisticated.

Second, although perception, memory and action are capacities that can be present in both animals and plants, they take different forms. Animals, insofar as they are heterotrophic organisms that require organic foodstuff to survive, exploit a number of mobility-related competencies in order to navigate in complex and contingent environments (Neumann 2006). Animals also appear to be better fitted to escape from predators or harmful environments. Plants, by contrast, do not require contractile muscles for fast responses to environmental contingencies. Insofar as plants are autotrophic organisms, they operate over slower timescales, since inorganic substrates can be synthesized into organic compounds while remaining stationary. This also means that the computational solutions found by both plants and animals can diverge even when they exhibit similar complex functions.

Someone could nevertheless argue that at the level of the architecture, a borderline should be drawn between the two kingdoms. Animal neural networks allow for intelligent behavior courtesy of hardware arranged in *parallel* (Rumelhart et al. 1986), rather than in series. However, in our view, it would be a mistake to demand the same type of computational architecture in plants and animals to accommodate constraints 3 and 4. Different architectures can serve to approximate the same function. Nonlinearly separable functions may be computed, for instance, by inserting a layer of processing units in-between the sensory and the motor layers. In this way, the metric relations of sensory similarity are recoded in terms of functional relations that allow for the approximation of the sensorimotor function, as the logical operation *exclusive disjunction* (aka "the XOR problem") has taught cognitive scientists. However, there are many different ways in which a nonlinearly separable function can be computed, and inserting a layer in-between is not a prerequisite. A two-layer network can equally well approximate the function assuming that the right sort of activation function is chosen for the problem at hand, such as the cosine or sine transfer functions to solve the XOR problem (Rosen et al. 1990). The relevance of this to our current concern is thus that in order for plants to exhibit sophisticated forms of computation, a parallel operating neural network may not be necessary.

Thus, differences in speed and architecture have fostered the idea that plant behavior, compared to animal behavior, is strictly determinate and invariant under a variety of conditions. However, animal and plant avoidance responses are both graded as a function of the stimulus strength and both involve modifications in cellular morphology. Plant behavioral invariance is in the eye of the anthropocentric

—or perhaps *heterotrophicentric*—observer. However, once we approach the study of (minimal) cognition nonanthropocentrically, perception, memory, and action are common currencies across phylogeny.

Summing up, differences in speed and form do not serve to exclude a cognitive interpretation of plant behavior. In our view, the issue is not whether plants can move about, acting intelligently in this way, but rather whether plants, being autotrophic organisms, integrate information, have memory, can make decisions, in such a way that their adaptive coupling to their environment can be called "cognitive." Light, gravity, moisture, and touch are signals that plants integrate and respond to in complex, nonlinear ways. Roots make decisions in particular contexts as to which type of signal(s) to honor (Li and Zhang 2008). Furthermore, phenotypic plasticity in ever-changing environments requires the exploitation of memory resources. Plants integrate exogenous and endogenous information channels in an attempt to phenotypically adapt to environmental contingencies; a sophisticated form of competency that, we believe, is comparable with animals' predictive behavior. In the remainder of this chapter, we will further highlight the similarities between plant growth and animal memory, and try to show why the behavior of plants, as they are coupled to their environment, allows for a cognitive interpretation of their adaptive responses.

6 Similarities Between Growth and Memory

Let us take the carnivorous plants *D. muscipula* and *A. vesiculosa* as examples. In the case of *D. muscipula*, an action potential (AP) is generated whenever an upper trap hair is bent. Crucially, a single stimulation of the hair does not trigger the closure of the trap. For the trap to close, a second AP that takes place only when another hair is bent within 40 s after the first AP has been generated is necessary (Baluška et al. 2006). This setup comprises a basic form of memory, similar to the TCST system in bacteria (Di Primio et al. 2000), as well as basic forms of animal memory.

Now consider the avoidance responses of plants in relation to drought (Trewavas 2003). Drought avoidance behavior results in a reduction in the rate of cell growth that involves, on the one hand, changes in cytosolic Ca^{2+}, $[Ca^{2+}]_i$ and in other secondary messengers, and, on the other hand, phosphorylation changes in ATPases and associated ion channels related to turgor (Palmgren 2001). Trewavas (2003) compares drought avoidance responses in plants with the avoidance behavior of *Aplysia*. The pattern of avoidance of this marine slug involves a form of short-term memory whose mechanism includes Ca^{2+} and, in addition to the second messengers, cyclic nucleotides and several protein kinases that operate as a temporary memory (Greengard 2001) by phosphorylating ion channels.

Phenotypic plasticity underlies the ability to memorize. The genesis of dendrites delivers the goods in animal brains by effectively altering the architectural features of the network. Different patterns of connectivity permit the network to acquire new functions. Plant networks cannot exploit phenotypic plasticity with the computational resources of animal networks. For one thing, plants lack neuron-like cells as computational building blocks where new dendrites allow for new functional profiles overall. However, by contrast, plant cell divisions continue at any stage of development. This means that the method of acquiring different functionalities throughout the life of the plant will involve architectural changes somehow. Fortunately, as we saw earlier, this need not be an insurmountable hurdle, since intelligence should not be univocally linked to a specific type of architecture. Trewavas (2003) points out where the architectural divergence may lie: "Just as the process of learning in a brain could be represented as a time series, a set of snapshots of developing brain connections, in plants, each snapshot may possibly be represented by developing plasmodesmatal connections or equally, successive new tissues. So, instead of changing dendrite connections, plants form new networks by creating new tissues, a series of developing brains as it were" (p 14). In this view, it is not the modification of patterns of connectivity that allows the plant network to remain competent. Rather, as new tissue accumulates, new networks with different computational resources are stored on top of each other. Note that newer tissue networks do not replace former ones; instead, we have a succession of operative serial networks that are obtained as cells continue to divide throughout the life of the plant.

The key issue when attempting a nonbiased approach to intelligence, as Trewavas (2003) points out, resides in the "delays in the transfer of information between the sensory system and the motor tissues acting upon the signals" (p 1). Note that these delays cancel out a rendering of plant forms of memory with some class of developmental progression, where previous cellular states determine future outcomes *linearly* (Firn 2004). As Bose and Karmakar (2003) point out, animal neuronal networks and plant calcium signaling systems are not that different in terms of nonlinearities. In the case of plants, nonlinearities are obtained by means of the succession of signaling networks. Chakrabarti and Dutta (2003) have put forward an electrical network that models plant calcium signaling systems and approximates Boolean functions, including XOR. Open/closed ion channels play the role of neurons in networks of plants. As calcium ions are released, diffusion across nearby channels gives rise to further calcium release, ultimately giving rise to a calcium wave that flows throughout the whole network. This modeling of calcium signaling networks, Bose and Karmakar (2003) argue, illustrates how memory mechanisms can be implemented in plants. The dynamics of the calcium wave are governed by nonlinear equations, opening up the possibility of integrating and computing all incoming data (Trewavas 2002) in a way that functionally resembles animal-based computation.

If animal memory is characterized by processes that are nonlinear, the same goes for plants. For a case in point of transfer delay in sensorimotor terms, consider the orchestrated behavior of roots and shoots. Here, there is genuine emergent behavior that cannot be accounted for by summing the local behavior of roots and shoots. Put bluntly, root–shoot interactions can only be understood synergistically (Corning 2003). Or, to put it in computational terms, the XOR function exemplifies the basis of complex behavior insofar as it exploits the delays in the transfer of information between sensory and motor processing tissues, as the metric relations in sensory space are altered in order for the problem to be approximated. We therefore have the means to overcome sensorimotor organization as a restriction (constraint 3, above). In what follows, we illustrate from a behavioral perspective how plants have offline capabilities (constraint 4) that allow them to do more than reacting to immediate stimuli.

7 Offline Cognition: Leaf Heliotropism

Someone may argue that the issue is not whether the function to be computed is linearly or nonlinearly separable, embodied or not, but rather whether the recoding of the metric relations that are obtained as a result of the approximation of the function can be interpreted in representational terms. In other words, the issue is whether plants, however complex the calcium wave networks happen to be, manipulate representational states or not. Take the case of the stilt palm. In order to avoid competition for light, the stilt palm (Allen 1977) "walks" away from shade and into sunlight. The stilt palm grows new roots in the direction of sunlight, allowing older roots to die. Although Trewavas (2003) interprets this as an *intentional* form of light-foraging behavior, it is usually assumed that decoupled, offline modeling tasks are what distinguish sophisticated forms of behavior from merely tropistic online routines. Put bluntly, the sine qua non of representation-based competency is offline adaptive behavior (Clark 1997).

Generally speaking, tropisms involve a directional change such as growth or movement in response to a given environmental stimulation (Stanton and Galen 1993). Nevertheless, plant tropisms can vary substantially as a function of the type of stimulus that the plant is responsive to, and the part of the plant that responds to the stimulation. Complex—although still online—tropistic reactions are frequent. Roots that manifest gravitropism stop developing downwards as they encounter a physical obstacle, and grow horizontally instead. However, they are able to assess the state of affairs online, and periodically try to move downwards, remaining horizontal if unable to respond gravitropically (Massa and Gilroy 2003). Likewise, we can think of thigmotropism, hydrotropism or thermotropism as other sophisticated—although possibly still online—tropistic responses.

However, some plant tropisms do constitute a form of offline anticipatory behavior. In relation to light-related tropisms, we can distinguish phototropisms, which may represent directed responses to a static light source, from heliotropism, a more complex response that involves a correlated response to changes in sunlight orientation as the day changes from sunrise to sunset. We may also differentiate between flower heliotropism and leaf heliotropism. In the case of flower heliotropism, no "memory" mechanisms seem to be required for flowers to keep track of the position of the sun. Unless flowers are exposed to light in the morning they will fail to reorient to sunrise, remaining in a random orientation throughout the night.

On the other hand, offline nocturnal reorientation by plant leaves represents a qualitative change with regard to stimulus-controlled online behavior. Leaf laminas of *Lavatera cretica* can not only anticipate the direction of the sunrise but they also allow for this anticipatory behavior to be retained for a number of days in the absence of solar tracking. That is, the laminas reorient during the night and keep facing the direction of the sunrise even after a few days without tracking the sun, and without sensing the position of sunset. Schwartz and Koller (1986) report a series of experiments that clearly show that this is a complex offline response. Three groups of plants were taken at sunset to three different cabinets. One was kept in darkness; another was illuminated during daylight hours, and the other group was kept illuminated vertically throughout the experiment. On the first day, the plants in cabinets 1 and 2 (darkness and daylight hours, respectively) could anticipate sunrise and leaves were oriented towards that direction. The plants in cabinet 3, on the other hand, were horizontal, with laminas facing upwards, as they had been kept under constant vertical artificial illumination. The nocturnal reorientation behavior exhibited by the plants in cabinets 1 and 2 lasted for as long as three days under the same experimental conditions; that is, in the absence of daytime solar tracking. In their study, the nocturnal reorientation could not be explained by the sunrise of the day before, since plants were prevented from tracking the sun for 3–4 days.

The explanation for nocturnal reorientation involves the internal modeling of environmental rhythms. Circadian clocks allow time to be estimated through the synchronization of endogenously generated activity with exogenous cyclic periods such as day–night planetary patterns. Circadian clocks can mimic biological rhythms on a 24-h cycle, and this explains nocturnal reorientation in plants for up to four days in the absence of sunrise stimulation. Plant genetics point towards underlying shared molecular components that explain day-length estimations and the operation of light receptors (see Pruitt et al. 2003, and references therein). In the case of time estimation, recent research in genomics has unearthed the molecular mechanisms of overt plant behavior, with the result that both plants and animals draw on the very same molecular networks (Cashmore 2003) in their adaptive exploitation of circadian clocks. Thus, a single level explains the origins of the anticipatory behavior as exemplified by circadian clocks.

As we saw earlier, the types of restrictions imposed by constraints 3 and 4 serve to stress the importance of having a sensorimotor organization on the one hand, and improving the offline and information processing capacities of the system on the other. Offline plant behavior may thus be considered to be minimally cognitive insofar as information is processed flexibly and adaptively in accordance with the aforementioned restraints. Constraint 5, nonetheless, remains an open empirical question, although there are reasons to be optimistic, as we point out in the closing section.

8 Concluding Remarks

The notion of cognition is currently in flux. Within embodied cognition, the meaning of cognition has been changed from human thought to a broader interpretation centered on perception and action. From this perspective, free-moving animals are part of the cognitive domain but sessile plants, that lack clear perception and action features and a dedicated sensorimotor organization, form a more difficult case. Given that plants are definitely capable of many forms of complex behavior, they provide an interesting domain for both questioning and clarifying the focus of embodied cognition on free-moving agents. At the same time, the wide reading of cognition allowed by embodied cognition can reinforce the study of intelligence in plants, such as that now occurring within the field of plant neurobiology.

We discussed how embodied cognition interpreted cognitive phenomena, and summarized this view in terms of five constraints on cognition (Sect. 5). Two of these constraints (3 and 5) emphasize being free-moving as a definite requirement for minimal forms of cognition. This seems to contradict the possibility of plant cognition. We used the analysis put forward by Hans Jonas to articulate the requirements for being considered "free-moving" and to assess possible reasons for its importance. Subsequently, we argued that these two constraints should be readjusted, allowing the possibility that plants fulfill these constraints in ways that differ from those in free-moving creatures. We then discussed work in the plant sciences to see whether it is plausible that plants can fulfill the five constraints formulated above.

Plants fulfill the first two of these constraints: they have a metabolic organization that provides a normative context, and exploit environmental spatiotemporal structure to enhance metabolic functioning. Our discussion centered on evidence for constraints 3 (sensorimotor organization), 4 (offline control), and 5 (acting as a global unit). Interestingly, while it is relatively easy to establish the presence of offline control in plants, which is generally considered the key notion of cognition, evidence concerning the role and organization of motility and perception in plants remains more equivocal and difficult to interpret.

There is a large gap between the facts and possible interpretations here. New developments in plant neurobiology have made it clear that plants are capable of much more complex behavior than many of us previously tended to attribute to them, including Jonas. We have argued that it is plausible that plants can circumvent the requirement for an animal sensorimotor organization. They are capable of organizing their behavior in ways that are different but still highly complex and adaptive, making this behavior cognitive in the general and minimal sense used within embodied cognition. It remains to be seen to what extent plants can really fulfill constraint 5. Plants integrate information simultaneously provided by many different sensors on a variety of parameters in real time, and it may very well be that this provides them with their own equivalent of the sensitivity provided by animal sensory surfaces. We expect that plant neurobiology will clarify this issue further in the near feature, given its emphasis and focus on the integration of information present in different parts of the plant.

Concluding, we hope that we have shown here that, in a number of specific issues relating to cognition, animals and plants do not differ fundamentally, and that plants are cognitive in a minimal, embodied sense that also applies to many animals and even bacteria. The scientific target in both cases is to understand the continuous interplay of animals and plants in relation to the environmental contingencies that impinge upon them. Plant cognition is, from this viewpoint, not a contradiction at all, but an empirical issue that requires much more attention, not only from plant scientists but also more generally from cognitive scientists.

Acknowledgments Fred Keijzer wants to thank Pamela Lyon, Marc van Duijn and Daan Franken for their helpful comments and discussion. The preparation of this chapter was supported in part by DGICYT Project HUM2006-11603-C02-01 (Spanish Ministry of Science and Education and Feder Funds) to Paco Calvo.

References

Adams F, Aizawa K (2001) The bounds of cognition. Philos Psychol 14:43–64
Allen PH (1977) The rain forests of Golfo Dulce. Stanford University Press, Stanford
Alpi A, Amrhein N, Bertl A, et al. (2007) Plant neurobiology: no brain, no gain? Trends Plant Sci 12:135–136
Baluška F, Hlavacka A, Mancuso S, Volkmann D, Barlow PW (2006) Neurobiological view of plants and their body plan. In: Communication in Plants: Neuronal Aspects of Plant Life, Baluška F, Mancuso S, Volkmann D (eds), Springer Verlag, 19–35
Barandiaran X (2008) Mental life: a naturalized approach to the autonomy of cognitive agents. Dissertation, University of the Basque Country, San Sebastian
Barlow PW (2008) Reflections on 'plant neurobiology.' BioSystems 99:132–147
Beer R (1995) A dynamical systems perspective on agent–environment interaction. Artific Int 72:173–215
Beer RD (2000) Dynamical approaches to cognitive science. Trends Cogn Sci 4:91–99
Beer RD (2003) The dynamics of active categorical perception in an evolved model agent. Adapt Behav 11:209–243
Bekoff M, Allen C, Burghardt GM (eds) (2002) The cognitive animal: empirical and theoretical perspectives on animal cognition. MIT, Cambridge, MA
Berg HC (2000) Motile behavior of bacteria. Phys Today 53:24
Bickhard M (2008) Is embodiment necessary? In: Calvo P, Gomila T (eds) Handbook of cognitive science: an embodied approach. Elsevier, Amsterdam, pp 29–40
Bose I, Karmakar R (2003) Simple models of plant learning and memory. Physica Scripta T106:9–12
Brenner ED, Stahlberg R, Mancuso S, Vivanco J, Baluška F, Van Volkenburgh E (2006) Plant neurobiology: an integrated view of plant signaling. Trends Plant Sci 11:413–419
Brooks R (1999) Cambrian intelligence. MIT, Cambridge, MA
Calvo P (2007) The quest for cognition in plant neurobiology. Plant Signal Behav 2:208–211
Calvo, Gomila (eds.) (2008) Handbook of Cognitive Science: An Embodied Approach, Elsevier
Carruthers P (2004) On being simple minded. Am Philos Quart 41:205–220
Cashmore AR (2003) Cryptochromes: enabling plants and animals to determine circadian time. Cell 114:537–543
Chakrabarti BK, Dutta O (2003) An electrical network model of plant intelligence. Ind J Phys 77A:549–551
Clark A (1997) Being there: putting brain, body, and world together again. MIT, Cambridge, MA
Corning P (2003) Natures magic. Synergy in evolution and the fate of humankind. Cambridge University Press, Cambridge
Darwin C (1875) Insectivorous plants. John Murray, Edinburgh
Darwin C (1880) The power of movements in plants. John Murray, Edinburgh
Dennett DC (1984) Elbow room: the varieties of free will worth wanting. Bradford, Cambridge
Dennett DC (1996) Kinds of minds. Basic, New York

Di Paolo EA (2005) Autopoiesis, adaptivity, teleology, agency. Phenomen Cogn Sci 4:97–125
Di Primio F, Müller BS, Lengeler JW (2000) Minimal cognition in unicellular organisms. In: Meyer JA, Berthoz A, Floreano D, Roitblat HL, Wilson SW (eds) SAB2000 Proceedings Supplement. International Society for Adaptive Behavior, Hawaii, pp 3–12
Firn R (2004) Plant intelligence: an alternative viewpoint. Ann Bot 93:345–351
Godfrey-Smith P (1996) Complexity and the function of mind in nature. Cambridge University Press, Cambridge
Godfrey-Smith P (2001) Environmental complexity and the evolution of cognition. In: Sternberg R, Kaufman J (eds) The evolution of intelligence. Lawrence Elrbaum, London
Gould JL, Gould CG (1998) Reasoning in animals. Sci Am Presents 9:52–59
Greengard P (2001) The neurobiology of slow synaptic transmission. Science 294:1024–1030
Greenspan RJ, van Swinderen B (2004) Cognitive consonance: complex brain functions in the fruit fly and its relatives. Trends Neurosci 27:707–711
Heinrich B (2000) Testing insight in ravens. In: Heyes C, Huber L (eds) The evolution of cognition. MIT, Cambridge, MA
Hofstadter DR (1985) On the seeming paradox of mechanizing creativity. In: Metamagical themas: questing for the essence of mind and pattern. Penguin, London, pp 526–546
Hurley SL (1998) Consciousness in action. Harvard University Press, Cambridge, MA
Jonas H (1966) The phenomenon of life: toward a philosophical biology. Harper and Row, New York
Jonas H (1968) Biological foundations of individuality. Int Philos Quart 8:231–251
Keijzer FA (2001) Representation and behavior. MIT, Cambridge
Keijzer FA (2003) Making decisions does not suffice for minimal cognition. Adapt Behav 11:266–269
Keijzer FA (2006) Differentiating animality from agency: towards a foundation for cognition. In: Sun R, Miyake N (eds) Proc CogSci/ICCS 2006. Sheridan Printin, Alpha, pp 1593–1598
Lengeler JW, Müller BS, di Primio F (2000) Neubewertung kognitiver Leistungen im Lichte der Fähigkeiten einzelliger Lebewesen. Kognitionswissenschaft 8:160–178
Li X, Zhang WS (2008) Salt-avoidance tropism in *Arabidopsis thalania*. Plant Signal Behav 3:351–353
Lyon P (2006a) The agent in the organism (dissertation). Australian National University, Canberra
Lyon P (2006b) The biogenic approach to cognition. Cogn Process 7:11–29
Massa G, Gilroy S (2003) Touch modulates gravity sensing to regulate the growth of primary roots of *Arabidopsis thaliana*. Plant J 33:435–445
Merleau Ponty M (1963) The structure of behaviour. Duquesne University Press, Pittsburgh
Moreno A, Etxeberria A (2005) Agency in natural and artificial systems. Artif Life 11:161–176
Moreno A, Umerez J, Ibañez J (1997) Cognition and life. The autonomy of cognition. Brain Cogn 34:107–129
Müller BS, di Primio F, Lengeler JW (2001) Contributions of minimal cognition to flexibility. In: Callaos N, Badawy W, Bozinovski S (eds) SCI 2001 Proceedings of the 5th World Multi-Conference on Systemics, Cybernetics and Informatics, Volume XV, Industrial Systems: Part II. International Institute of Informatics and Systemics, Orlando, pp 93–98
Neisser U (1967) Cognitive psychology. Appleton-Century Crofts, New York
Neumann PM (2006) The role of root apices in shoot growth regulation: support for neurobiology at the whole plant level? In: Baluška F, Mancuso S, Volkmann D (eds) Communication in plants: neuronal aspects of plant life. Springer, Berlin
Palmgren MG (2001) Plant plasma membrane H^+-ATPases; powerhouses for nutrient uptake. Annu Rev Plant Physiol 52:817–845
Pfeifer R, Scheier C (1999) Understanding intelligence. MIT, Cambridge, MA
Pickard BG (1973) Action potentials in higher plants. Bot Rev 39:172–201
Prescott TJ, Redgrave P, Gurney K (1999) Layered control architectures in robots and vertebrates. Adapt Behav 7:99–127
Pruitt R, Bowman J, Grossniklaus U (2003) Plant genetics: a decade of integration. Nat Genet 33:294–304

Rosen BE, Goodwin JM, Vidal JJ (1990) Transcendental functions in backward error propagation. Proc IEEE Int Conf Syst Man Cybern 4–7:239–241
Roth G, Wullimann MF (eds) (2001) Brain evolution and cognition. Wiley, New York
Rumelhart DE, McClelland JL, PDP Research Group (1986) Parallel distributed processing: explorations in the microstructure of cognition, vol. 1. MIT Press, Cambridge, MA
Schwartz A, Koller D (1986) Diurnal phototropism in solar tracking leaves of *Lavatera cretica*. Plant Physiol 80:778–781
Shettleworth SJ (1998) Cognition, evolution, and behavior. Oxford University Press, New York
Smirnova AA, Lazareva OF, Zorina ZA (2003) Prototype symbolization in hooded crows. Neurosci Behav Physiol 33:335–348
Stanton ML, Galen C (1993) Blue light controls solar tracking by flowers of an alpine plant. Plant Cell Environ 16:983–989
Sterelny K (2001) The evolution of agency and other essays. Cambridge University Press, Cambridge
Trewavas AJ (2002) Mindless mastery. Nature 415–841.
Todar K (2004) Todar's online textbook of bacteriology. Department of Bacteriology, University of Wisconsin-Madison, Madison (see http://www.textbookofbacteriology.net/, accessed 27 Sept 2005)
Trewavas A (2003) Aspects of plant intelligence. Ann Bot 92:1–20
Trewavas A (2005) Green plants as intelligent organisms. Trends Plant Sci 10:413–419
Van Duijn M, Keijzer F, Franken D (2006) Principles of minimal cognition: casting cognition as sensorimotor coordination. Adapt Behav 14:157–170
Varela FJ, Thompson JE, Rosch E (1991) The embodied mind. MIT Press, Cambridge, MA

Memorization of Abiotic Stimuli in Plants: A Complex Role for Calcium

Camille Ripoll, Lois Le Sceller, Marie-Claire Verdus, Vic Norris, Marc Tafforeau, and Michel Thellier

Abstract Plants are sensitive to various abiotic stimuli (including electromagnetic radiations), to which they respond by modifying their development. The response is sometimes delayed, relative to the reception of the stimulus, which implies that the corresponding information is memorized. A few cases of such behavior are described, including that controlling the induction of meristems in the hypocotyl of flax seedlings. Using this model system, calcium has been shown to play a key role not only in stimulus sensing and the possible storage of that information, but also in its final expression. Modifications of genome expression, as well as posttranslational modifications (e.g., phosphorylation) of proteins, are involved in signal transduction and the possible memorization of information. SIMS methodology provides us with interesting experimental possibilities. The process of "ion condensation," which has been practically ignored by biologists so far, may be involved in the memorization mechanism. A few cases of the application of plant sensitivity and information memorization to agronomical or research problems are described. The role of memorization mechanisms in the elaboration of an integrated response from plants to the many, varied stimuli that they receive permanently from their environment is discussed.

Abbreviations ppb; Parts per billion, ppm; Parts per million, SIMS; Secondary ion mass spectrometry

1 Introduction

Plants are sensitive to many different abiotic signals (mechanical stimuli, cold shock, osmotic stress, etc.). They react to such stimuli in a few cases by rapid movements (e.g., *Mimosa pudica*, Venus fly trap), but most generally by modifying

---

C. Ripoll, L. Le Sceller, M.-C. Verdus,V. Norris, M. Tafforeau, and M. Thellier (✉)
Laboratoire AMMIS, Faculté des Sciences de l'Université de Rouen, F-76821
Mont-Saint-Aignan, France
e-mail: Michel.Thellier@univ-rouen.fr

F. Baluška (ed.), *Plant-Environment Interactions*, Signaling and Communication in Plants,
DOI: 10.1007/978-3-540-89230-4_14,

their metabolism, including genome expression and mRNA translation (Davies and Schuster 1981; Davies 1987; Braam and Davis 1990; Henry Vian et al. 1995a; Braam et al. 1996; Knight et al. 2004; Kaplan et al. 2006), growth and organogenesis (Jaffé and Forbes 1993). Sometimes the modifications occur in the very organ that has been stimulated. For instance, gently rubbing the terminal, actively growing internode of a shoot of *Bryonia dioica* causes the elongation of this internode to be inhibited without appreciably modifying the growth of the neighboring internodes (Boyer et al. 1979). In other cases, the modifications take place at a distance from the stimulated tissue (Gautheret 1935; Desbiez et al. 1978, 1983; Davies and Schuster 1981). This implies the existence of a transfer of information from the stimulated to the reactive tissue in the plant. The rate of this transfer of information, as evaluated in the different cases under consideration, usually lies in the range of 0.1–1 mm s^{-1} (Desbiez et al. 1991a). The interpretation of the mechanisms involved in that process of information transfer has been the subject of intense debate; for a review and discussion, see for instance Davies (1987, 1993) and Malone and Alarcon (1995). Apart from that spatial disconnection between the stimulated and the responding tissues, there may also be a temporal disconnection between stimulus and response. This means that the information corresponding to the reception of the stimulus can be memorized within the plant. Sometimes the memorized information remains latent, not taking effect until the application of another signal makes the plant able to recall the memorized message and allow it to take effect. A few examples of cases of the memorization of a signal in a plant are summarized in Sect. 3.

The chain of events between the sensing of an abiotic stimulus by a plant and the final (metabolic and/or morphogenetic) response of that plant is rather complex; for reviews, see, e.g., Davies (1987), Desbiez et al. (1992), Bowler and Chua (1994), Bowles (1995), Knight and Knight (2001), Plieth (2005) and Kaplan et al. (2006). However, there is common agreement that plants react to such stimuli by an almost immediate elevation of the free calcium in the cell cytosol (Knight et al. 1991, 1992, 1995; Bush 1995; Knight 2000; Plieth 2001). Especially by working with plants genetically transformed to express the Ca^{2+}-dependent luminescent protein aequorin (Knight et al. 1991; Knight and Knight 1995), and possibly by using calcium chelators and calcium channel blockers, it was inferred that this elevation of free cytosolic calcium derived from the uptake of external calcium and/or release from internal Ca^{2+} stores (Klüsener et al. 1995; Polisensky and Braam 1996; Cessna et al. 1998). Moreover, these calcium changes show enormous variability in their nature (transient, sustained or oscillatory), amplitude, kinetics and spatial distribution (Bush 1995; Knight et al. 1998). Some authors have suggested that the time course of changes in the cytosolic calcium contributes to the identification of the type and intensity of the stimulus (McAinsh and Hetherington 1998; Sanders et al. 2002; Ng and McAinsh 2003). For instance, the cytosolic Ca response of *Arabidopsis* seedlings to gravistimulation exhibits kinetics that are very different from those induced by movement and wind (Plieth and Trewavas 2002). There is also some evidence that calcium can act as a protective agent against ionic stresses or as a chemical switch (Plieth 2005).

2 Examples of the Memorization of Signals in Plants

2.1 *Breaking the Symmetry of the Growth of Opposite Buds*

To our knowledge, the possible existence of some sort of plant memory was first hypothesized for the control of the relative growth of the cotyledonary buds of seedlings of *Bidens pilosa* L. at the age of 2–3 weeks (Thellier et al. 1981, 1982, 2000; Desbiez et al. 1984, 1986, 1991c). When the seedlings were subjected to an asymmetrical stimulus such as pricking or rubbing one of the two seedling cotyledons, or even when a drop of an appropriate aqueous solution was deposited on one cotyledon (Desbiez et al. 1991b), the growth of the bud at the axil of the unstimulated cotyledon was favored. However, depending on the reception of a variety of other, complementary stimuli, the seedlings were either in a state in which their cotyledonary buds actually started to grow asymmetrically (responding seedlings) or in a state in which no asymmetry of bud growth could be observed (nonresponding seedlings). In the second case, the seedling development was practically indistinguishable from that of the seedlings that were not stimulated asymmetrically; however, when the nonresponding seedlings were finally given the same complementary stimuli as those used to obtain the responding seedlings, bud-growth asymmetry rapidly became apparent. This means that the initial, asymmetrical stimulus caused information on asymmetrical growth to be stored (memorized) within the seedlings in all cases (STO function), but that the complementary stimuli made these seedlings either able (responding seedlings) or unable (nonresponding seedlings) to recall the stored information and allow it to take effect (RCL function). The combined effect of the STO and RCL functions on the final response of the plants is illustrated in Table 1. Memorization lasting up to 15 days was observed in these experiments. Various theoretical modeling approaches were developed to cope with the rather complex phenomenology of this memorization process (Kergosien et al. 1979; Desbiez et al. 1994; Thellier et al. 2004; Demongeot et al. 2006). By using a model based on the theory of dynamic systems, the storage of information is interpreted as corresponding to the formation of limit cycles in the state space of the system, while the recall of that information depends on each particular pathway that an appropriate perturbation can impose to the point representing the system in its state space according to each particular experimental condition (Demongeot et al. 2006).

In the above experiments, it was also observed that the pricking of one cotyledon of unresponsive seedlings had a rapid, specific and asymmetric effect on the renewal of the cell cycle in the meristematic cells of the cotyledonary buds (Desbiez et al. 1998; Tafforeau et al. 2006). Namely, the meristematic cells were approximately half and half in the states G1 and G2 of the cell cycle in the cotyledonary buds of unstimulated, control seedlings. After the pricking of one cotyledon, practically all of the meristematic cells in G2 underwent division almost immediately in the bud at the axil of the pricked cotyledon, whereas this was the case for only a small proportion of the cells in G2 in the bud opposite to the pricked cotyledon; the situation then remained virtually unchanged for the next five days. It is not known whether this

Table 1 Storage and recall of signals controlling the asymmetry of bud growth in *Bidens* seedlings[a]

Pricking treatments[b]	STO function[c]	RCL function[d]	Asymmetrical bud growth[e]
Nonpricked control	Off	Off	No
2A–2B	Off	On	No
2A	On	Off	No
2A(1h)2A–2B	On	On	Yes
2A(1h)2A–2B(3h)2A–2B	On	Off	No
2A(1h)2A–2B(3h)2A–2B(5h)2A–2B	On	On	Yes

[a]This table is a selection of data originating from various publications by M.O. Desbiez and collaborators (see the list of references)

[b]A and B are the two seedling cotyledons; 2A is an asymmetrical treatment consisting of the administration of two needle pricks to cotyledon A; 2A–2B is a symmetrical treatment consisting of the simultaneous administration of two needle pricks to both cotyledons A and B; the time elapsed between two successive treatments is indicated between parentheses.

[c]The asymmetrical treatment 2A switches the STO function "on." This is apparently an all-or-nothing and irreversible process.

[d]When applied at the correct time intervals, the symmetrical treatment 2A–2B switches the RCL function reversibly "on" then "off" then "on" again, etc.

[e]Bud growth is asymmetrical only when both STO and RCL are "on." Note that the "asymmetry information" stored in the seedlings as a consequence of treatment 2A can be evoked (recalled) at least twice depending on the status of the RCL function (see the last four lines in this table)

asymmetrical renewal of the cell cycle is somehow related to the storage of the symmetry-breaking information controlling bud growth (see the above paragraph). In any event, this shows that the relative abundances of the G2 cells in the meristems of the two cotyledonary buds can retain the memory of the administration of the asymmetrical pricking stimulus for at least five days. The discovery of such an effect of a pricking stimulus on cell division is consistent with the well-known fact (Jonak et al. 1996) that mitogen-activated protein (MAP) kinases are involved in differentiation, cell division and stress response.

2.2 Inhibition of Hypocotyl Growth

Again with seedlings of *B. pilosa*, it is well known that the hypocotyl elongates shortly after seed germination. At that age, the development of the seedlings is practically indifferent to them being grown in a conventional nutrient solution or in distilled water. Pricking the seedling cotyledon at the beginning of hypocotyl elongation, either asymmetrically (pricking a single cotyledon) or symmetrically (pricking the two cotyledons simultaneously), tends to inhibit the elongation of the hypocotyl, but the inhibition is very intense or almost insignificant according to whether the seedlings are grown in distilled water or in a nutrient solution. Using seedlings pricked while they were grown in a nutrient solution (hypocotyl elongating almost normally) then transferred to distilled water a few days later, inhibition of hypocotyl elongation occurred immediately after the transfer of the seedlings to distilled water (Desbiez et al. 1983, 1987a). Again, this means that the pricking treatment caused "elongation inhibition" information to be stored within the seedlings, but that this information could take effect (by controlling hypocotyl development) only after the transfer of the seedlings to distilled water had rendered them able to recall the stored information.

2.3 Inhibition of Internode Elongation

As already stated in the "Introduction," gently rubbing the terminal internode of a shoot of *B. dioica* causes a severe inhibition of its elongation. This inhibition is associated with an increase of several enzyme activities, including different peroxidase activities (Boyer et al. 1979; De Jaegher et al. 1985). By preparing tissue cultures of stimulated internodes, the same enzyme activities were found to exist in a few successive callus subcultures, and then (after a number of subcultures, which varied according to the different enzymes involved) these activities decreased progressively back to the level of the unstimulated controls (Bourgeade et al. 1989). In this case, the information initiated by rubbing thus remained stored within the plant tissues for a period of up to several months but which varied according to the type of enzyme activity studied.

2.4 Kinetics of the Effect of Wind Stimulation on Calcium Signaling

With nonstimulated *Nicotiana plumbaginifolia* seedlings, wind stimulation causes an immediate increase in cytosolic calcium (as per usual); but repeated wind stimulation makes the plant cells transiently (i.e., for 45–60 s) refractory to further calcium signaling (Knight et al. 1992). This means that the plants under study have encoded a memory of the previous wind stimuli, which is revealed by an alteration of the calcium response to a new wind stimulus.

2.5 Effect of Stress History on Drought Calcium Signaling Pathways

The elevation of cytosolic calcium concentration as a result of hyperosmotic stress (treatment with mannitol) in *Arabidopsis thaliana* was reduced by oxidative stress pretreatment, while it was increased by hyperosmotic stress pretreatment (Knight et al. 1998). Different combinations of environmental stress can thus produce novel calcium signal responses. Again, this means that the plants under study have encoded a memory of previous stress encounters, which is revealed by an alteration of the calcium response to a new stimulus.

2.6 *Temperature Sensing*

With *A. thaliana*, it was observed that cold elicited an immediate rise in the cytosolic concentration of free calcium, which was dependent on the cooling rate rather than on the absolute temperature values. Prolonged or repeated cold treatments were shown to attenuate the cytosolic calcium response to subsequent episodes of cooling (Plieth et al. 1999). Plants are thus able to encode a memory of previous cold treatments (which is revealed by an alteration of the Ca response to a new cold stimulus), as was the case with other types of stimuli (see Sects. 2.4 and 2.5).

2.7 *Effect of the Preceding Phosphate Supply on Phosphate Uptake*

When a population of the cyanobacterium *Anabaena variabilis*, previously grown under conditions of phosphate deficiency, was subjected to pulsewise increases in the external phosphate concentration, the phosphate-uptake system was transformed within a few minutes from a very active to a less active state. Moreover, novel adaptation was influenced in a distinct way by the pattern of previous phosphate fluctuations to which the cyanobacterium population had been exposed (Falkner and Falkner 2003; Plaetzer et al. 2005). This is a suggestion that the *Anabaena* cells have the potential to retain a memory of previous phosphate nutrition conditions.

2.8 *Plant Electrical Memory*

An electrical stimulus given above a threshold value is as efficient as the mechanical stimulation of the trigger hairs to cause the closure of the upper leaf of the Venus flytrap. The plant can accumulate small, subthreshold charges, and when the threshold value is reached, the leaf closes. This accumulative character of electrical stimuli reveals the existence of an electrical memory in this plant system (Volkov et al. 2008)."

3 Our Model System of the Induction of Meristems in Flax Hypocotyls

In addition to the cases described in Sect. 2, a model system of plant memory that is well adapted to easy experimental study has been developed by our group. Combining an abiotic stimulus with a transient (1–3 days) depletion of calcium resulted in the induction of numerous epidermal meristems in flax-seedling hypocotyls during the following three weeks (Verdus et al. 1996, 1997; Tafforeau et al. 2006; Verdus et al. 2007). When unstimulated seedlings were calcium depleted, or when stimulated seedlings were not calcium depleted, usually no more than two meristems were produced. When calcium depletion was delayed relative to the abiotic stimulus, the production of meristems was correspondingly delayed (Table 2). This means that the meristem production information induced by the abiotic stimulus was stored within the seedlings, without any apparent effect, until calcium depletion finally allowed this stored information to be recalled and to take effect (meristem formation). For storage periods of up to eight days, no loss of potency of the stored information was observed.

We did not find any treatment other than calcium depletion that had an effect on the recall of the stored meristem production information. Nevertheless, a variety of abiotic stimuli were able to induce the storage of a meristem production signal. These include mechanical signals, such as wind or manipulation stimulus (transferring the seedlings from one nutrient medium to another), and nonmechanical stimuli, such as cold shock (roots bathing for 1 min in a medium at 4°C), slow cold treatment (seedlings transferred from the normal growth room to a cold room, then brought back to the normal room after 24 h), drought stress (seedlings left for 2 h with their roots in open air), and irradiation with electromagnetic radiation emitted at nonthermal levels at either 0.9 GHz by a GSM (Global System for Mobile communications) telephone or 105 GHz by a Gunn oscillator (Verdus et al. 1996, 1997; Tafforeau 2002; Tafforeau et al. 2002a, b, 2004). That plants were sensitive to electromagnetic radiation was fully confirmed by tomato seedlings that were irradiated with a 900 MHz electromagnetic field (Vian et al. 2006; Roux et al. 2006). Coming back to the flax seedlings, the production of epidermal meristems was inhibited by the addition to the nutrient medium of pharmacological agents (EGTA, ruthenium red, lanthanum or gadolinium) that are known to affect calcium availability or calcium transport. The use of these agents revealed a period of

Table 2 Correlation of the delay in the application of the calcium depletion treatment with the delay in the appearance of the meristems[a]

Delay in the calcium depletion treatment (days)	Delay in the production of the meristems[b] (days)
4	3.4
8	6.7

[a]In this experiment, flax seedlings were subjected to a manipulation stress then to a calcium depletion treatment (2 days), beginning either at the moment of the manipulation stress or after delays of 4 or 8 days. The curves representing the time course of the production of meristems (each experimental point is the mean value calculated from 10 seedlings) reached a plateau value of approximately 14 meristems per seedling in the three cases. The delay in the production of meristems was measured at the mid-height of the plateau value, i.e., when the mean production of meristems was equal to 7. Given the fluctuations in the measurements, the delay in the production of the meristems is not appreciably different from the delay in the application of the calcium depletion treatment.

[b]Computed from experimental data in Verdus et al. (1997)

vulnerability in information processing that was less than 2 min for mechanical stimuli and over 5 min for other abiotic stimuli (Verdus et al. 2007). This was consistent with information about mechanical stimuli being stored particularly rapidly. This means that (1) the initial reaction of the flax seedlings to abiotic stimuli (including the transduction/storage of the information for meristem production) involves uptake of external calcium, elevating the cytosolic calcium, with a variability that depends on the type of stimulus, as was the case with other systems (see the "Introduction"), and (2) a temporary depletion of external calcium is needed for the actual production of meristems.

4 Gene Expression and Proteome Modifications

Wounding (e.g., by excision, abrasion or puncture) elicited the massive, rapid and enduring formation of polysomes in aged pea stems and other mature tissues (Davies and Schuster 1981). In response to a variety of abiotic stimuli (e.g., water spray, wind, touch, wounding) *Arabidopsis* plants increased the transcription of several touch-induced (tch), calmodulin-related genes (Braam and Davis 1990). When *Bidens* seedlings were subjected to noninjurious stimuli, calmodulin-related mRNA accumulated only in the stimulated region, whereas mRNA accumulation took place in both wounded and distant, unwounded tissues when the stimulus was injurious (Vian et al. 1996). Again with *Bidens* seedlings (see the experimental conditions in Sect. 2.2), it was shown that the accumulation of transcripts for tch and other (e.g., hsp) genes was involved in the final biochemical events causing the inhibition of hypocotyl growth rather than in the initial storage of the growth inhibition information (Henry Vian et al. 1995b).

When tomato plants were irradiated with 900 MHz electromagnetic radiation, the accumulation of the mRNA encoding the stress-related bZIP transcription factor occurred in a manner similar to that evoked by mechanical stimulation (Vian et al. 2006). Other stress-related transcripts (calmodulin, protease inhibitor and chloroplast mRNA-binding protein) increased four- to sixfold 15 min after the end of electromagnetic stimulation, dropped almost as far as the initial level by 30 min, and then increased again at 60 min (Roux et al. 2006). It is not known whether some sort of signal memorization was involved in this case.

When specific cytosolic calcium transients were generated in *A. thaliana* seedlings, a thorough analysis of transcriptome changes revealed the existence of 230 Ca-responsive genes, of which 162 were upregulated and 68 were downregulated (Kaplan et al 2006). Moreover, the upregulated genes were shown to possess a consensus sequence comprising two abscisic acid-specific *cis* elements. Some of these genes were associated with one or several types of stress (e.g., drought, salt stress, cold stress, touch). The function of a significant proportion of these genes was identified, but no investigation was carried out, as far as we know, to determine which of these genes (if any) were involved in the storage of stress-induced morphogenetic information.

With the induction of meristems in flax hypocotyls (see Sect. 3), gel electrophoresis has shown that some protein spots were slightly displaced (thus corresponding to posttranslational modifications), and that a few new spots appeared or preexisting spots disappeared (modification of transcriptional or translational activities) as a consequence of the application of the treatments that store or recall meristem production information (Tafforeau et al. 2002a, 2006). Some of these protein modifications were only associated with the application of an abiotic stimulus, while others took place only after a transient deprivation of calcium. Among the protein changes of the first group, some were common to several abiotic signals, while others were specific to each particular type of abiotic stimulus. Some of these protein modifications were transient, while others were long-lasting. Attempts to identify the proteins under consideration were not very successful. Identifying the responding genes (see the above paragraph) might aid the identification of these proteins in future work.

5 Potential of the SIMS Methodology as an Experimental Approach

In the SIMS (secondary ion mass spectrometry) method (Castaing and Slodzian 1962; Slodzian et al. 1990), the sample is bombarded with a beam of medium-weight or heavy ions (primary ions), and the ions that are sputtered from the sample surface (secondary ions) are collected, sorted by mass spectrometry and used for analytical and imaging purposes. For a description of the principle and performances of the SIMS method and a comparison with other methods of analytical imaging, see, e.g., Thellier et al. (1993, 2001) and Guerquin-Kern et al. (2005). Particular advantages of the SIMS technique are that it can detect all of the isotopes of almost all chemical elements, its lateral resolution is of the order of 0.3 μm with classical SIMS machines and even better (Hillion et al. 1997) with the nanoSIMS50 instrument (Fig. 1), its mass resolution is so good that almost all mass interference issues can be solved, and that its limit of detection is usually in the ppm and sometimes in the ppb range. When studying mobile substances (e.g., inorganic ions), sample preparation must be accomplished in a way that minimizes the mobilization of these substances. The adaptation by our group (Dérue et al. 2006) of a SIMS machine to the study of frozen-hydrated specimens is a promising way to overcome these difficulties.

The SIMS study of the calcium deprivation step related to epidermal meristem production induced in flax seedlings by abiotic stimuli (see Sect. 3) highlighted decreases in calcium, sodium and potassium and an increase in magnesium that did not substantially alter the overall ratio of divalent to monovalent cations in the seedling tissues (Tafforeau et al. 2002b). This is consistent with early observations that not only calcium but also other inorganic cations are implicated in plant sensitivity to abiotic stimuli and possible memorization of the corresponding information (Desbiez and Thellier 1975; Desbiez et al. 1987a, b,1991a, b; Bowles 1995).

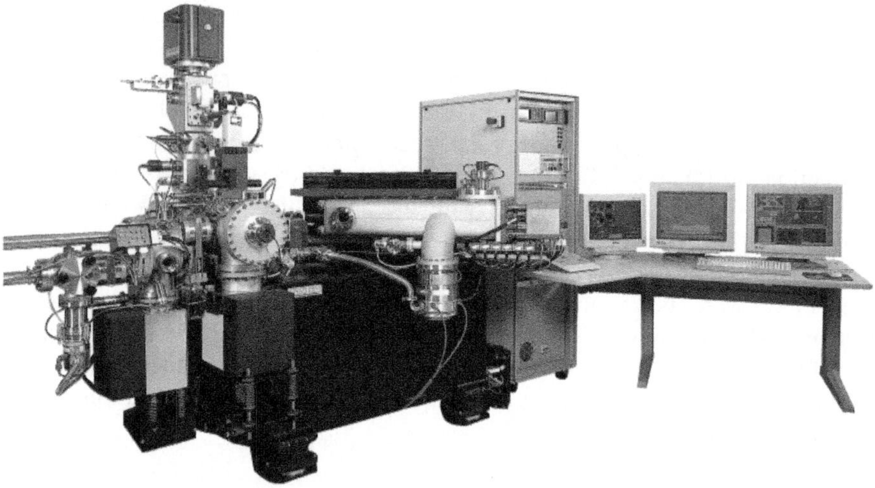

Fig. 1 The nanoSIMS50 instrument

The SIMS method has also been applied to the study of the involvement of inorganic ions in types of signaling and plant response other than those described in Sects. 2 and 3. For instance, in *Fagus sylvatica*, at the end of the period of winter quiescence, a strong temporary increase in calcium concentration was observed to take place in cambium and phloem, but not in xylem cells (Folley-Gueye et al. 1998); in flax plants grown under normal (i.e., not excessively saline) conditions, a dramatic increase in the Na/Ca concentration ratio was observed to strictly parallel the differentiation of the secondary walls of the fibers (Ripoll et al. 1993).

Taking advantage of the remarkable sensitivity of the SIMS methodology, it was possible to measure $^{31}P/^{12}C$ ratios in protein spots that were displaced in gel electrophoreses as a consequence of subjecting flax seedlings to abiotic stimuli (cf. fourth paragraph of Sect. 4 for more detail). In one of the protein spots corresponding to seedlings subjected to a manipulation stimulus only (no calcium depletion), it was observed that the $^{31}P/^{12}C$ ratio (1) was not significantly different from zero in the control seedlings, while (2) it increased up to approximately 4×10^{-4} and 6×10^{-4} in the fifth and tenth minutes after the manipulation stimulus, and was back to zero after the thirtieth min. This time course of the $^{31}P/^{12}C$ ratio was identical to that observed for the pI shift of this protein spot in the gel. This is a direct confirmation of the suggestion by previous authors (Bowles 1995; Bögre et al. 1996; Jonak et al. 1996; Takahashi et al. 1997) that the activation of kinases (e.g., those of the mitogen-activated protein kinase pathway) and the transient phosphorylation of a few proteins are among the early events that occur after abiotic stimulation.

6 A Possible Role of Ion Condensation in Signal Transduction

Biologists usually assume that (1) the cell calcium is either free or bound, and (2) the changes of cytosolic calcium that have been shown above to occur when a plant has received an abiotic stimulus are a consequence of an uptake of free calcium from the external medium or of a mobilization of calcium from internal, organelle sources (see the second paragraph of the "Introduction"). We have pointed out that this assumption neglects the fact that the cytosol, as well as other cell compartments, contains many electrically charged (usually negatively charged), filamentous structures, as reviewed by, e.g., Davies et al. (1996, 2001), and that such structures may be involved in a process of "counterion condensation" (Ripoll et al. 2004).

Briefly, consider a system in which charged filamentous structures, characterized by their mean linear charge density, β, are bathing in a saline medium. This medium thus contains co-ions and counterions, i.e., small, mobile ions with charges that are identical or opposite, respectively, to those of the charges on the filamentous structures. When the charge density of these structures exceeds a critical threshold value, β_c, the counter-ions "condense" onto the filamentous structures, thus decreasing β to $\beta = \beta_c$. This condensation process (Manning 1969; Oosawa 1971) is very different from a chemical reaction binding the counterions to the filamentous structures, in the sense that (1) the condensed counterions are delocalized and can diffuse along the filamentous structures, and (2) the process is an all-or-nothing process (resembling a phase transition) that does not obey the law of mass action. The counterions with the largest valence are the first to condense, i.e., Ca^{2+} ions will condense before Na^+ or K^+ ions on negatively charged filamentous structures.

Since the network of filamentous cellular structures has physicochemical characteristics that enable counterion condensation, we have proposed a model in which the feedback relationships between the condensation/decondensation of calcium and the activation of calcium-dependent kinases and phosphatases control the charge density of this filamentous network (Ripoll et al. 2004). In this model, the condensation process contributes to mediating the levels of free calcium, and calcium condensation/decondensation on the cellular filamentous network creates coherent patterns of protein phosphorylation that are well suited to signal integration. This raises the exciting possibility that the process of ion condensation can have an integrative role in signal transduction, and that the cell network of filamentous charges may serve as an integrative receptor.

7 Practical Applications

Some of the data summarized in the previous sections have yielded or may yield valuable practical applications in rather diverse domains.

Shaking or flexing has been used as means for reducing undesirable excessive stem elongation of greenhouse chrysanthemums without the use of chemicals

(Hammer et al. 1974). An automated, mechanical oscillatory shaking was developed, permitting mechanical height control of diverse greenhouse crops for a large number of plants treated simultaneously (Beyl and Mitchell 1977). A sufficiently strong water spray exerted an effect similar to that of shaking in reducing the growth of tomato plants; this made it possible to produce shorter, sturdier plants in the greenhouse, albeit with the disadvantage of a slight reduction in fresh fruit yield (Wheeler and Salisbury 1979).

There is increasing concern about possible adverse effects of the electromagnetic radiation emitted by mobile telephones on human health (Elwood 2003). The debate is no longer restricted to the scientific community and has become the subject of intense and emotional confrontations in the media and the general public. Since plants have been shown to be sensitive to such radiation (Tafforeau et al. 2002b, 2004; Vian et al. 2006; Roux et al. 2006) (see the second paragraph in Sect. 3), we propose that plants should be used as much as possible in order to perform a less emotionally charged study of the basic biochemical events involved in cell responses to electromagnetic radiation. It is also worth noting that no pathologic reaction has been observed so far in the plant responses to such radiation.

It has been observed that an increase in the Na/Ca concentration ratio paralleled the differentiation of the secondary walls of the fibers in flax seedlings. Moreover, sodium deprivation delayed and specifically altered the fiber secondary-wall differentiation, while, in the field, the flax plants with the best fiber quality were those exposed to salt water spray close to the seashore (Ripoll et al. 1993). It is thus worth studying whether flax could be adapted to be grown on excessively saline soils (as encountered for instance in northern Africa) that are deleterious to most other crops.

Thermal reflectance and fluorescence imaging can be used to detect stress-related changes in the pattern of light emission from plant leaves (Chaerle and Van der Straeten 2000). These techniques allow nondestructive, presymptomatic monitoring of changes in the plant state. Their utilization for crop monitoring would allow one to alleviate stress at an early stage, thus avoiding irreversible damage and consequently reducing yield losses.

We noted above that there are cases in which the information induced by an abiotic stimulus was stored within a plant but could only be recalled and take effect (by modifying the metabolism and organogenesis) after that plant had been subjected to a second, appropriate stimulus (see Sects 2.1, 2.2 and 3). This has been interpreted as being due to the existence of two functions: STO (ability to store morphogenetic information) and RCL (ability to recall the stored information and allow it to take effect), with the RCL function being turned "on" (or "off") by applying (or not) the appropriate second stimulus (Desbiez et al. 1984, 1986, 1987b). In plants that have been subjected to a potentially deleterious, stressing stimulus (e.g., temperature shock, drought) and that have stored the corresponding information, it would be possible to alleviate the adverse effect of that stress by finding a way to manipulate the RCL function such that it precludes the recall of the adverse stored information and thus prevents it from taking effect.

The elementary forms of memory that we have observed to exist in plants may provide an experimental system that is simpler that the animal brain that can be used to test the validity of the theoretical models for interpreting important processes such as memory storage and evocation (Demongeot et al. 2000a, b; Thellier et al. 2000).

8 What is the Purpose of Plant Memory?

In contrast to animals, plants have no means to search for nutrients or escape from predators by moving from a place to another. They can only optimize their conditions of life by adapting their growth and organogenesis to the main features of their environment. It is generally accepted that the ability of plants to perceive and transduce abiotic and biotic signals, and eventually produce appropriate biochemical, physiological and morphological responses, plays a key role in successful adaptation and acclimation; see, e.g., Smith (1990), Bowler and Chua (1994) and Knight (2000) for reviews. Since plants have to adapt to all of the stimuli (and their fluctuations) they receive over time, and not just to one at a time, it is not very surprising that they have developed an ability to register and store information about the different stimuli to which they are exposed for long or short periods of time, finally integrating this information in order to produce a coherent response at the right time (Knight et al. 1998; Thellier et al. 2000; Verdus et al. 2007).

The difficulty involved in understanding how each of the different aspects of memory described above (see Sects. 2 and 3) contributes to this integration of information remains, however. For example, what is the evolutionary advantage for flax seedlings (see Sect. 3) of memorizing information about abiotic signals, when this information will only take effect in the production of epidermal meristems if the seedlings are then subjected to transient calcium depletion (an event which is very unlikely to occur under natural conditions)? The answer may be that only a small part of the picture has been studied: (1) the ability of plants to memorize and integrate information received from the environment is based on a complicated set of interconnected functions, possibly involving nonlocalized processes (see third paragraph of Sect. 6), but (2) experimental access to this set is limited by the highly artificial conditions under which the plants are kept. This answer would at least be consistent with evidence that the signaling pathways leading to the responses to abiotic stress constitute a network that is interconnected at many levels (Knight and Knight 2001).

References

Beyl CA, Mitchell CA (1977) Automated mechanical stress. Application for height control of greenhouse chrysanthemum. HortSci 12:575–577

Bögre L, Ligterink W, Heberle-Bors E, Hirt H (1996) Mechanosensors in plants. Nature 383:489–490

Bourgeade P, Boyer N, De Jaegher G, Gaspar T (1989) Carry-over of thigmomorphogenesic characteristics in calli derived from *Bryonia dioica* internodes. Plant Cell Tissue Organ Cult. 19:199–211

Bowler C, Chua NH (1994) Emerging themes of plant signal transduction. Plant Cell 6:1529–1541
Bowles DJ (1995) Signal transduction in plants. Trends Cell Biol 5:404–408
Boyer N, Gaspar T, Lamand M (1979) Modification des isoperoxydases et de l'allongement des entre-noeuds de bryone à la suite d'irritations mécaniques. Z Pflanzenphysiol 93:459–470
Braam J, Davis RW (1990) Rain-, wind- and touch-induced expression of calmodulin and calmodulin-related genes in *Arabidopsis*. Cell 60:357–364
Braam J, Sistrunk ML, Polisensky DH, Xu W, Purugganan MM, Antosiewicz DM, Campbell P, Johnson K (1996) Life in a changing world: *TCH* gene regulation of expression and responses to environmental signals. Physiol Plant 98:909–916
Bush DS (1995) Calcium regulation in plant cells and its role in signalling. Annu Rev Plant Physiol Plant Mol Biol 46:95–122
Castaing R, Slodzian G (1962) Microanalyse par émission secondaire. J Microsc 1:395–414
Cessna SG, Chandra S, Low PS (1998) Hypo-osmotic shock of tobacco cells stimulates Ca^{2+} fluxes deriving first from external and then internal Ca^{2+} stores. J Biol Chem 42:27286–27291
Chaerle L, Van Der Straeten D (2000) Imaging techniques and the early detection of plant stress. Trends Plant Sci 5:495–501
Davies E (1987) Plant responses to wounding. In: Davies DD (ed) The biochemistry of plants, vol. 12. Academic, London, pp 243–264
Davies E (1993) Intercellular and intracellular signals and their transduction via the plasma membrane–cytoskeleton interface. Cell Biol 4:139–147
Davies E, Schuster A (1981) Intercellular communication in plants: evidence for a rapidly generated, bidirectionally transmitted wound signal. Proc Natl Acad Sci USA 78:2422–2426
Davies E, Fillingham BD, Abe S (1996) The plant cytoskeleton. In: Hesketh JE, Pryme IM (eds) The cytoskeleton: role in specialised tissues, vol 3. JAI, New York, pp 405–449
Davies E, Stankovic B, Azama K, Shibata K, Abe S (2001) Novel components of the plant cytoskeleton: a beginning to plant "cytomics." Plant Sci 160:185–196
De Jaegher G, Boyer N, Gaspar T (1985) Thigmomorphogenesis in *Bryonia dioica*, changes in soluble and wall peroxidases, phenylalanine ammonia lyase activity, cellulose, lignin content and monomeric constituents. Plant Growth Regul 3:133–148
Demongeot J, Kaufman M, Thomas R (2000a) Positive feedback circuits, regulation circuits and memory. C R Acad Sci Paris (Sciences de la Vie/Life Sciences) 323:69–79
Demongeot J, Thomas R, Thellier M (2000b) A mathematical model for storage and recall functions in plants. C R Acad Sci Paris (Sciences de la Vie/Life Sciences) 323:93–97
Demongeot J, Thellier M, Thomas R (2006) Storage and recall of environmental signals in a plant: modelling by use of a differential (continuous) formulation. C R Biol 329:971–978
Dérue C, Gibouin D, Demarty M, Verdus MC, Lefebvre F, Thellier M, Ripoll C (2006) Dynamic-SIMS imaging and quantification of inorganic ions in frozen-hydrated plant samples. Microsc Res Tech 69:53–63
Desbiez MO, Thellier M (1975) Lithium inhibition of the mechanically induced precedence between cotyledonary buds. Plant Sci Let 4:315–321
Desbiez MO, Champagnat P, Thellier M (1978) Décapitation, traumatismes et variations nycthémérales des corrélations entre cotylédons et bourgeons axillaires chez le *Bidens pilosus* L. C R Acad Sci Paris (série D) 286:1289–1292
Desbiez MO, Champagnat P, Boyer N, Frachisse JM, Gaspar T, Thellier M (1983) Inhibition corrélative de la croissance de l'hypocotyle de *Bidens pilosus* L. par des traumatismes cotylédonaires légers. Bull Soc Bot Fr [Actual Bot] 130:67–77
Desbiez MO, Kergosien Y, Champagnat P, Thellier M (1984) Memorization and delayed expression of regulatory messages in plants. Planta 160:392–399
Desbiez MO, Champagnat P, Thellier M (1986) Mécanisme de "mise en mémoire" et de "rappel de mémoire" de messages morphogènes chez *Bidens pilosus* L. C R Acad Sci Paris (série III) 302:573–578
Desbiez MO, Gaspar T, Crouzillat D, Frachisse JM, Thellier M (1987a) Effect of cotyledonary prickings on growth, ethylene metabolism and peroxidase activity in *Bidens pilosus*. Plant Physiol Biochem 25:137–143

Desbiez MO, Thellier M, Champagnat P (1987b) Storage and retrieval of morphogenetic messages in plantlets of *Bidens pilosus* L. In: Wagner E (ed) The cell surface in signal transduction, NATO ASI series, vol. H12. Springer, Berlin, pp 189–203

Desbiez MO, Boyer N, de Jaegher G, Frachisse JM, Ripoll C, Demarty M, Thellier M (1991a) Fast signalling involved in the morphogenetic responses of plants to mechanical stimuli, Rev Cytol Biol Veg 14:185–200

Desbiez MO, Ripoll C, Pariot C, Thellier M (1991b) Elicitation of developmental processes in higher plants by hexoses or myo-inositol, in the presence of K^+ or Ca^{2+}. Plant Physiol Biochem 29:457–462

Desbiez MO, Tort M, Thellier M (1991c) Control of a symmetry-breaking process in the course of the morphogenesis of plantlets of *Bidens pilosa* L. Planta 184:397–402

Desbiez MO, Boyer N, Thellier M (1992) Les messages de croissance chez les plantes. La Recherche 240:188–196

Desbiez MO, Mikulecky D, Thellier M (1994) Growth messages in plants: principle of a possible modeling and further experimental characteristics. J Biol Syst 2:127–136

Desbiez MO, Tort M, Monnier C, Thellier M (1998) Asymmetrical triggering of the cell cycle in opposite buds of a young plant, after a slight cotyledonary wound. C R Acad Sci Paris (Sciences de la Vie/Life Sciences) 321:403–407

Elwood JM (2003) Epidemiological studies of radio-frequency exposures and human cancer. Bioelectromagnetics 24(Suppl 6):S63–S73

Falkner R, Falkner G (2003) Distinct adaptability during phosphate uptake by the Cyanobacterium *Anabaena variabilis* reflects information processing about preceding phosphate supply. J Trace Microprobe Tech 21:363–375

Folley-Gueye ML, Verdus MC, Demarty M, Thellier M, Ripoll C (1998) Cambium pre-activation in beech correlates with a strong temporary increase of calcium in cambium and phloem but not in xylem cells. Cell Calcium 24:205–211

Gautheret RJ (1935) Recherches sur la culture des tissus végétaux. Rev Cytol Cytophysiol Veg 1:1–279

Guerquin-Kern JL, Wu TD, Quintana C, Croisy A (2005) Progress in analytical imaging of the cell by dynamic secondary ion mass spectrometry (SIMS microscopy). Biochim Biophys Acta 1724:228–238

Hammer PA, Mitchell CA, Weiler TC (1974) Height control in greenhouse chrysanthemum by mechanical stress. HortScience 9:474–475

Henry Vian C, Vian A, Dietrich A, Ledoigt G, Desbiez MO (1995a) Change in the polysomal mRNA population upon wound signal expression or storage in *Bidens pilosa*. Plant Plysiol 33:337–344

Henry Vian C, Vian A, Davies E, Ledoigt G, Desbiez MO (1995b) Wounding regulates polysomal incorporation of hsp70 and tch1 transcripts during signal storage and retrieval. Physiol Plant 95:387–392

Hillion F, Daigne B, Girard F, Slodzian G (1997) The Cameca nanoSIMS50: experimental results. In: Benninghoven A, Hagenhoff B, Werner HW (eds) Secondary ion mass spectrometry SIMS X. Wiley, Chichester, pp 979–982

Jaffé MJ, Forbes S (1993) Thigmomorphogenesis: the effect of mechanical perturbations on plants. Plant Growth Regul 12:313–324

Jonak C, Kiegerl S, Ligterink W, Barker PJ, Huskisson NS, Hirt H (1996) Stress signaling in plants: a mitogen-activated protein kinase pathway is activated by cold and drought. Proc Natl Acad Sci USA 93:11274–11279

Kaplan B, Davydov O, Knight H, Galon Y, Knight MR, Fluhr R, Fromm H (2006) Rapid transcriptome changes induced by cytosolic Ca^{2+} transients reveal ABRE-related sequences as Ca^{2+}-responsive *cis* elements in *Arabidopsis*. Plant Cell 18:2733–2748

Kergosien Y, Thellier M, Desbiez MO (1979) Préséances entre bourgeons axillaires chez *Bidens pilosus* L.: modélisation au niveau macroscopique en termes de catastrophes, ou au niveau microscopique en termes de "pompes et fuites" cellulaires. In: Delattre P, Thellier M (eds) Elaboration et justification des modèles: applications en biologie, part I. Maloine, Paris, pp 323–343

Klüsener B, Boheim G, Liss H, Engelberth J, Weiler EW (1995) Gadolinium-sensitive, voltage-dependent calcium release channels in the endoplasmic reticulum of a higher plant mechanoreceptor organ. EMBO J 14:2708–2714

Knight H (2000) Calcium signalling during abiotic stress in plants. Int Rev Cytol 195:269–324

Knight H, Knight MR (1995) Recombinant aequorin methods for intracellular calcium measurement in plants. Methods Cell Biol 49:201–216

Knight H, Knight MR (2001) Abiotic stress signalling pathways: specificity and cross-talk, Trends Plant Sci 6:262–267

Knight MR, Campbell AK, Smith SM, Trewavas AJ (1991) Transgenic plant aequorin reports the effect of touch and cold-shock and elicitors on cytoplasmic calcium. Nature 352:524–526

Knight MR, Smith SM, Trewavas AJ (1992) Wind-induced plant motion immediately increases cytosolic calcium. Proc Natl Acad Sci USA 89:4967–4971

Knight MR, Knight H, Watkins J (1995) Calcium and the generation of plant form. Phil Trans R Soc Lond B 350:83–86

Knight H, Brandt S, Knight MR (1998) A history of stress alters drought calcium signalling pathways in *Arabidopsis*. Plant J 16:681–687

Knight H, Zarka DG, Okamoto H, Thomashow MF, Knight MR (2004) Abscisic acid induces CBF gene transcription and subsequent induction of cold-regulated genes via the CRT protomer element. Plant Physiol 135:1710–1717

Malone M, Alarcon JJ (1995) Only xylem-born factors can account for systemic wound signalling in the tomato plant. Planta 196:740–746

Manning GS (1969) Limiting laws and counterion condensation in polyelectrolyte solutions. I. Colligative properties. J Chem Phys 51:924–933

McAinsh MR, Hetherington AM (1998) Encoding specificity in Ca^{2+} signalling systems. Trends Plant Sci 3:32–36

Ng CKY, McAinsh MR (2003) Encoding specificity in plant calcium signalling: hot-spotting the ups and downs and waves. Ann Bot 92:477–485

Oosawa F (1971) Polyelectrolytes. Marcel Dekker, New York

Plaetzer K, Thomas SR, Falkner R, Falkner G (2005) The microbial experience of environmental phosphate fluctuations: an essay on the possibility of putting intentions into cell biochemistry. J Theoret Biol 235:540–554

Plieth C (2001) Plant calcium signaling and monitoring: pro and cons and recent experimental approaches. Protoplasma 218:1–23

Plieth C (2005) Calcium: just another regulator in the machinery of life. Ann Bot 96:1–8

Plieth C, Trewavas AJ (2002) Reorientation of seedlings in the Earth's gravitational field induces cytosolic calcium transients. Plant Physiol 129:786–796

Plieth C, Hansen UP, Knight H, Knight MR (1999) Temperature sensing by plants: the primary characteristics of signal perception and calcium response. Plant J 18:491–497

Polisensky DH, Braam J (1996) Cold-shock regulation of the *Arabidopsis* TCH genes and the effect of modulating intracellular calcium levels. Plant Physiol 111:1271–1279

Ripoll C, Pariot C, Jauneau A, Verdus MC, Catesson AM, Morvan C, Demarty M, Thellier M (1993) Involvement of sodium in a process of cell differentiation in plants. C R Acad Sci Paris (Sciences de la Vie/Life Sciences) 316:1433–1437

Ripoll C, Norris V, Thellier M (2004) Ion condensation and signal transduction. BioEssays 26:549–557

Roux D, Vian A, Girard S, Bonnet P, Paladian F, Davies E, Ledoigt G (2006) Electromagnetic fields (900 MHz) evoke consistent molecular responses in tomato plants. Physiol Plant 128:283–288

Sanders D, Pelloux J, Brownlee C, Harper JF (2002) Calcium at the crossroads of signalling. Plant Cell 14:S401–S417

Slodzian G, Daigne B, Girard F, Boust F, Hillion F (1990) Cartographie parallèle de plusieurs éléments ou isotopes par balayage avec une sonde ionique submicronique: premiers résultats. C R Acad Sci Paris (série II) 311:57–64

Smith H (1990) Signal perception, differential expression within multigene families and the molecular basis of phenotypic plasticity. Plant Cell Environ 13:585–594

Tafforeau M (2002) Etude des phases précoces de la transduction des signaux environnementaux chez le lin : une approche protéomique. Doctorate Thesis of the University of Rouen, France, pp. 1–207

Tafforeau M, Verdus MC, Charlionet R, Cabin-Flaman A, Ripoll C (2002a) Two-dimensional electrophoresis investigation of short term response of flax seedlings to cold shock. Electrophoresis 23:2534–2540

Tafforeau M, Verdus MC, Norris V, White G, Demarty M, Thellier M, Ripoll C (2002b) SIMS study of the calcium-deprivation step related to epidermal meristem production induced in flax by cold shock or radiation from a GSM telephone. J Trace Microprobe Tech 20:611–623

Tafforeau M, Verdus MC, Norris V, White GJ, Cole M, Demarty M, Thellier M, Ripoll C (2004) Plant sensitivity to low intensity 105 GHz electromagnetic radiation. Bioelectromagnetics 25:403–407

Tafforeau M, Verdus MC, Norris V, Ripoll C, Thellier M (2006) Memory processes in the response of plants to environmental signals. Plant Signal Behav 1:9–14

Takahashi K, Isobe M, Knight MR, Trewavas AJ, Muto S (1997) Hypoosmotic shock induces increases in cytosolic Ca^{2+} in tobacco suspension-culture cells. Plant Physiol 113:587–594

Thellier M, Desbiez MO, Kergosien Y, Champagnat P (1981) Mise en mémoire de signaux morphogènes chez *Bidens pilosus* L. C R Acad Sci Paris (série III) 292:1187–1190

Thellier M, Desbiez MO, Champagnat P, Kergosien Y (1982) Do memory processes also occur in plants? Physiol Plant 56:281–284

Thellier M, Ripoll C, Quintana C, Sommer F, Chevallier P, Dainty J (1993) Physical methods to locate metal atoms in biological systems. Methods Enzymol 227:535–586

Thellier M, Le Sceller L, Norris V, Verdus MC, Ripoll C (2000) Long-distance transport, storage and recall of morphogenetic information in plants: the existence of a primitive plant "memory". C R Acad Sci Paris (Sciences de la Vie/Life Sciences) 323:81–91

Thellier M, Dérue C, Tafforeau M, Le Sceller L, Verdus MC, Massiot P, Ripoll C (2001) Physical methods for in vitro analytical imaging in the microscopic range in biology, using radioactive or stable isotopes. J Trace Microprobe Tech 19:143–162

Thellier M, Demongeot J, Norris V, Guespin J, Ripoll C, Thomas R (2004) A logical (discrete) formulation for the storage and recall of environmental signals in plants. Plant Biol 6:590–597

Verdus MC, Cabin-Flaman A, Ripoll C, Thellier M (1996) Calcium dependent storage/retrieval of environmental signals in plant development. C R Acad Sci Paris (Sciences de la Vie/Life Sciences) 319:779–782

Verdus MC, Thellier M, Ripoll C (1997) Storage of environmental signals in flax: their morphogenetic effect as enabled by a transient depletion of calcium. Plant J 12:1399–1410

Verdus MC, Le Sceller L, Norris V, Thellier M, Ripoll C (2007) Pharmacological evidence for calcium involvement in the long-term processing of abiotic stimuli in plants. Plant Signal Behav 2:212–220

Vian A, Henry-Vian C, Schantz R, Ledoigt G, Frachisse JM, Desbiez MO, Julien JL (1996) Is membrane potential involved in calmodulin gene expression after external stimulation in plants? FEBS Lett 380:93–96

Vian A, Roux D, Girard S, Bonnet P, Paladian F, Davies E, Ledoigt G (2006) Microwave irradiation affects gene expression in plants. Plant Signal Behav 1:67–70

Volkov AG, Carrell H, Adesina T, Markin VS, Jovanov VS (2008) Plant electrical memory. Plant Signal Behav 3:490-492

Wheeler RM, Salisbury FB (1979) Water spray as a convenient means of imparting mechanical stimulation to plants. HortScience 14:270–271

Plants and Animals: Convergent Evolution in Action?

František Baluška and Stefano Mancuso

Abstract The Aristotelian–Linnean heritage of our current sciences is tightly associated with a view of automata-like passive plants lacking active sensory-driven lifestyles. Charles Darwin made the first attempt to escape from this "Aristotelian trap." Although his work on plants stimulated lively research into plant tropisms and hormones, his unconventional view of plants was largely ignored by the mainstream of plant sciences until recently. Darwin witnessed early studies on electrical signaling in plants, including plant action potentials. Nevertheless, this important and well-developed field of plant sciences was almost wiped out following the publication of the controversial book The Secret Life of Plants in the 1970s. The resulting "esoteric stigma" hindered the further development of this branch of plant sciences. Recently, advances in cell and molecular biology as well as in ecology led to the birth of plant neurobiology, which aims to study plants in their full sensory and communicative complexity. New concepts are needed and new questions must be asked in order to advance our still rudimentary understanding of plants.

> It is hardly an exaggeration to say that the tip of the radicle thus endowed [with sensitivity] and having the power of directing the movements of the adjoining parts, acts like the brain of one of the lower animals; the brain being seated within the anterior end of the body, receiving impressions from the sense-organs, and directing the several movememnts.
> —Charles Darwin in *The Power of Movement in Plants* (1880; John Murray, London, p 573)

1 Introduction

One particularly puzzling problem in biology is the sensory and communicative complexity of higher plants (Baluška et al. 2006a), which is linked to the high sensitivity and inherent excitability of plant cells (Wayne 1994). This branch of plant biology sciences

F. Baluška (✉)
Rheinische Friedrich-Wilhelms-Universität Bonn, Bonn, Germany
e-mail: baluska@uni-bonn.de

S. Mancuso
University of Florence, Florence, Italy

F. Baluška (ed.), *Plant-Environment Interactions*, Signaling and Communication in Plants, 285
DOI: 10.1007/978-3-540-89230-4_15, © Springer-Verlag Berlin Heidelberg 2009

is not new, as it can be traced back to Charles Darwin and his three books devoted to plant life (Darwin 1875a, b, 1880). Nevertheless, the Aristotelian–Linnean approach to reasoning and thinking, which dominates contemporary scientific concepts, is based on an almost dogmatic idea that sessile plants are passive organisms that do not need rapid electrical signals, and so these, when recorded, should be considered for some kind of biological oddity that do not require further detailed studies (Baluška and Mancuso 2006). Moreover, these features and aspects of higher plants are associated with the negative stigma of *sentient plants* and their links to parapsychology, paranormal properties and even esoteric ideas (Backster 1968; Tompkins and Bird 1973). As a result, attempts to improve our understanding of plants have been limited to hunting for more proteins and genes. The ultimate goal of current research is to improve the agricultural properties and stress adaptability of plants. However, we will not achieve this goal until we unravel the communicative, sensory and cognitive aspects of these organisms (Trewavas 2003, 2005a, b, 2007; Calvo Garzon 2007; Barlow 2008).

Recent advances in chemical and sensory ecology, including the discovery of plant communication via volatile and allelochemical chemical messengers, have placed all of these so-called oddities into a new perspective (Bais et al. 2003; Dicke et al. 2003; Baldwin et al. 2006). Moreover, advances in plant cell biology have revealed that plant endocytosis, despite negative mechanistic predictions based on high turgor pressure of plant cells, is a cellular process that affects almost all biological processes (Šamaj et al. 2005, 2006). These breakthrough discoveries will force a reassessment of almost all metabolic, physiological, and sensory processes from the perspective of endocytosis, endosomal compartments, and endocytic vesicle recycling. In addition, it is becoming clear that cell–cell communication in plants is (similar to the situation in animal cells) based on synaptic cell–cell adhesion domains enriched with the actin cytoskeleton and specialized for the endocytic vesicle recycling-based exchange of signaling molecules among communicating cells (Baluška et al. 2005a, b; Kwon et al. 2008a, b).

In this chapter, we will analyze the historical background and the newest results in this area of research, as well as their relevance to not only the plant sciences in particular but to our understanding of communicative organisms in general. Finally, we will place the emerging field of plant neurobiology into a broader philosophical context of adaptive and convergent biological evolution, focusing on the position of humans in living systems. Our point is that humans still are, and will continue to be, fully dependent on plants since they, together with unicellar photosynthetic organisms, are the only primary source of oxygen and organic matter on this planet. Our future is dependent on a full understanding of higher plants in their whole complexity. If we can gain an understanding of how plant communicate using their volatile-based "language" (Dicke et al. 2003; Baldwin et al. 2006), then we should be able to access the huge amount of knowledge these fascinating organisms have gathered during their evolution.

2 Historical Excursion: Charles Darwin Versus Julius Sachs

You may be quite surprised to learn that there are several notes of ancient Greek philosophers that take a rather sensitive and active view of plants; in particular, Empedocles (495–435 BC) was very much in favor of plants. However, our current sciences are based mostly on Aristoteles (384–322 BC), and he was less favorable to plants, claiming that they are fully passive creatures, insensitive, and incapable of any movements (Radl 1909; Ingensiep 2001). Aristotelian philosophy led to the passive view of plants, which, to a greater or lesser degree, persists even today. The sixteenth century saw the first botanists to question this Aristotelian view of plants (Webster 1966). The most interesting figure in this respect is Erasmus Darwin (Darwin 1899; King-Hele 1974), who was the grandfather of Charles Darwin and encouraged him to take an interest to plants. Despite being rather amateur, especially in comparison to German botanists and plant physiologists of that era (Darwin 1899; Heslop-Harrison 1980), Charles Darwin devoted much time, effort, and energy to plant research. It was possibly his fresh perspective on plants, which was not biased by any preconceived views (Darwin 1899), that allowed him to propose his novel—indeed neurobiological—views about plants, eliciting fiercely antagonistic responses, especially from Julius Sachs (Heslop-Harrison 1980). Charles Darwin was not afraid to make controversial statements if these were backed up by his very careful observations and experiments. For example, his last book, *The Power of Movement in Plants* (Darwin 1880), closes with a statement (see the quotation in the "Abstract") that Darwin clearly believed to be important enough to place at the very end of his book: that the root apex has similar features to the brains of lower animals (Baluška et al. 2006b; Barlow 2006; Trewavas 2007). This book is highly cited in the contemporary plant biology literature, but only due to its discussion of coleoptiles and light; the part of the book that deals with roots has largely been ignored (see, e.g., Whippo and Hangarter 2006). We will come back to this issue later when discussing the complex nature of the polar cell–cell transport of auxin, which mediates sensory perceptions and underlies the adaptive and coordinated motor behavior of plant organs (Esmon et al. 2005; Whippo and Hangarter 2006). It is quite obvious that all three of these books by Charles Darwin represent strong founding pillars for the emerging field of plant neurobiology (Brenner et al. 2006). The negative responses from several leading plant scientists of that time, especially those from Julius Sachs (Heslop-Harrison 1980), resemble the current opposition in some areas of plant science (Alpi et al. 2007) to this more complex neurobiological view of higher plants (Baluška and Mancuso 2006; Baluška et al. 2006a, b; Trewavas 2003, 2005a, b, 2007; Brenner et al. 2006, 2007; Calvo Garzon 2007; Barlow 2006, 2008).

Charles Darwin was aware that his book *The Power of Movement in Plants* (Darwin 1880), and especially his conclusion about the root apex acting similarly to the brains of lower animals, would elicit strongly negative responses and opposition, especially in Germany (Heslop-Harrison 1980). The root apex was central to his ideas about plants as, in some plants such as maize, the root cap can be removed completely without interfering too much with root growth per se. This important finding was published by Theophil Ciesielski in 1871, but Julius Sachs failed to repeat these experiments two years later (Heslop-Harrison 1980), as he removed part of the root apex meristem (Darwin 1880; Heslop-Harrison 1980). On the other hand, Darwin confirmed Ciesielski's findings (Darwin 1880), and this work, as well as subsequent experiments, apparently yielded useful data that enabled important conclusions to be drawn regarding the brain-like properties of the root apex (Darwin 1880; Barlow 2006, 2008; Baluška et al. 2006b; Trewavas 2007). However, the last sentence from Darwin's book, cited in full above, contains another important message: that the root apex represents the anterior pole of the plant body (Darwin 1880; Baluška et al. 2006b). This correlates plant body organization (Fig. 1) with that of the animal body, which has sensory organs and brain tissue, or brain-like tissue, at the anterior pole of the body, while organs and openings for excretion as well as sexual organs occur at the posterior pole of the body (Baluška et al. 2006b).

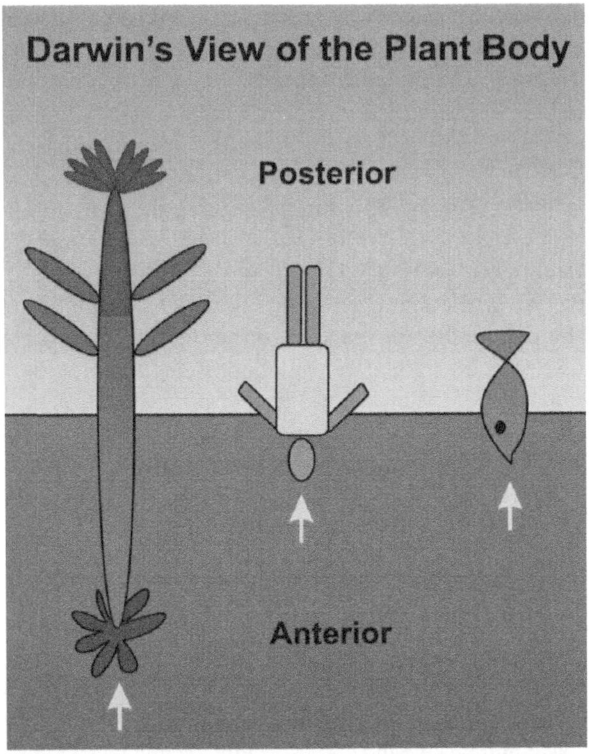

Fig. 1 Highly schematic view of plants from the Darwin's perspective. Root apices are specialized for anterior-like activities, including nutrient uptake and integration of sensory information into adaptive behavior. Shoots and their apices are specialized for posterior-like activities, including metabolic support for the whole organism (photosynthesis), exclusion of byproducts of this process via openings (carbon dioxide is released via stomata, secretory glands), and sexual reproduction (flowers). Darwin's *neurobiological* view of the plant body harmonizes plants with animals and removes the 'schizophrenic' apical–basal dichotomy (Baluška et al. 2005b; Friml et al. 2006) that is currently present in plant sciences

3 Sensory Biology in Plants and Animals: Bioelectricity Underlies Sensorimotor Circuits

The survival of an organism depends on its ability to respond to the environment through its senses when neuronal systems translate sensory information into electrical impulses—so-called neural code (DeWeese and Zador 2006). This enables multisensory integration that in turn leads to adaptive motor responses and behavior (Stein 1998; Benda et al. 2007). In order to survive, all biological systems continuously retrieve information from their environment and use this information in order to effectively adapt to it (Kovac 2007). In animals, neurons and associated neurobiological systems translate sensory information obtained from the environment into bioelectrical impulses, which are then transformed into biological signals that induce motor responses (DeWeese and Zador 2006). Similarly, in plants, numerous physical environmental factors, especially light and gravity, are continuously monitored (Baluška et al. 2006a, b; Wayne 1994; Esmon et al. 2005; Whippo and Hangarter 2006; Brenner et al. 2006). Specialized plant cells have been optimized by evolution to translate sensory information obtained from the physical environment into motor responses known as tropisms (Esmon et al. 2005; Whippo and Hangarter 2006), with root gravitropism representing one of the most intensively studied plant organ tropisms (Baluška et al. 2007, 2008a). Electrical signals are induced by all known physical factors in plants, suggesting that electricity mediates physical–biological communication in plants too (Baluška et al. 2006a, b; Wayne 1994; Volkov 2006; Fromm and Lautner 2007; Felle and Zimmermann 2007). Root gravitropism is a particularly instructive example in this respect. Sensory perception of gravity is accomplished at the very tip of the root apex, at the root cap (Barlow 1993), which includes a vestibular-like gravisensing organ composed of statocytes. On the other hand, the motor responses, which begin almost immediately after root cap sensory events, are accomplished in remote growth zones of the root apex (Baluška et al. 2007, 2009). Therefore, root gravitropism represents a nice example of a neuronal sensorimotor circuit in plants. Importantly, there are several other examples in plants, such as the leaf traps of carnivorous plants (Volkov et al. 2007a, b, 2008), or the mechanosensitive leaves of *Mimosa* (Braam 2005).

4 Plant Action Potentials, Synapses, Neurons, Neuronal Molecules, and Transmitters

The occurrence of action potentials (APs) in plants was established more than 150 years ago (Shepherd 2005; Stahlberg 2006; Volkov 2006; Wayne 1994). Although it is well known that these rapid electric signals modulate and control metabolic processes such as respiration and photosynthesis (Wayne 1994; Volkov 2000, 2006; Davies 2004; Fromm and Lautner 2007; Felle and Zimmermann 2007), the dominant view in plant biology at the moment is that plants do not need rapid electrical signals, and so these potentials are considered to be an evolutionary oddity. Claims that plants do not have classical neurons and synapses have been raised (Alpi et al. 2007). This is undoubtedly true—plant cells are different to animal cells; their cell walls and plasmodesmata in particular pose challenges to neuronal cell–cell communication. However, it is clear that plants do initiate APs, which propagate rapidly through the whole plant body. These trains of electrical activity are energetically extremely costly (Lennie 2003; Crotty et al. 2006). As evolution should "weed out" all of the nonprofitable activities, it is not surprising that plant APs regulate growth as well as physiological and metabolic processes (Wayne 1994; Volkov 2000, 2006; Volkov et al. 2007a, b, 2008; Davies 2004; Fromm and Lautner 2007; Felle and Zimmermann 2007). It should be noted that plant cells are organized into regular cell files which converge at "organizing centers" at organ apices. The arrangement of root apex cells, which shows a good correspondence with higher neuronal activities, is particularly regular (Baluška et al. 2006b).

The ultimate question is: how can walled cells support the propagation of action potentials for several meters if they lack synapses? One possibility is that the only tissue suitable for this task is the acellular phloem, which forms axon-like supracellular channels (Volkov 2000; Barlow 2008) spanning the whole plant body (Baluška et al. 2006b). However, plants support a variety of AP types, from slow ones to ultrafast ones resembling animal APs (Wayne 1994; Volkov 2000, 2006; Volkov et al. 2007a, b, 2008; Davies 2004; Fromm and Lautner 2007; Felle and Zimmermann 2007). Our immunolabelings, performed over ten years ago, revealed dense F-actin meshworks at the nongrowing end-poles in root apices (Baluška et al. 1997). More recent work has revealed that these F-actin-enriched end-poles represent domains that are specialized for active endocytosis and endocytic vesicle trafficking of not only plasma membrane proteins but also cell wall pectins (Baluška et al. 2002, 2003; Dhonukshe et al. 2006). Moreover, these end-poles of root apex cells accomplish the secretion of auxin via endocytic vesicle recycling (Baluška et al. 2005a, b, 2009; Schlicht et al. 2006; Mancuso et al. 2007). Although auxin is mainly considered a plant hormone by the mainstream plant sciences, there are several inconsistencies in such a simple interpretation. Besides having properties that are very close to those of morphogens (Bhalerao and Bennett 2003), auxin also resembles a transmitter, as it is secreted from root cells and induces electrical responses after "landing" on the outside surfaces of adjacent cells of the plasma membrane (Barbier-Brygoo et al. 1991; Felle et al. 1991; Steffens et al. 2001). Unfortunately, these electrical responses of plant cells to extracellular, apparently secreted, auxin have not been studied extensively yet. Nevertheless, it is known that polar auxin transport is both sensitive to and contributes to electric fields generated around the growing plant cells and root apices. In the transition zone, the electrical current inherent to the growing root apices is at its highest value, and it is also sensitive to inhibitors of polar auxin transport (Collings et al. 1992). In this context, it is interesting that the same root apex zone shows a peak in the synaptic activity driving the polar auxin transport (Baluška et al. 2002; Mancuso et al. 2005, 2007). The synaptic activity underlying the auxin secretion is extremely sensitive to gravitational stimulation, and we have proposed that these auxin-secreting plant synapses act as mechanosensitive gravisensing domains (Baluška et al. 2005a, b, 2007, 2008b, 2009).

Besides auxin-secreting neuron-like plant synapses, we have also predicted the existence of "plant immunological synapses" (Baluška et al. 2005a, b), which would coordinate communication between plant cells and pathogens as well as symbionts. In accordance with our prediction, plant immunological synapses have recently been identified (Kwon et al. 2008a, b). Importantly, plant cells are inherently excitable and sensory (Wayne 1994; Brenner et al. 2006), and are equipped with very robust molecular systems capable of supporting neuron-like complexity in their complex vesicle trafficking pathways (Baluška et al. 2006b). As discussed above, plant cells are closely apposed in plant tissues, especially at root apices, so they do not need to extend long axons to find their communicative cell partners (Baluška et al. 2006b). Therefore, most plant cells fulfill criteria originally reserved for neurons. In addition, recent advances in cell biology have revealed that most of the features that were thought to be unique to neurons, and which form the basis of the "neuron doctrine," are in fact characteristics of other cell types too, meaning that this doctrine is now not only valid for neurons (Guillery 2005).

Another issue that is relevant to our currently changing views of plants is the status of plant hormones. Auxin (discussed above) is the prime example, but new aspects of other so-called plant hormones such as abscisic acid, ethylene and cytokinins are also accumulating. Many of the recent findings suggest that these signaling molecules tend to act in a transmitter mode; they are released from cells and act on neighboring cells. Therefore, plant hormones rather resemble neurohormones, which are produced by neurons and affect other neurons. Surprisingly, plant hormones are also used in interorganism communication, and many of them are even produced by bacteria and fungi. Certain plant hormones are also biologically active in lower and higher animals. For example, abscisic acid acts as a cytokine in brains (Bruzzone et al. 2007), as well as a signaling molecule in sponges (Puce et al. 2004), and cytokinins induce sporulation in *Dictyostelium* (Anjard and Loomis 2008). Thus, plant hormones represent multipurpose signaling molecules that are active in not only plants but in almost all organisms. Not surprisingly, some of them are also very promising agents in cancer therapy (Rotem et al. 2005; Kim et al. 2006).

Plants synthesize large amounts of neuroactive substances and also express almost all known neurotransmitters (Barlow 2008; Baluška et al. 2003, 2006b). There are indications that several of these classical neuronal molecules are indeed used for neuron-like cell–cell communication in plants (Baluška et al. 2007, 2009). In addition, plants synthesize pain-modulating and pain-relieving substances. Interestingly, ethylene, which is a wound-induced plant hormone, also acts as a general anesthetic (Campagna et al. 2003), and roots are sensitive to anesthetics (Grant et al. 1974). In fact, anesthetics and ethylene induce similar responses in roots (Powell et al. 1973). Moreover, several other neuronal transmitters have been found to act in a similar neuronal mode in plants too, including glutamate, glycine, acetylcholine, ATP, GABA, histamine, dopamine, melatonin, and serotonin (Baluška et al. 2006a, 2007, 2009; Brenner et al. 2006; Sagane et al. 2005; Stephens et al. 2008). Moreover, surprisingly, GABA is (similar to auxin) also active in interorganism communication (Shelp et al. 2006, 2008).

5 Sensitive and Communicative Plants: Lessons from Root Apices

Plants continuously monitor numerous parameters from the environment, and the sensory information obtained is integrated into adaptive responses and complex plant behavior (Baluška and Mancuso 2006; Trewavas 2003, 2005a, b, 2007; Barlow 2008; Bais et al. 2003; Dicke et al. 2003; Baldwin et al. 2006; Brenner et al. 2006; Braam 2005). Plants are able to store information and use it later to achieve adaptive behavior (Trewavas 2003, 2005a, b, 2007). These plant memories allow them to make fairly good predictions about their future circumstances (Trewavas 2005a, b; Tafforeau et al. 2006; Bruce et al. 2007; Ruuhola et al. 2007; Ripoll et al. 2009). Similar to animal memories, these plant memories are based on electricity, chemicals and calcium (Verdus et al. 2007; Volkov et al. 2008).

Plants are known to use touch genes to mechanically sense their environment (Trewavas and Knight 1997). A new touch-sensing gene from *Arabidopsis* roots expressing a plasma membrane resident protein that regulates calcium levels was reported recently (Nakagawa et al. 2007). Moreover, we report the sound-induced genes rbcS and Ald, the most recent additions to the plant sensory repertoire, here. Expression of these genes is specifically induced by frequencies of 125 and 250 Hz (Jeong et al. 2007). Similar to the case of magnetoperception (see below), these two genes are also sensitive to light, indicating that the association of light with the sensing of other physical parameters by plants may be generally importance. Even high-frequency low-amplitude electromagnetic fields induced by wireless devices induce rapid changes in plant cells (Vian et al. 2007). On top of this, plants are much more sensitive to low-amplitude electromagnetic fields than animals (Beaubois et al. 2007; Roux et al. 2008).

The root apex has been known for many years to act as the principal sensory organ of plants (Barlow 1993, 2008; Baluška et al. 2006b). Besides gravisensing, which is accomplished in "vestibular-like" root cap statocytes (Baluška et al. 2007, 2009; Barlow 1993) there are a number of other parameters that are sensed by root apices. They sense and integrate multiple signals for thigmotropism, hydrotropism, phototropism, rheotropism, electrotropism, and magnetotropism. Of course, this list is still not complete and will grow further. Recently, two breakthrough findings further advanced this view of plants as sensory organisms that are able to perceive even more physical parameters associated with the environment and to translate them into biologically relevant information. Firstly, a gene responsible for plant root hydrotropism was discovered, and its protein product was localized to gravisensing root cap statocytes (Kobayashi et al. 2007). Secondly, low-phosphate sensing, which is very important for the architecture and physiology of the whole plant, was also localized to the root apex cells (Svistoonoff et al. 2007). In addition, plants sense magnetic fields (Galland and Pazur 2005), and recent studies revealed that cryptochromes are behind this magnetic sensing by plants (Ahmad et al. 2007; Solov'yov et al. 2007), similar to those discovered in birds recently (Johnsen et al. 2007). These findings suggest that there are intriguing parallels between the light-sensing-related magnetoreception of animals and plants, which both appear to be based on some common physical properties of photoexcited cryptochromes. So this last "holy grail" of sensory biology (Johnsen et al. 2007) may reveal further surprising parallels between plants and animals.

Roots show very active behavior, as they continuously search for water and nutrition. However, roots can do much more: they can avoid dangerous soil patches even before they encounter them, they are capable of recognizing kin as well as discriminating self from nonself roots (Gruntman and Novoplansky 2004). For example, roots can recognize roots from the same plant or species and can be aggressive towards roots from other species (Bais et al. 2003; Rudrappa et al. 2007; Rudrappa and Bais 2008). Root apices of invasive plants can, in fact, kill other roots (even the whole plants) by exuding toxic allelochemicals (Rudrappa et al. 2007; Rudrappa and Bais 2008). Alternatively, root apices of parasitic plants can recognize prey root apices at distance, perform active chemotropism to reach them, and then generate haustoria (root hair-like protrusions), which then invade the prey roots to steal the nutritive substances from their vascular tissues (Keyes et al. 2001).

Recently, dramatic salinity-induced modifications of the root growth direction were been observed (Sun et al. 2007), which represent a new salt-avoidance tropism of root apices (Li and Zhang 2008). This salinity-avoidance behavior of growing root apices represents an active adaptive mechanism for plants grown under saline conditions. Importantly, root apices turn away from a dangerous salt-stress-inducing medium (Sun et al. 2007; Li and Zhang 2008). These authors demonstrated the disintegration of gravity-sensing statoliths and the degradation of PIN2 auxin efflux transporter, allowing them to perform negative gravitropism as part of the salinity-avoidance tropism (Sun et al. 2007; Li and Zhang 2008). It was also shown that growing root apices can recognize dangerous substrate (soil) patches with high aluminum levels in advance, and can then avoid them using a similar active avoidance root tropism (Hawes et al. 2000). Furthermore, high salinity, which is experienced primarily at the roots, stimulates the alteration of the gravitropic growth of shoots (Sun et al. 2007; Li and Zhang 2008).

As vascular cells are the most active neuronal cells of plants (Volkov 2000, 2006; Baluška et al. 2006a; Barlow 2008), it is not surprising that aboveground organs also often show very prominent animal-like behavior. For example, parasitic plants of the genus *Cuscuta* are unable to make their own roots and rely fully on their host plants. *Cuscuta* plants show many intelligent features as aspects of their parasitic lifestyle (Trewavas 2002, 2003, 2005a, b, 2007). Their sprouts move in a circular fashion when searching for a suitable host, which is identified using host-emitted volatiles. When given a choice between volatiles released by the preferred host (tomato) and the nonhost (wheat), the parasite grows toward the former (Runyon et al. 2006).

All of these reports document that plants have a fine and sophisticated neuronal system which encompasses both sensory and communicative aspects that enable them to efficiently cope with rapidly changing environments (Darwin 1875a,b, 1880; Baluška et al. 2006a; Barlow 2008).

6 Plant Intelligence: Oddity or Convergent Evolution?

Tony Trewavas introduced the concept of plant intelligence (Trewavas 2003, 2005a, b, 2007), which, although controversial, is very attractive and explains many mysteries and complexities in plant sensory biology as well as the animal-like behavior of plant roots. As we mentioned above, roots are able to perform navigation driven by the search for nutrition, and even avoidance tropisms (Hawes et al. 2000; Sun et al. 2007; Li and Zhang 2008), indirectly implicating some negative perceptive experiences which roots try to avoid. In fact, sensory-driven plant behavior fulfills criteria associated with cognition (Calvo Garzon 2007). This is perhaps not quite so surprising if we recall that even bacteria are capable of both cognition and communication (Stock and Levit 2000; Jenal et al. 2005; Baker and Stock 2007; Vermeij 2006). As multicellularity evolved independently in plants and animals (Meyerowitz 2000), it is tempting to hypothesize that synaptic cell–cell communication as well as neuronal computation and integration of sensory perceptions evolved via convergent evolution (Conway Morris 2003, 2006; Vermeij 2006). Besides plant neurobiology and intelligence, there are other examples of convergent evolution between plants and animals (Table 1). Thus, plant intelligence (Trewavas 2003, 2005a, b, 2007) and cognition (Calvo Garzon 2007), as well as the whole emerging field of plant neurobiology, may prove to be important phenomena as we attempt to improve our understanding of the elusive processes that "channel" biological evolution towards similar states despite different outcome situations (Conway Morris 2003,

Table 1 Some major examples of the convergent evolution between plants and animals

Convergent evolution: animals and plants
1. Action potentials
2. Multicellularity (homeobox transcription factors, retinoblastoma protein)
3. Sexual reproduction
4. Embryo
5. Placenta versus endosperm
6. Vivipary
7. Anterior–posterior plant body organization
8. Innate immunity
9. Self and nonself
10. Programmed cell death
11. Circadian clock
12. Dormancy and sleep
13. Stem cells
14. Genomic imprinting
15. Hormones, morphogens, and transmitters
16. Cell–cell channels
17. Cell walls and cuticle
18. Synapses and synaptotagmins
19. Intelligence–adaptive behavior
20. Plant–plant communication and plant sociality

2008; Vermeij 2006). It may well be that for complex multicellular organisms, neurobiological integration of sensory information and communicative organism behavior is the only path to survival.

Recently, Struik et al. (2008) questioned the validity of the concept of plant neurobiology by arguing against the parsimony principle, also known as "Occam's razor," which suggests that the most plausible concept is the one that is based on the simplest ideas and requires the fewest assumptions. However, Francis Crick has commented on the potential limitations of Occam's razor in biology (Crick 1988). Because biological systems are the products of (ongoing) natural selection, the mechanisms are not necessarily optimal in an obvious sense. He cautions that "while Occam's razor is a useful tool in the physical sciences, it can be a very dangerous implement in biology. It is thus very rash to use simplicity and elegance as a guide in biological research" (Crick 1988).

7 Unicellular "Neurons" and Plant Neurobiology: Unifying the Plant and Animal Kingdoms?

On the other hand, one can also propose that the discovery of convergent neuronal processes in animals and plants allows us to unite these two kingdoms. Such a unified view would conform with Occam's razor better than the two contrasting kingdoms of *Animalia* and *Plantae*, which we inherited from Aristoteles and Carl von Linné. For instance, the complex behavior of *Paramecium* combined with its sensory and signal integration complexities have resulted in surprising proposals that this unique organism represents a "swimming neuron," which resulted in the inclusion of this protozoan in neuroscience (Greenspan 2007). Interestingly, there are very close similarities between plant cells and unicellular swimming neurons such as the animal-like *Paramecium* and the plant-like *Chlamydomonas*. Eric R. Kandel won the Nobel Prize for his work with *Aplysia* entitled *In Search of Memory* (Kandel 2006), and his definition of neuron is as follows:

> The fundamental unit of any nervous system. Neurons are similar to other cells in having common molecular machinery for cellular function, but they have the unique ability to communicate rapidly with one another over great distances and with great precision.

Neurons are the only cells in animals/humans that lack centrioles, thus closely resembling plant cells. Also, neurons are the only cells in animals/humans that do not bathe directly in blood, but are instead protected and nourished by the blood–brain barrier (Rubin and Staddon 1999), again closely resembling plant cells. Finally, the latest evo–devo studies on lower animals such as *Hydra* reveal that neurons are the oldest (from an evolutionary perspective) cells of animals/humans. Hans Meinhardt recently proposed that the opening on *Hydra* is not the mouth but the anus, while the whole body represents the brain (Meinhardt 2002), leading to the "brain with anus" concept of *Hydra* body organization (Holland 2003). Interestingly in this respect, all sessile lower animals are anchored in substrate via the anterior poles of their bodies (Dawkins 2005; Smith 2008). Similar to plants, the settled anterior pole performs filter feeding, resembling the solute-based root nutrition of plants. It appears that plants harmonize well with animals if we consider which body pole is anchored into the substrate.

8 Conclusions and Outlook

Why do plants invest so much energy in transporting auxin via synaptic-like processes? Since all plant cells appear to be capable of auxin biosynthesis, the polar transport of this multipurpose signaling molecule along the whole plant body must have some crucial importance for plants. Advances made in the last few years have revealed that the cell–cell transport of auxin is inherently linked to the sensory perception of light and gravity that allows the adaptive shaping of plant bodies via sensorimotor circuits (Baluška et al. 2007, 2009). Root apices are particularly complex in this respect and represent anterior-like neuronal organs that are specialized for both nutrient uptake and neuronal activities (Fig. 1), in accordance with the prediction made by Charles and Francis Darwin in 1880 (Darwin 1880). Intriguingly, auxin transport is much more complex in the root apex than in the shoot apex (Baluška et al. 2007, 2009).

Similarly, we should ask why plants invest so much energy in the synthesis of large amounts of secondary metabolites, many of which have pain-relieving sedative or hallucinogenic effects on humans and animals. Most of these substances are synthesized in response to stressful situations, and, aside from its plant-specific actions, the "stress hormone" ethylene acts as a general anesthetic. Is this mere coincidence? Or could it be that plants can locally or even systematically lower their putative "pain-like" sensations by deliberately regulating their complex biochemistry? We simply do not know. This is precisely why we should invest energy in trying to answer these difficult but important questions. Interestingly, sessile lower animals also synthesize large amounts of secondary metabolites, so it may be that this is a general feature of sessile multicellular organisms, irrespective of whether they are plants or animals.

Sleep may be relevant to plants, as many plant organs perform "sleep movements," a phenomenon that was studied over a century ago by Wilhelm Pfeffer and Erwin Bünning (Brenner et al. 2007; Stahlberg 2006). The chemical basis for these leaf movements has recently been discovered (Shoji et al. 2006; Ueda and Nakamura 2007), allowing us to chemically manipulate these plant movements. Interestingly, those leaves which are chemically prevented from performing sleep movements eventually die (Shoji et al. 2006). In its extreme form, the sleep-like state can progress into long-term dormancy, which allows some perennial plants to live for several thousands of years (Ueda and Nakamura 2007; Rohde and Bhalerao 2007; Munne-Bosch 2008; Flanary and Kletetschka 2005). Thus, plants are also very important objects for biologists studying senescence. As pain, sleep and senescence are still rather elusive and perplexing phenomena in animal biology and medicine, plants may prove to be useful in unraveling these last mysteries of biology.

In conclusion, more than 120 years after Charles Darwin, together with his son Francis Darwin, first stirred our interest in sensory plants with his book *Power of Movement in Plants*, plants have been recognized in sensory biology to be highly sensible organisms. Plants are far more complex than we normally consider them to be. In order to interpret molecular, genetic, and physiologic data correctly, we need to understand why this sensory complexity has been evolved by sessile plants. New concepts are needed, and new questions must be asked, to advancing our still rudimentary understanding of the communicative nature of sensory plants.

References

Ahmad M, Galland P, Ritz T, Wiltschko R, Wiltschko W (2007) Magnetic intensity affects cryptochrome-dependent responses in *Arabidopsis thaliana*. Planta 225:615–624

Alpi A, Amrhein N, Bertl A, et al. (2007) Plant neurobiology: no brain, no gain? Trends Plant Sci 12:135–136

Anjard C, Loomis WF (2008) Cytokinins induce sporulation in *Dictyostelium*. Development 135:819–837

Backster C (1968) Evidence of a primary perception in plant life. Int J Parapsychol 10:329–348

Bais HP, Park S-W, Weir TL, Callaway RM, Vivanco J (2003) How plants communicate using the underground information superhighway. Trends Plant Sci 9:26–32

Baker MD, Stock JB (2007) Signal transduction: networks and integrated circuits in bacterial cognition. Curr Biol 17:R1021–R1024

Baldwin IT, Halitschke R, Paschold A, von Dahl CC, Preston CA (2006) Volatile signaling in plant–plant interactions: "talking trees" in the genomics era. Science 311:812–815

Baluška F, Mancuso S (2006) Plant neurobiology as a paradigm shift not only in the plant sciences. Plant Signal Behav 2:205–207

Baluška F, Vitha S, Barlow PW, Volkmann D (1997) Rearrangements of F-actin arrays in growing cells of intact maize root apex tissues: a major developmental switch occurs in the postmitotic transition region. Eur J Cell Biol 72:113–121

Baluška F, Hlavacka A, Šamaj J, Palme K, Robinson DG, Matoh T, McCurdy DW, Menzel D, Volkmann D (2002) F-actin-dependent endocytosis of cell wall pectins in meristematic root cells: insights from brefeldin A-induced compartments. Plant Physiol 130:422–431

Baluška F, Šamaj J, Menzel D (2003) Polar transport of auxin: carrier-mediated flux across the plasma membrane or neurotransmitter-like secretion? Trends Cell Biol 13:282–285

Baluška F, Volkmann D, Menzel D (2005a) Plant synapses: actin-based domains for cell-cell communication. Trends Plant Sci 10:106–111

Baluška F, Barlow PW, Baskin TI, Chen R, Feldman L, Forde BG, Geisler M, Jernstedt J, Menzel D, Muday GK, Murphy A, Samaj J, Volkmann D (2005b) What is apical and what is basal in plant root development? Trends Plant Sci 10:409–411

Baluška F, Mancuso S, Volkmann D (2006a) Communication in plants: neuronal aspects of plant life. Springer, Berlin

Baluška F, Hlavacka A, Mancuso S, Barlow PW (2006b) Neurobiological view of plants and their body plan. In: Baluška F, Mancuso S, Volkmann D (eds) Communication in plants: neuronal aspects of plant life. Springer, Berlin, pp 19–35

Baluška F, Barlow PW, Volkmann D, Mancuso S (2007) Gravity related paradoxes in plants: plant neurobiology provides the means for their resolution. In: Witzany G (ed) Biosemiotics in transdisciplinary context. Umweb, Helsinky, pp 9–35

Baluška F, Schlicht M, Wan Y-L, Burbach C, Volkmann D (2009) Intracellular domains and polarity in root apices: from synaptic domains to plant neurobiology. Nova Acta Leopold (in press)

Baluška F, Schlicht M, Volkmann D, Mancuso S (2008) Vesicular secretion of auxin: evidences and implications. Plant Signal Behav 3:254–256
Barbier-Brygoo H, Ephritikhine G, Klämbt D, Maurel C, Palme K, Schell J, Guern J (1991) Perception of the auxin signal at the plasma membrane of tobacco mesophyll protoplasts. Plant J 1:83–93
Barlow PW (1993) The root cap: cell dynamics, cell differentiation and cap function. J Plant Growth Regul 21:261–286
Barlow PW (2006) Charles Darwin and the plant root apex: closing the gap in living systems theory as applied to plants. In: Baluška F, Mancuso S, Volkmann D (eds) Communication in plants: neuronal aspects of plant life. Springer, Berlin, pp 37–51
Barlow PW (2008) Reflections on plant neurobiology. Biosystems 92:132–147
Benda J, Gollisch T, Machens CK, Herz AVM (2007) From response to stimulus: adaptive sampling in sensory physiology. Curr Opin Neurobiol 17:430–436
Beaubois E, Girard S, Lallechere S, Davies E, Paladian F, Bonnet P, Ledoigt G, Vian A (2007) Intercellular communication in plants: evidence for two rapidly transmitted systemic signals generated in response to electromagnetic field stimulation in tomato. Plant Cell Environ 30:733–744
Bhalerao RP, Bennett MJ (2003) The case for morphogens in plants. Nat Cell Biol 5:939–941
Braam J (2005) In touch: plant responses to mechanical stimuli. New Phytol 165:373–389
Brenner ED, Stahlberg R, Mancuso S, Vivanco J, Baluška F, Van Volkenburgh E (2006) Plant neurobiology: an integrated view of plant signaling. Trends Plant Sci 11:413–419
Brenner ED, Stahlberg R, Mancuso S, Vivanco J, Baluška F, Van Volkenburgh E (2007) Response to Alpi et al.: plant neurobiology—the gain is more than the name. Trends Plant Sci 12:285–286
Bruce TJA, Matthes MC, Napier JA, Pickett JA (2007) Stressful "memories" of plants: evidence and possible mechanisms. Plant Sci 173:604–608
Bruzzone S, Moreschi I, Usai C, Guida L, Damonte G, Salis A, Scarfi S, Millo E, De Flora A, Zocchi E (2007) Abscisic acid is an endogenous cytokine in human granulocytes with cyclic ADP-ribose as second messenger. Proc Natl Acad Sci USA 104:5759–5764
Calvo Garzon F (2007) The quest for cognition in plant neurobiology. Plant Signal Behav 2:208–211
Campagna JA, Miller KW, Forman SA (2003) Mechanisms of actions of inhaled anesthetics. N Engl J Med 348:2110–2124
Collings DA, White RG, Overall RL (1992) Ionic current changes associated with the gravity-induced bending response in roots of *Zea mays* L. Plant Physiol 100:1417–1426
Conway Morris S (2003) Life's solution. Inevitable humans in a lonely universe. Cambridge University Press, Cambridge
Conway Morris S (2006) Evolutionary convergence. 16:R826–R827
Crick FHC (1988) What mad pursuit: a personal view of scientific discovery. Basic, New York
Crotty P, Sangrey T, Levy WB (2006) Metabolic energy cost of action potential velocity. J Neurophysiol 96:1237–1246
Darwin CR (assisted by Darwin F) (1880) The power of movement in plants. John Murray, London
Darwin CR (1875a) Insectivorous plants. John Murray, London
Darwin CR (1875b) The movements and habits of climbing plants. John Murray, London
Darwin F (1899) The botanical work of Darwin. Ann Bot 13:626
Davies E (2004) New functions for electric signals in plants. New Phytol 161:607–610
Dawkins R (2005) The ancestor's tale: a pilgrimage to the dawn of evolution. Mariner, New York
DeWeese MR, Zador A (2006) Efficiency mesures. Nature 439:920–921
Dhonukshe P, Baluška F, Schlicht M, Hlavacka A, Šamaj J, Friml J, Gadella TWJ Jr (2006) Endocytosis of cell surface material mediates cell plate formation during plant cytokinesis. Dev Cell 10:137–150
Dicke M, Agrawal AA, Bruin J (2003) Plants talk, but are they deaf? Trends Plant Sci 8:403–405
Esmon CA, Pedmale UV, Liscum E (2005) Plant tropisms: providing the power of movement to a sessile organism. Int J Dev Biol 49:665–674
Felle HH, Zimmermann MR (2007) Systemic signalling in barley through action potentials. Planta 226:203–214
Felle H, Peters W, Palme K (1991) The electrical response of maize to auxins. Biochim Biophys Acta 1064:199–204

Flanary BE, Kletetschka G (2005) Analysis of telomere length and telomerase activity in tree species of various life-spans, and with age in the bristelcone pine *Pinus longaeva*. Biogerontology 6:101–111
Friml J, Benfey P, Benková E, Bennett M, Berleth T, Geldner N, Grebe M, Heisler M, Hejátko J, Jürgens G, Laux T, Lindsey K, Lukowitz W, Luschnig C, Offringa R, Scheres B, Swarup R, Torres-Ruiz R, Weijers D, Zazímalová E (2006). Apical–basal polarity: why plant cells don't stand on their heads. Trends Plant Sci 11:12–14
Fromm J, Lautner S (2007) Electrical signals and their physiological significance in plants. Plant Cell Environ 30:249–257
Galland P, Pazur A (2005) Magnetoreception in plants. J Plant Res 118:371–389
Grant CJ, Powell JN, Radford SG (1974) Effects of halothane on DNA synthesis and mitosis in root tip meristems of *Vicia faba*. Br J Anaesth 46:653–657
Greenspan RJ (2007) An introduction to nervous systems. Cold Spring Harbor Laboratory, New York
Gruntman M, Novoplansky A (2004) Physiologically mediated self/non-self discrimination in roots. Proc Natl Acad Sci USA 10:3863–3867
Guillery RW (2005) Observations of synaptic structures: origins of the neuron doctrine and its current status. Phil Trans R Soc B 360:1281–1307
Hawes MC, Gunawardena U, Miyasaka S, Zhao X (2000) The role of root border cells in plant defense. Trends Plant Sci 5:128–133
Heslop-Harrison J (1980) Darwin and the movement of plants: a retrospect. In: Skoog F (ed) Plant growth substances 1979. Springer, Berlin, pp 3–14
Holland ND (2003) Early central nervous system evolution: an era of skin brains? Nat Rev Neurosci 4:617–627
Ingensiep HW (2001) Geschichte der Pflanzenseele: Philosophische und biologische Entwürfe von der Antike bis zur Gegenwart. Kröner, Stuttgart
Jenal U, Silversmith RE, Sogaard-Andersen L, Sockett L (2005) Sense and sensibility in bacteria. EMBO Rep 6:615–619
Jeong MJ, Shim CK, Lee JO, Kwon HB, Kim YH, Lee SK, Byun MO, Park SC (2007) Plant gene responses to frequency-specific sound signals. Mol Breed 21:217–226
Johnsen S, Mattern E, Ritz Th (2007) Light-dependent magnetoreception: quantum catches and opponency mechanisms of possible photosensitive molecules. J Exp Biol 210:3171–3178
Kandel ER (2006) In search of memory. The emergence of new theory of mind. Norton, New York
Keyes WJ, Taylor JV, Apkarian RP, Lynn DG (2001) Dancing together. Social controls in parasitic plant development. Plant Physiol 127:1508–1512
Kim DS, Kim SY, Jeong YM, Jeon SE, Kim MK, Kwon SB, Na JI, Park KC (2006) Light-activated indole-3-acetic acid induces apoptosis in g361 human melanoma cells. Biol Pharm Bull 29:2404–2409
King-Hele D (1974) Erasmus Darwin, master of many crafts. Nature 247:87–91
Kobayashi A, Takahashi A, Kakimoto Y, Miyazawa Y, Fujii N, Higashitani A, Takahashi H (2007) A gene essential for hydrotropism in roots. Proc Natl Acad Sci USA 104:4724–4729
Kovac L (2007) Information and knowledge in biology: time for reappraisal. Plant Signal Behav 2:65–73
Kwon C, Neu C, Pajonk S, Yun HS, Lipka U, Humphry M, Bau S, Straus M, Kwaaitaal M, Rampelt H, El Kasmi F, Jürgens G, Parker J, Panstruga R, Lipka V, Schulze-Lefert P (2008a) Co-option of a default secretory pathway for plant immune responses. Nature 451:835–840
Kwon C, Panstruga, Schulze-Lefert P (2008b) Les liaisons dangereuses: immunological synapse formation in animals and plants. Trends Immunol 29:159–166
Lennie P (2003) The cost of cortical computation. Curr Biol 13:493–497
Li X, Zhang WS (2008) Salt-avoidance tropism in *Arabidopsis thaliana*. Plant Signal Behav 3:351–353
Mancuso S, Marras AM, Volker M, Baluška F (2005) Non-invasive and continuous recordings of auxin fluxes in intact root apex with a carbon-nanotube-modified and self-referencing micro-electrode. Anal Biochem 341:344–351
Mancuso S, Marras AM, Mugnai S, Schlicht M, Zarsky V, Li G, Song L, Hue HW, Baluška F (2007) Phospholipase Dζ2 drives vesicular secretion of auxin for its polar cell-cell transport in the transition zone of the root apex. Plant Signal Behav 2:240–244

Meinhardt H (2002) The radial-symmetric hydra and the evolution of the bilateral body plan: an old body became a young brain. Bioessays 24:185–191
Meyerowitz EM (2000) The plant plan: multicellular life in the other Kingdom. Harvey Lect 96:51–72
Munne-Bosch S (2008) Do perennials really senesce? Trends Plant Sci 13:216–220
Nakagawa Y, Katagiri T, Shinozaki K, Qi Z, Tatsumi H, Furuichi T, Kishigami A, Sokabe M, Kojima I, Sato S, Kato T, Tabata S, Iida K, Terashima A, Nakano M, Ikeda M, Yamanaka T, Iida H (2007) Arabidopsis plasma membrane protein crucial for Ca^{2+} influx and touch sensing in roots. Proc Natl Acad Sci USA 104:3639–3644
Powell JN, Grant CJ, Robinson SM, Radford SG (1973) A comparison with halothane of the hormonal and anaesthetic properties of ethylene in plants. Br J Anaesth 45:682–690
Puce S, Basile G, Bavestrello G, Bruzzone S, Cerrano C, Giovine M, Arillo A, Zocchi W (2004) Abscisic acid signaling through cyclic ADP-ribose in hydroid regeneration. J Biol Chem 17:39783–39788
Radl E (1909) Die Geschichte der biologischen Theorien. Wilhelm Engelmann, Leipzig
Ripoll C, Le Sceller L, Verdus MC, Norris V, Tafforeau M, Thellier M (2009) Memorization of abiotic stimuli in plants. A complex role for calcium. In: Baluška F (ed) Plant–environment interactions. Springer, Berlin
Rohde A, Bhalerao RP (2007) Plant dormancy in the perennial context. Trends Plant Sci 12:217–223
Rotem R, Heyfets A, Fingrut O, Blickstein D, Shaklai M, Flescher E (2005) Jasmonates: novel anticancer agents acting directly and selectively on human cancer cell mitochondria. Cancer Res 65:1984–1993
Roux D, Faure C, Bonnet P, Girard S, Ledoigt G, Davies E, Gendraud M, Paladian F, Vian A (2008) A possible role for extra-cellular ATP in plant responses to high frequency, low amplitude electromagnetic field. Plant Signal Behav 3:383–385
Rubin LL, Staddon JM (1999) The cell biology of the blood–brain barrier. Annu Rev Neurosci 22:11–28
Rudrappa T, Bais HP (2008) Genetics, novel weapons and rhizospheric microcosmal signaling in the invasion of *Phragmites australis*. Plant Signal Behav 3:1–5
Rudrappa T, Bonsall J, Gallagar J, Seliskar DM, Bais HP (2007) Root secreted allelochemical in noxious weed *Phragmites australis* deploys reactive oxygen species response and microtubule assembly disruption to execute rhizotoxicity. J Chem Ecol 33:1898–1918
Runyon JB, Mescher MC, De Moraes CM (2006) Volatile chemical cues guide host location and host selection by parasitic plants. Science 313:1964–1967
Ruuhola T, Salminen JP, Haviola S, Yang S, Rantala MJ (2007) Immunological memory of mountain birches: effects of phenolics on performance of the autumnal moth depend on herbivory history of trees. J Chem Ecol 33:1160–1176
Sagane Y, Nakagawa T, Yamamoto K, Michikawa S, Oguri S, Momonoki YS (2005) Molecular characterization of maize acetylcholinesterase. A novel enzyme family in the plant kingdom. Plant Physiol 138:1359–1371
Šamaj J, Read ND, Volkmann D, Menzel D, Baluška F (2005) Endocytic network in plants. Trends Cell Biol 15:425–433
Šamaj J, Baluška F, Menzel D (2006) Plant endocytosis. Springer, Berlin
Schlicht M, Strnad M, Scanlon MJ, Mancuso S, Hochholdinger F, Palme K, Volkmann D, Menzel D, Baluška F (2006) Auxin immunolocalization implicates vesicular neurotransmitter-like mode of polar auxin transport in root apices. Plant Signal Behav 1:122–133
Shelp BJ, Bown AW, Faure D (2006) Extracellular γ-aminobutyrate mediates communication between plants and other organisms. Plant Physiol 142:1350–1352
Shelp BJ, Allan WL, Faure D (2008) Role of γ-aminobutyrate and γ-hydroxybutyrate in plant communication. In: Baluška F (ed) Plant–environment interactions. Springer, Berlin
Shepherd VA (2005) From semi-conductors to the rhythms of sensitive plants: the research of J.C. Bose. Cell Mol Biol 51:607–619
Shoji M, Kato E, Nakamura Y, Fujii T, Manabe Y, Ueda M (2006) Bioorganic studies on plant movement, from natural products to its receptor. Chem Rec 6:344–355

Smith AB (2008) Deuterostomes in a twist: the origins of a radical new body plan. Evol Dev 10:493–503

Solov'yov IA, Chandler DE, Schulten K (2007) Magnetic field effects in *Arabidopsis thaliana* cryptochrome-1. Biophys J 92:2711–2726

Stahlberg R (2006) Historical overview on plant neurobiology. Plant Signal Behav 1:6–8

Steffens B, Feckler C, Palme K, Christian M, Böttger M, Lüthen H (2001) The auxin signal for protoplast swelling is perceived by extracellular ABP1. Plant J 27:591–599

Stein BE (1998) Neural mechanisms for synthesizing sensory information and producing adaptive behaviors. Exp Brain Res 123:124–135

Stephens NR, Qi Z, Spalding EP (2008) Glutamate receptor subtypes evidenced by differences in desensitization and dependence on the GLR3.3 and GLR3.4 genes. Plant Physiol 146:529–538

Stock J, Levit M (2000) Signal transduction: hair brains in bacterial chemotaxis. Curr Biol 10:R11–R14

Svistoonoff S, Creff A, Reymond M, Sigoillot-Claude C, Ricaud L, Blanchet A, Nussaume L, Desnos T (2007) Root tip contact with low-phosphate media reprograms plant root architecture. Nat Genet 39:792–793

Struik PC, Yin X, Meinke H (2008) Plant neurobiology and green plant intelligence: science, metaphors and nonsense. J Sci Food Agric 88:363–370

Sun F, Zhang W, Hu H, Li B, Wang Y, Zhao Y, Li K, Liu M, Li X (2007) Salt modulates gravity signaling pathway to regulate growth direction of primary roots in *Arabidopsis thaliana.* Plant Physiol 146:178–188

Tafforeau M, Verdus MC, Norris V, Ripoll C, Thellier M (2006) Memory processes in the response of plants to environmental signals. Plant Signal Behav 1:9–14

Tompkins P, Bird CO (1973) The secret life of plants. Avon, New York

Trewavas A (2002) Mindless mastery. Nature 415:841

Trewavas A (2003) Aspects of plant intelligence. Ann Bot 92:1–20

Trewavas A (2005a) Plant intelligence. Naturwissenschaften 92:401–413

Trewavas A (2005b) Green plants as intelligent organisms. Trends Plant Sci 10:413–419

Trewavas A (2007) Response to Alpi et al.: plant neurobiology—all metaphors have value. Trends Plant Sci 12:231–233

Trewavas A, Knight M (1997) Mechanical signalling, calcium and plant form. Plant Mol Biol 26:1329–1341

Ueda M, Nakamura Y (2007) Chemical basis of plant leaf movement. Plant Cell Physiol 48:900–907

Verdus MC, Sceller LL, Norris V, Thellier M, Ripoll C (2007) Pharmacological evidence for calcium involvement in the long-term processing of abiotic stimuli in plants. Plant Signal Behav 2:212–220

Vermeij GJ (2006) Historical contigency and the purported uniqueness of evolutionary innovations. Proc Natl Acad Sci USA 103:1804–1809

Vian A, Faure C, Girard S, Davies E, Hallé F, Bonnet P, Ledoigt G, Paladian F (2007) Plants do respond to GSM-like radiations. Plant Signal Behav 2:522–524

Volkov AG (2006) Plant electrophysiology. Springer, Berlin

Volkov AG, Adesina T, Jovanov E (2007a) Closing of Venus flytrap by electrical stimulation of motor cells. Plant Signal Behav 2:139–145

Volkov AG, Adesina T, Markin VS, Jovanov E (2007b) Kinetics and mechanism of *Dionaea muscipula* trap closing. Plant Physiol 146:694–702

Volkov AG, Carrell H, Adesina T, Markin VS, Jovanov E (2008) Plant electrical memory. Plant Signal Behav 3:490–492

Wayne R (1994) The excitability of plant cells: with emphasis on characean internodal cells. Bot Rev 60:265–335

Webster C (1966) The recognition of plant sensitivity by English botanists in the seventeenth century. Isis 57:5–23

Whippo CW, Hangarter RP (2006) Phototropism: bending towards enlightenment. Plant Cell 18:1110–1119

Index

GPSR Compliance

The European Union's (EU) General Product Safety Regulation (GPSR) is a set of rules that requires consumer products to be safe and our obligations to ensure this.

If you have any concerns about our products, you can contact us on ProductSafety@springernature.com

In case Publisher is established outside the EU, the EU authorized representative is:

Springer Nature Customer Service Center GmbH
Europaplatz 3
69115 Heidelberg, Germany

Batch number: 10761599

Printed by Printforce, the Netherlands